T0178528

Communications
in Computer and Information Science   1354

More information about this series at http://www.springer.com/series/7899

Samson Lasaulce · Panayotis Mertikopoulos ·
Ariel Orda (Eds.)

Network Games, Control and Optimization

10th International Conference, NetGCooP 2020
France, September 22–24, 2021
Proceedings

 Springer

Editors
Samson Lasaulce
CRAN – CO2 Team ENSEM
Vandoeuvre-lès-Nancy, France

Panayotis Mertikopoulos
CNRS Researcher Laboratoire
d'Informatique de Grenoble (LIG)
Saint-Martin d'Hères, France

Ariel Orda
Technion – Israel Institute of Technology
Haifa, Israel

ISSN 1865-0929   ISSN 1865-0937 (electronic)
Communications in Computer and Information Science
ISBN 978-3-030-87472-8   ISBN 978-3-030-87473-5 (eBook)
https://doi.org/10.1007/978-3-030-87473-5

This Springer imprint is published by the registered company Springer Nature Switzerland AG
The registered company address is: Gewerbestrasse 11, 6330 Cham, Switzerland

Preface

This volume in the Communications in Computer and Information Science series (CCIS, volume 1354) is a collection of the papers accepted for the 10th International Conference on NETwork Games, COntrol and OPtimization (NetGCooP 2020). The event was originally planned to take place in Cargèse, Corsica, France during March 18–20, 2020, but due to the recent COVID-19 pandemic the conference was postponed to September 22–24, 2021. Because of the exceptional circumstances, and to ensure timely dissemination of the contributions by the authors that would normally have been presented last March, we decided to publish the accepted papers before the event.

Networks form the backbone of many complex systems, ranging from the Internet to social interactions. The proper design and control of networks have been long-standing issues in various engineering and science disciplines. The vision of the conference is to provide a platform for researchers to share novel ideas and network applications in the areas of control and optimization. From an application point of view, the 2020 edition focused on resource allocation, energy markets, and opinion dynamics. And from a theoretical point of view, we received and accepted papers concerning learning in games, the value of information in games, and how to design robust defense strategies.

The program for NetGCooP 2020 comprised 14 papers (3 short papers and 11 regular papers) selected from submissions received through open calls, and an additional 15 papers selected from submissions received via invitation. We would like to thank the authors for having done their best to produce works of high quality and for having supported the difficult decisions that the organizers were forced to make because of the challenging situation. Moreover, we would like to thank the reviewers and the Technical Program Committee (TPC) members for providing reviews of quality in spite of the challenges created by the pandemic.

June 2021

Samson Lasaulce
Panayotis Mertikopoulos
Ariel Orda

Organization

General Co-chairs

Tamer Başar   University of Illinois Urbana-Champaign, USA
Merouane Debbah   Huawei, France
Alexandre Reiffers   IMT Atlantique, France
Corinne Touati   Inria, France

Technical Program Committee Co-chair

Samson Lasaulce   CRAN, CNRS, University of Lorraine, France
Panayotis Mertikopoulos   CNRS and Laboratoire d'Informatique de Grenoble,
                          France
Ariel Orda   Technion, Israel

Publication Chair

Essaid Sabir   ENSEM, Hassan II University of Casablanca, Morocco

Technical Program Committee

Samson Lasaulce   CRAN, CNRS, University of Lorraine, France
Panayotis Mertikopoulos   CNRS and Laboratoire d'Informatique de Grenoble,
                          France
Alexandre Reiffers   IMT Atlantique, France
Yezekael Hayel   LIA, University of Avignon, France
K. Clay McKell   California Polytechnic State University, USA
Eli Meirom   Technion, Israel
Essaid Sabir   ENSEM, Hassan II University of Casablanca, Morocco
Ramakrishnan   Indian Institute of Technology, Madras, India
Ariel Orda   Technion, Israel
E. Veronica Belmega   ETIS, ENSEA, Université Cergy Pontoise, CNRS,
                          France
Balakrishna Prabhu   LAAS-CNRS, France
Swapnil Dhamal   Chalmers University of Technology, Sweden
Vijay Kamble   University of Illinois at Chicago, USA
Francesco De Pellegrini   University of Avignon, France
Isabel Amigo   IMT Atlantique, France
Salah Eddine   Elayoubi CentraleSupélec, France
Bruno Gaujal   Inria and Laboratoire d'Informatique de Grenoble,
                          France

Iriniel-Constantin   CRAN, Université de Lorraine, France
  Morarescu
Eitan Altman   Inria, France
Romain Negrel   ESIEE Paris and LIGM, France

Contents

Scheduling and Resource Allocation Problems in Networks

Advance in Game Theory

Social Networks

Electrical Networks

Game Theory and Iterative Algorithms Applied to Wireless Communication

On the Existence and Uniqueness of Nash Equilibria in MIMO Communication Games with a Jammer

K. Clay McKell[1]([⊠])[iD] and Gürdal Arslan[2][iD]

[1] California Polytechnic State University, San Luis Obispo, CA, USA
kmckell@calpoly.edu
[2] University of Hawai'i at Mānoa, Honolulu, HI, USA
gurdal@hawaii.edu

Abstract. This paper models multiple users of a multiple-input multiple-output communications channel and a jammer as players in a competitive game. When users maximize their quality of service and the jammer strives to minimize the aggregate quality of service in the network with all players subject to a power constraint, we show that a pure Nash equilibrium always exists. Under assumptions on channel structure, including no user-user interference, we show that there can be at most one Nash equilibrium where a user transmits on all of its subchannels. A similar but weaker result holds for channels with limited amounts of user-user interference.

Keywords: Multi-user MIMO · Jammer · Nash equilibrium

1 Introduction and Game Theoretic Model

We consider a single communication channel that is shared by r users, $\mathcal{R} = \{1, \ldots, r\}$, each comprising a transmit/receive pair. When the transmitter and receiver have multiple antennae, N_t and N_r, respectively, the channel is said to be multiple-input multiple-output (MIMO). The multiple paths that information streams can travel are referred to as subchannels. When $N_t = N_r$ and if there is no interference between subchannels, they are said to be orthogonal. In general, however, the channel inputs may interfere with each other (constructively or destructively).

Regarding notation: we use $\mathbb{H}^n$ to denote the set of n-dimensional Hermitian matrices. If $X \in \mathbb{H}^n$, then $X \succeq 0$ ($X \succ 0$) means that X is positive semidefinite (definite). The trace of a matrix is denoted $\operatorname{Tr} X$ and the conjugate transpose is $X^\dagger$. We use $\overline{\sigma}(Y)$ to denote the maximum singular value of Y and $\underline{\lambda}(X)$ for the minimum eigenvalue of Hermitian X.

We model the channel users' objectives as their information throughput subject to an average power constraint. A *utility game* is then a set of agents, $\mathcal{P} = \mathcal{R} \cup \{0\}$, a set of channel matrices, $\{H_{j,k}\}_{\mathcal{P} \times \mathcal{R}}$, and a vector of positive

S. Lasaulce et al. (Eds.): NetGCooP 2021, CCIS 1354, pp. 3–7, 2021.
https://doi.org/10.1007/978-3-030-87473-5_1

power constraints $c \in \mathbb{R}^{r+1}$. Gaussian interference implies that a zero-mean Gaussian codebook is the optimal encoding scheme [5], so the strategy set for each agent $k \in \mathcal{P}$ is the set of positive semidefinite covariance matrices with bounded power: $\mathcal{S}_k = \{Q_k \in \mathbb{H}^{N_t} : Q_k \succeq 0, \operatorname{Tr} Q_k \leq c_k\}$. A joint action is then a vector of all agent actions: $Q = \{Q_k\}_{k \in \mathcal{P}}$. Let each user k have a utility function equal to the mutual information between its transmitter and receiver (as in [8]):

$$u_k(Q) = \log \left| \left(I + R_k^{-1/2}(Q) H_{k,k} Q_k H_{k,k}^{\dagger} R_k^{-1/2}(Q) \right) \right|, \tag{1}$$

where $R_k(\cdot)$ is the noise-plus-interference from agents other than k:

$$R_k(Q) = I + \sum_{j \in \mathcal{P} \setminus \{k\}} H_{k,j} Q_j H_{k,j}^{\dagger}. \tag{2}$$

The jammer, however, strives to minimize the total mutual information across the network. This is equivalent to maximizing $J^u(Q) = -\sum_{k \in \mathcal{R}} u_k(Q)$.

To apply Rosen's theorems on concave games played over real strategy sets [6], we will frequently view N_t-dimensional Hermitian matrices as real $2N_t^2$ vectors using the reshaping for $X \in \mathbb{H}^{N_t}$, $\overrightarrow{X} = [\operatorname{vec}^{\mathsf{T}}(\operatorname{Re} X), \operatorname{vec}^{\mathsf{T}}(\operatorname{Im} X)]^{\mathsf{T}} \in \mathbb{R}^{2N_t^2}$, where $\operatorname{vec}(\cdot)$ is the standard vectorization operator that vertically concatenates the columns of its argument.

The existence and uniqueness of equilibria in this game with full-rank channel matrices and small interference levels in the absence of a jammer was shown in [1]. Restricted to square nonsingular channel matrices, [7] provided sufficient conditions for uniqueness—including explicit bounds on interference levels. Similarly, [3] gave conditions for equilibrium uniqueness in the presence of a jammer but restricted agents to orthogonal subchannels. This paper aims to bridge the gaps between these works.

2   Existence of Equilibria

Not all games have pure Nash equilibria. In this section, we use results on concave games [6] to guarantee the existence of pure equilibria for any channel condition.

Theorem 1. *For any vector of costs and any set of channel matrices, the utility game possesses a pure Nash equilibrium.*

Proof. For all $k \in \mathcal{P}$, the strategy set $\mathcal{S}_k$ is compact, convex, and nonempty. Furthermore, $u_k(Q)$ is continuous in Q, concave in Q_k, and convex in Q_0 for all $k \in \mathcal{R}$. As the negative sum of these functions, $J^u(Q)$ is continuous in Q and concave in Q_0. Thus the existence result for concave games in [6] applies.

3 Uniqueness of Equilibria

The classical tool for showing the uniqueness of an equilibrium in concave games is to employ Rosen's sufficient condition for diagonal strict concavity which requires the symmetrized pseudohessian matrix of a game be negative definite when evaluated at any point in the joint action set. The utility game does not meet this (rather restrictive) sufficient condition. The pseudohessian matrix, $G(Q)$, for a utility game comprises blocks $G_{k,j}(Q)$ where $G_{k,j}(Q) = \nabla_{\overrightarrow{Q_j}}\nabla_{\overrightarrow{Q_k}}u_k(Q)$, for all $(k, j) \in \mathcal{R} \times \mathcal{P}$ and $G_{0,j}(Q) = \nabla_{\overrightarrow{Q_j}}\nabla_{\overrightarrow{Q_0}}J^u(Q)$, for all $j \in \mathcal{P}$. Even utility games with "nice" channel structure—invertible user and jammer-user matrices with no user-user interference—may fail to meet this condition. Under certain conditions, the set of joint actions in a utility game that satisfy the KKT optimality conditions outlined in Rosen's proof is a singleton. There is a natural split in this analysis between the case were users do not interfere with each other—termed *zero cross-talk*—and the case in which each user sees all manner of interference.

We define a zero cross-talk channel to be one which for every $k \in \mathcal{R}$, $H_{k,j} = 0$ for all $j \in \mathcal{R}\backslash\{k\}$. In games played on these channels, each user sees ambient noise and interference from the jammer but is unaffected by the other users.

Theorem 2. *In addition to being a zero cross-talk channel, assume the channel matrices meet the following conditions for all $k \in \mathcal{R}$:*

$$\text{rank } H_{k,k} = N_t \tag{3a}$$
$$\text{rank } H_{k,0} = N_t. \tag{3b}$$

Then for any budget c the set of equilibria of the utility game will have one of the following properties:

1. *It is a singleton, $\hat{Q}$, in which $\hat{Q}_k \succ 0$ for at least one $k \in \mathcal{R}$, or*
2. *It may contain multiple equilibria, but no equilibrium will have positive definite user action for any user.*

Proof. Assume there are two equilibria, Q^0 and Q^1 and that Q^0 possesses a nonempty set of users with positive definite equilibrium actions. Define $Q(\theta) = \theta Q^1 + (1-\theta)Q^0$. The KKT conditions that must hold at both of these equilibria require that

$$\int_0^1 \left(\overrightarrow{Q^1} - \overrightarrow{Q^0}\right)^\mathsf{T} [G(Q(\theta)) + G^\mathsf{T}(G(Q(\theta))]\left(\overrightarrow{Q^1} - \overrightarrow{Q^0}\right) d\theta = 0. \tag{4}$$

To produce a contradiction, it suffices to show that $\bar{G}(Q(\theta)) = G(Q(\theta)) + G^\mathsf{T}(Q(\theta))$ is a negative real quadratic form over vectors drawn from $\{\overrightarrow{Z} : Z \in \prod_{k\in\mathcal{P}}\mathbb{H}^{N_t}\}$. Under the zero cross-talk assumption, all off-diagonal blocks of $\bar{G}(\cdot)$ are zero. As in [1], for $k \in \mathcal{R}$ the k-th diagonal block of the pseudohessian is negative definite for all joint actions.

The directional derivative of the jammer's component of the pseudogradient when Q_0 is perturbed in the direction $Z_0 \in \mathbb{H}^{N_t}$ is

$$
\begin{aligned}
&\mathcal{D}\left(\mathcal{D}J^u(Q; Z_0); Z_0\right) = \\
&\quad -\sum_{k \in \mathcal{R}} \operatorname{Tr} H_{k,0}^{\dagger} \left(R_k^{-1} H_{k,0} Z_0 H_{k,0}^{\dagger} R_k^{-1} - A_k^{-1} H_{k,0} Z_0 H_{k,0}^{\dagger} A_k^{-1} \right) H_{k,0} Z_0,
\end{aligned}
\tag{5}
$$

where R_k is defined in (2) and $A_k = R_k + H_{k,k} Q_k H_{k,k}^{\dagger}$. From this summation, select term $\hat{k}$ where $Q_{\hat{k}}^0 \succ 0$ and write it as

$$
\operatorname{Tr} A_{\hat{k}}^{-1/2} \tilde{Z}_0 \left(A_{\hat{k}}^{-1} - R_{\hat{k}}^{-1} \right) \tilde{Z}_0 A_{\hat{k}}^{-1/2} + \operatorname{Tr} R_{\hat{k}}^{-1/2} \tilde{Z}_0 \left(A_{\hat{k}}^{-1} - R_{\hat{k}}^{-1} \right) \tilde{Z}_0 R_{\hat{k}}^{-1/2}, \tag{6}
$$

where we have substituted the Hermitian $\tilde{Z}_0$ for $H_{k,0} Z_0 H_{k,0}^{\dagger}$. Since rank $H_{k,k} = N_t$ for all users, $A_{\hat{k}}(Q(\theta)) \succ R_{\hat{k}}(Q(\theta))$ over $0 \leq \theta < 1$. Inversion reverses this ordering so the $\hat{k}$ term in the sum is strictly negative. Thus we can conclude that $\left(\overrightarrow{Z_0} \right)^{\mathsf{T}} G_{0,0}(Q(\theta)) \overrightarrow{Z_0} < 0$ for all $0 \leq \theta < 1$, which means the two implications of the theorem are mutually exclusive.

We now endeavor to generalize Theorem 2 to instances with nonzero cross-talk between the users. For any collection of channel matrices the pseudohessian can be decomposed: $\bar{G}_{\mathrm{CT}}(Q) = \bar{G}_{\mathrm{ZCT}}(Q) + \tilde{G}(Q)$, where $\bar{G}_{\mathrm{ZCT}}(\cdot)$ is the pseudohessian from Theorem 2. The cross-talk channel matrices do not change the diagonal blocks of the perturbation: $\tilde{G}_{k,k}(Q) = 0$ for all $k \in \mathcal{P}$. We will bound the effect on $\bar{G}_{\mathrm{CT}}(Q)$ of the user-user interference channels with the largest maximum singular value of those matrices: $\bar{\sigma}_c = \max_{(k \neq j) \in \mathcal{R} \times \mathcal{R}} \bar{\sigma}(H_{k,j})$.

Theorem 3. *For any budget c, assume the channel matrices of the utility game obey (3). For any $\epsilon > 0$, there is a $\hat{\sigma}_c > 0$ such that if $\bar{\sigma}_c < \hat{\sigma}_c$, there can be at most one equilibrium Q with the property that $\underline{\lambda}(Q_k) > \epsilon$ for some $k \in \mathcal{R}$.*

Proof. Assume that there are two Nash equilibria, Q^0 and Q^1, that have nonempty sets of users with minimum eigenvalues greater than ϵ, and again define $Q(\theta)$ as the convex combination of Q^0 and Q^1. As in Theorem 2, $\bar{G}_{\mathrm{ZCT}}(Q(\theta))$ is negative definite with respect to Hermitian matrices for all $0 \leq \theta \leq 1$. To generate a contradiction with the KKT conditions in the nonzero cross-talk case, it suffices to provide a tight enough bound on $\left(\overrightarrow{Z} \right)^{\mathsf{T}} \tilde{G}(Q(\theta)) \overrightarrow{Z}$, provided Z is drawn from a compact set. To this end, we use [2] to bound

$$
\bar{\sigma}(\tilde{G}(Q(\theta))) \leq \max_{k \in \mathcal{P}} \sum_{j \in \mathcal{P} \setminus \{k\}} \bar{\sigma}(\tilde{G}_{k,j}(Q(\theta))). \tag{7}
$$

For row $k \in \mathcal{R}$ and column $j \in \mathcal{R} \setminus \{k\}$ we have

$$
\bar{\sigma}(\tilde{G}_{k,j}(Q(\theta))) \leq \bar{\sigma}^2 (H_{k,k}) \bar{\sigma}^2 (H_{k,j}) + \bar{\sigma}^2 (H_{j,j}) \bar{\sigma}^2 (H_{j,k}), \tag{8}
$$

and the jammer's block in this row can be bounded

$$\overline{\sigma}(\tilde{G}_{k,0}(Q(\theta))) \leq \sum_{\substack{j \in \mathcal{R} \\ j \neq k}} \overline{\sigma}^2 (H_{j,0}) \, \overline{\sigma}^2 (H_{j,k}) . \tag{9}$$

If all user-user channel matrices have singular values no larger than $\overline{\sigma}_c$, then

$$\left(\overrightarrow{Q^1} - \overrightarrow{Q^0}\right)^{\mathsf{T}} \tilde{G}(Q(\theta)) \left(\overrightarrow{Q^1} - \overrightarrow{Q^0}\right) \tag{10}$$

can be resticted in magnitude without knowledge of Q^0 and Q^1 by appropriate choice of $\hat{\sigma}_c$. This guarantees $\bar{G}_{\mathrm{CT}}(Q(\theta))$ remains negative definite with respect to Hermitian matrices for all $0 \leq \theta \leq 1$, which produces a contradiction on the assumption that two Nash equilibria exist with user actions more than ϵ away from singularity.

Theorems 2 and 3 are consistent with the literature. In [3], the authors observe that for their generalized iterative water-filling algorithm to converge in a jammed multi-user MIMO network (limited to orthogonal subchannels), the users should not play singular actions. In future work, we shall attempt to provide distributed algorithm convergence results in the vein of [4] when the equilibrium is unique.

References

1. Arslan, G., Demirkol, M.F., Song, Y.: Equilibrium efficiency improvement in MIMO interference systems: a decentralized stream control approach. IEEE Trans. Wireless Commun. **6**(8), 2984–2993 (2007). https://doi.org/10.1109/TWC.2007.051043
2. Feingold, D.G., Varga, R.S.: Block diagonally dominant matrices and generalizations of the Gershgorin circle theorem. Pac. J. Math. **12**(4), 1241–1250 (1962). https://doi.org/10.2140/pjm.1962.12.1241
3. Gohary, R.H., Huang, Y., Luo, Z.Q., Pang, J.S.: A generalized iterative water-filling algorithm for distributed power control in the presence of a jammer. IEEE Trans. Signal Process. **57**(7), 2660–2674 (2009). https://doi.org/10.1109/TSP.2009.2014275
4. Mertikopoulos, P., Belmega, E.V., Moustakas, A.L., Lasaulce, S.: Distributed learning policies for power allocation in multiple access channels. IEEE J. Sel. Areas Commun. **30**(1), 96–106 (2012). https://doi.org/10.1109/JSAC.2012.120109
5. Neeser, F.D., Massey, J.L.: Proper complex random processes with applications to information theory. IEEE Trans. Inf. Theory **39**(4), 1293–1302 (1993). https://doi.org/10.1109/18.243446
6. Rosen, J.B.: Existence and uniqueness of equilibrium points for concave N-person games. Econometrica **33**(3), 520–534 (1965). https://doi.org/10.2307/1911749
7. Scutari, G., Palomar, D.P., Barbarossa, S.: Competitive design of multiuser MIMO systems based on game theory: a unified view. IEEE J. Sel. Areas Commun. **26**(7), 1089–1103 (2008). https://doi.org/10.1109/JSAC.2008.080907
8. Telatar, E.: Capacity of multi-antenna Gaussian channels. Eur. Trans. Telecommun. **10**(6), 585–595 (1999). https://doi.org/10.1002/ett.4460100604

A Game-Theoretical Approach for Energy Efficiency in Multiuser MIMO System

Hang Zou[1(⊠)], Chao Zhang[1,2], Samson Lasaulce[3], Lucas Saludjian[4], and Patrick Panciatici[4]

[1] L2S, CNRS-CentraleSupelec-Univ. Paris Saclay, Gif-sur-Yvette, France
{hang.zou,chao.zhang}@centralesupelec.fr
[2] Telecom Paristech, Palaiseau, France
[3] CRAN, Nancy, France
samson.lasaucle@centralesupelec.fr
[4] RTE, Paris, France

Abstract. The game-theoretic analysis of Energy Efficiency (EE) game is known to be difficult due to the non-convexity of EE-utility, especially for tracing the Nash Equilibrium in MIMO system. In this paper, The existence and the uniqueness of the Nash Equilibrium (NE) is affirmed for a MIMO multiple access channel (MAC) communication system and a bisection search algorithm is designed to find this unique NE. Despite being sub-optimal for deploying approximate best response, the policy found by the proposed algorithm is shown to be more efficient than the classical allocation techniques. Simulation shows that even the policy found by proposed algorithm might not be the exact NE of the game, the deviation w.r.t. to the exact NE is small and the resulted policy actually Pareto-dominates the unique NE of the game.

Keywords: Energy efficiency · MIMO · Nash equilibrium

1 Introduction

Energy Efficiency (EE) is one of key performance of the next generation network system (5G and beyond) due to the exponential increase of connected devices. However, a proportional increasing of the power resulted by the goal of increasing the transmission rate 1000 times than 4G will lead to an unimaginable energy demand. Therefore energy-efficient design of the wireless system draws both the attention of industrial and academic researchers. One of the pioneer works of studying the maximization of EE in MIMO system is [6]. In [6], the optimal precoding scheme is studied and divided into different cases with different assumptions on the systems. Till now the optimal precoding matrix for general condition is merely conjectured and unproved. Then It is later widely

This work was funded by the RTE-CentraleSupelec Chair on "Digital Transformation of Electricity Networks".

© Springer Nature Switzerland AG 2021
S. Lasaulce et al. (Eds.): NetGCooP 2021, CCIS 1354, pp. 8–16, 2021.
https://doi.org/10.1007/978-3-030-87473-5_2

realized that the problem of EE maximization actually belongs to the category of fractional programming. Techniques such as Dinkelbach's algorithm (see [8]) is used to solve EE maximization in [10, 11]. The main difficulty of EE maximization OP is usually due to the non-convexity of the fractional structure of the EE function. The situation is even worse involving the decentralized EE problems or EE games. The main difficulty lies in the traceability of NE due to the high non-convexity of EE-type utility function. To the best knowledge of authors, NE is rarely considered and studied in EE games, especially for MIMO System. In [4], it is shown that there always exists an unique NE for scalar power allocation game in a relay-assisted MIMO systems due to the standard property of the best response dynamics. Similar results will be given in a MAC-MIMO system latter.

The contribution of this paper is twofold: 1) we first extend the state-of-the-art work from a scalar power situation to a more general situation where each user is allowed to choose its covariance matrix to maximize its individual EE. The existence and uniqueness of the NE is proved under some assumptions. 2) An algorithm is proposed to find the unique NE of this MIMO-MAC game. When the number of antennas of transmitter is equal the one of receiver, proposed algorithm leads to exact NE. Otherwise it only leads to approximated solution for replacing the exact best response dynamic by its linear approximation.

Notations: $(\cdot)^H$ and $(\cdot)^\dagger$ denote matrix transpose and Moore-Penrose inverse respectively. $\mathbf{I}_N$ stands for identity matrix of size N. $\det(\cdot)$ and $\mathrm{Tr}(\cdot)$ denote the determinant and the trace of a matrix respectively. Denote the natural number set inferior or equal than N as $[N] \triangleq \{1, \ldots, N\}$.

2   System Model

Consider a multiple access channel (MAC) with one base station (BS) and K users (players) to be served. BS is equipped with N_r receive antennas and each user terminal is equipped with N_t transmit antennas. We assume a block fading channel where the realization of channel remains a constant during the coherence time of transmission and randomly generated according to some statistic distribution from period to period. The received signal at BS is given by:

$$ \boldsymbol{y} = \sum_{k=1}^{K} \mathbf{H}_k \boldsymbol{x}_k + \boldsymbol{z}, \tag{1} $$

where $\mathbf{H}_k \in \mathbb{C}^{N_r \times N_t}$ is the channel transmit matrix of k-th user. Each entry is assumed to be i.i.d. complex Gaussian distributed according to $\mathcal{CN}(0,1)$. $\boldsymbol{x}_k = (x_{k,1}, \ldots, x_{k,N_t})^T$ is the transmit symbol of k-th user and $\boldsymbol{z}$ is the noise observed by the receiver with complex Gaussian distribution $\mathcal{CN}(\mathbf{0}, \sigma^2 \mathbf{I}_{N_r})$. For the sake of simplicity, we assume that single user decoding is implemented for each user. Then the capacity achieved by the k-th user is

$$ R_k = \log \frac{\det \left(\sigma^2 \mathbf{I}_{N_r} + \sum_{j=1}^{K} \mathbf{H}_j \mathbf{Q}_j \mathbf{H}_j^H \right)}{\det \left(\sigma^2 \mathbf{I}_{N_r} + \sum_{j \neq k}^{K} \mathbf{H}_j \mathbf{Q}_j \mathbf{H}_j^H \right)}, \tag{2} $$

where $\mathbf{Q}_k = \mathbb{E}\left[\boldsymbol{x}_k \boldsymbol{x}_k^H\right] \in \mathbb{C}^{N_t \times N_t}$ is the covariance matrix of symbol $\boldsymbol{x}_k$ and $P_c > 0$ is the power dissipated in transmitter's circuit to operate the devices. It is reasonable to assume that each user has perfect knowledge about its own channel, e.g., through downlink pilot training. Therefore user k is able to perform the singular value decomposition (SVD) of its own channel $\mathbf{H}_k$ and its covariance matrix $\mathbf{Q}_k$: $\mathbf{H}_k = \mathbf{U}_k \mathbf{\Lambda}_k \mathbf{V}_k^H$ and $\mathbf{Q}_k = \mathbf{W}_k \mathbf{P}_k \mathbf{W}_k^H$ respectively. To simplify the problem, we assume that user k always adapts its covariance matrix to $\mathbf{H}_k$, i.e., choosing $\mathbf{W}_k = \mathbf{V}_k$. $\mathbf{P}_k$ is a diagonal matrix with $\mathbf{P}_k = \mathrm{diag}\left(\boldsymbol{p}_k\right) = \mathrm{diag}\left(p_{k1}, \ldots, p_{kN_t}\right)$ where we use $\mathrm{diag}\left(\cdot\right)$ to generate a diagonal matrix from a vector or vice versa. Thus user k's only legal action is represented by $\boldsymbol{p}_k$ or $\mathbf{P}_k$ and the action set of k-th user is $\mathcal{P}_k = \left\{\boldsymbol{p}_k \left| \sum_{i=1}^{N_t} p_{ki} \leq \overline{P}_k, \; p_{ki} \geq 0 \right.\right\}$ with $\overline{P}_k$ is power budget of k-th user. Further more, we denote $\boldsymbol{p} = \left(\boldsymbol{p}_k, \boldsymbol{p}_{-k}\right)$ with $\boldsymbol{p}_{-k} \triangleq \left(\boldsymbol{p}_1, \ldots, \boldsymbol{p}_{k-1}, \boldsymbol{p}_{k+1} \ldots, \boldsymbol{p}_K\right) \in \mathcal{P}_{-k}$ and $\mathcal{P}_{-k} \triangleq \mathcal{P}_1 \times \cdots \times \mathcal{P}_{k-1} \times \mathcal{P}_{k+1} \times \cdots \times \mathcal{P}_K$. In this paper energy efficiency defined as the ratio between a benefit function and the power consumed by producing it has the following expression for user k after some simplifications:

$$u_k\left(\mathbf{P}_k, \mathbf{P}_{-k}\right) = \frac{\log \frac{\det\left(\sigma^2 \mathbf{I}_{N_r} + \sum_{j=1}^K \mathbf{U}_j \mathbf{\Lambda}_j \mathbf{P}_j \mathbf{\Lambda}_j^H \mathbf{U}_j^H\right)}{\det\left(\sigma^2 \mathbf{I}_{N_r} + \sum_{j \neq k}^K \mathbf{U}_j \mathbf{\Lambda}_j \mathbf{P}_j \mathbf{\Lambda}_j^H \mathbf{U}_j^H\right)}}{\mathrm{Tr}\left(\mathbf{P}_k\right) + P_c} \tag{3}$$

To this end, the MIMO MAC EE game is thus given by the following strategic form in triplet:

$$\mathcal{G} = \left(\mathcal{K}, \left(\mathcal{P}_k\right)_{k \in \mathcal{K}}, \left(u_k\right)_{k \in \mathcal{K}}\right) \tag{4}$$

3   Main Results

Some important properties satisfied by individual utility function are resumed in the following proposition:

Proposition 1. R_k *is a concave functions w.r.t.* $\boldsymbol{p}_k$ *and* u_k *is a pseudo-concave (quasi-concave) function w.r.t.* $\boldsymbol{p}_k$ *for* $\forall k \in \mathcal{K}$*; For any fixed* $\boldsymbol{p}_{-k} \in \mathcal{P}_{-k}$ *and* p_{kj} *with* $j \neq i$*, only one of following statements is true for all* $i \in [N_t]$*:*

i) $\exists\, p_{ki}^\star > 0$ s.t. u_k is an increasing function in $(0, p_{ki}^\star)$ and a decreasing function in $(p_{ki}^\star, +\infty)$ w.r.t. p_{ki}.
ii) u_k is a decreasing function in $(0, +\infty)$ w.r.t. p_{ki}.

Proof. The proof is omitted here for lack of space. For more details, see [1,3].

Before stating the best response dynamics of the game, we define the following boundary of set $\mathcal{P}_k$ indicated by an index subset $\mathcal{E} \subset [N_t]$: $\mathcal{P}_k[\mathcal{E}] \triangleq \{\boldsymbol{p}_k \in \mathcal{P}_k, p_{ki} = 0 \text{ for } i \in \mathcal{E}\}$ and the non-positive index set for a given action $\mathbf{P}_k$: $\mathcal{I}\left(\mathbf{P}_k\right) \triangleq \{i \in [N_t] \text{ s.t. } p_{ki} \leq 0\}$. We having the following proposition confirming the existence of NE in this game:

Proposition 2. *For any given* $\mathbf{P}_{-k}$ *and provided that the power budget* $\overline{P}_k$ *is sufficiently large, denote the unique solution of the following equation as* $\mathbf{P}_k^*$:

$$diag\left(\boldsymbol{\Lambda}_k^H\left(\boldsymbol{\Lambda}_k\mathbf{P}_k\boldsymbol{\Lambda}_k^H+\mathbf{F}_k+\sigma^2\mathbf{I}_r\right)^{-1}\boldsymbol{\Lambda}_k\right)=u_k\left(\mathbf{P}_k,\mathbf{P}_{-k}\right)\mathbf{I}_{N_t} \qquad (5)$$

with $\mathbf{F}_k=\sum_{j\neq k}\mathbf{S}_{j,k}\mathbf{P}_j\mathbf{S}_{j,k}^H$ *is the interference matrix of k-th user with* $\mathbf{S}_{j,k}=\mathbf{U}_k^H\mathbf{U}_j\boldsymbol{\Lambda}_j$. *Then the BR of* $\mathbf{P}_k$ *w.r.t.* $\mathbf{P}_{-k}$ *is standard and converges to the unique NE admitted by game (4); The BR is the unique solution of (5) restricted to the boundary of* $\mathcal{P}_k$ *indicated by* $\mathcal{I}\left(\mathbf{P}_k^*\right)$.

Proof. The proof is omitted here for lack of space. For more details, see [1,3].

4   Algorithm for Finding NE

Proposition 2 actually provides an approach for us to find the NE of the game (4). One can easily deduce an iterative equation according to (5):

$$\mathrm{diag}\left(\boldsymbol{\Lambda}_k^H\left(\boldsymbol{\Lambda}_k\mathbf{P}_k^{(t)}\boldsymbol{\Lambda}_k^H+\mathbf{F}_k^{(t-1)}+\sigma^2\mathbf{I}_r\right)^{-1}\boldsymbol{\Lambda}_k\right)=u_k\left(\mathbf{P}_k^{(t-1)},\mathbf{P}_{-k}^{(t-1)}\right)\mathbf{I}_{N_t} \quad (6)$$

However, due to Proposition 2, this stationary point might not be in the feasible action set. Nevertheless, one can design the following basic algorithm to find NE of the game (4) based on Proposition 2 summarized in Algorithm 1. However, Algorithm 1 is not a satisfatory way to find the NE of the game.

Algorithm 1. Basic Algorithm for finding NE of MAC-MIMO EE game

---

Initialization: $\mathbf{P}_k^{(0)}=\frac{1}{N_t}\mathbf{I}_{N_t},\forall k$. Choose T and ϵ
For $t=1$ to T, **do**
  For $k=1$ to K, **do**
    Compute $\mathbf{P}_k^{(t)}$ using (6)
    If $\mathcal{I}\left(\mathbf{P}_k^{(t)}\right)\neq\varnothing$
      Compute $\mathbf{P}_k^{(t)}$ using (6) restricted to $\mathcal{I}\left(\mathbf{P}_k^{(t)}\right)$
    End If
  End For
  If $\sum_k\left\|\mathbf{P}_k^{(t)}-\mathbf{P}_k^{(t-1)}\right\|<\epsilon$
    Break
  End If
End For
Output: $\mathbf{P}_k^{\mathrm{NE}}=\mathbf{P}_k^{(t)}$ for $\forall k$.

More precisely, to find the BR for given $\mathbf{P}_{-k}$, one actually need to solve an optimization problem. However, if $h = U(\mathbf{P}_{-k}) = \max_{\mathbf{P}_k \in \mathcal{P}_k} u_k(\mathbf{P}_k, \mathbf{P}_{-k})$ is known as *a priori* information, (6) can be transformed into following equation:

$$\text{diag}\left(\mathbf{\Lambda}_k^H \left(\mathbf{\Lambda}_k \mathbf{P}_k^{(t)} \mathbf{\Lambda}_k^H + \mathbf{F}_k^{(t-1)} + \sigma^2 \mathbf{I}_r\right)^{-1} \mathbf{\Lambda}_k\right) = h\mathbf{I}_{N_t} \tag{7}$$

Introducing an auxiliary parameter h, one obtains an iterative equation of $\mathbf{P}_k$. Without loss of generality, we assume that the solution of (6) belongs to the feasible action set for given $\mathbf{P}_{-k}$. Otherwise, similar analysis can applied for $\mathbf{P}_k$ but restricted on a boundary given by Proposition 2. For the sake of simplicity, we omit the discussion here and restrict ourselves to the situation where the BR is strictly included in the interior of the feasible action set. Therefore for all $i \in [N_t]$, there exists $p_{ki}^\star$ such that individual utility function $u_k(\mathbf{P}_k, \mathbf{P}_{-k})$ is an increasing function in $(0, p_{ki}^\star)$ and a decreasing function in $(p_{ki}^\star, +\infty)$ with respect to p_{ki}, where $p_{ki}^\star$ is the i-th component of user k's BR for given $\mathbf{P}_{-k}$. Then u_k is also an increasing function in $(0, U(\mathbf{P}_{-k}))$ and a decreasing function in $(U(\mathbf{P}_{-k}), +\infty)$ w.r.t. parameter h. In other words, to find $\mathbf{P}_k = \text{BR}(\mathbf{P}_{-k})$, it is sufficient to find $U(\mathbf{P}_{-k})$ by a bisection search due to the special monotonicity of the utility function.

However, it is worth mentioning that it is still difficult to directly find the solution of iterative Eq. (7). Because this solution is actually implicitly given. We would like to further simplify (7) to facilitate the calculation of BR or NE. To start with, we assume that $N_t = N_r$. Firstly, we remove the diagonal operator of LHS of (7). Therefore we have:

$$\mathbf{P}_k^{(t)} = \frac{1}{h}\mathbf{I}_{N_t} - \mathbf{\Lambda}_k^{-1}\left(\mathbf{F}_k^{(t-1)} + \sigma^2 \mathbf{I}_{N_r}\right)\mathbf{\Lambda}_k^{-1} \tag{8}$$

If $N_t > N_r$ or $N_t < N_r$ then $\mathbf{\Lambda}_k$ is not directly invertible, then we should consider the pseudo-inverse matrix of $\mathbf{\Lambda}_k$. Without loss of generality, we assume that $N_t > N_r$, denoting the right pseudo-inverse of $\mathbf{\Lambda}_k$ as $\mathbf{\Lambda}_k^\dagger$, one thus obtain:

$$\mathbf{\Lambda}_k^H \left(\mathbf{\Lambda}_k \mathbf{P}_k^{(t)} \mathbf{\Lambda}_k^H + \mathbf{F}_k^{(t-1)} + \sigma^2 \mathbf{I}_r\right)^{-1} \mathbf{\Lambda}_k = h\mathbf{I}_{N_t}$$

$$\left(\mathbf{\Lambda}_k \mathbf{P}_k^{(t)} \mathbf{\Lambda}_k^H + \mathbf{F}_k^{(t-1)} + \sigma^2 \mathbf{I}_r\right)^{-1} = h\left(\mathbf{\Lambda}_k^\dagger\right)^H \mathbf{\Lambda}_k^\dagger \tag{9}$$

However, it is generally impossible to have $\mathbf{\Lambda}_k^\dagger \mathbf{\Lambda}_k = \mathbf{I}_{N_t}$. Thus the equality does not always holds when we multiply $\mathbf{\Lambda}_k^\dagger$ on left and $\left(\mathbf{\Lambda}_k^\dagger\right)^H$ on the right on both sides of the equation. Nevertheless, this operation will yield a linear approximation of the BR dynamics:

$$\widehat{\mathbf{P}}_k^{(t)} = \frac{\mathbf{\Lambda}_k^\dagger \left[\left(\mathbf{\Lambda}_k^\dagger\right)^H \mathbf{\Lambda}_k^\dagger\right]^{-1} \left(\mathbf{\Lambda}_k^\dagger\right)^H}{h} - \mathbf{\Lambda}_k^\dagger \left(\mathbf{F}_k^{(t-1)} + \sigma^2 \mathbf{I}_{N_r}\right)\left(\mathbf{\Lambda}_k^\dagger\right)^H \tag{10}$$

Similarly, if $N_t < N_r$ we can obtain exactly same iterative equation as (10). This type of dynamics belongs to the so-called ε-approximate best response. To this end, we obtain a sub-optimal algorithm summarized in Algorithm 2 by using the iterative equation deduced in (10) instead of using (6).

Algorithm 2. Bisection Search Algorithm for find the NE of MAC-MIMO EE game

---

Initialization: $\mathbf{P}_k^{(0)} = \frac{1}{N_t}\mathbf{I}_{N_t}, \forall k.$ choose T, ϵ_1 and ϵ_2
For $t = 1$ to T, **do**
  For $k = 1$ to K, **do**
    Initialization: $\underline{h} = 0$ and $\overline{h} = h_{max}$
    Repeat Until $\overline{h} - \underline{h} \leq \epsilon_1$
    $h_M = \frac{\underline{h}+\overline{h}}{2}$, $h_L = \max\left\{0, h_M - \frac{\epsilon_1}{2}\right\}$ and $h_R = \min\left\{h_{max}, h_M + \frac{\epsilon_1}{2}\right\}$
    Compute $\mathbf{P}_k(h_i)$ using (10), $i \in \{L, M, R\}$
    $U_i = u_k\left(\mathbf{P}_k(h_i), \mathbf{P}_{-k}^{(t-1)}\right), i \in \{L, M, R\}$
    If $U_L < U_M < U_R$
      $\underline{h} = h_L$
      Else If $U_L > U_M > U_R$
        $\overline{h} = h_R$
      End If
    Else
      $\underline{h} = h_L$ and $\overline{h} = h_R$
    End If
    Compute $\mathbf{P}_k^{(t)}$ by (10) with $h = h_M$
  End For
  If $\sum_k \left\|\mathbf{P}_k^{(t)} - \mathbf{P}_k^{(t-1)}\right\| < \epsilon_2$
    Break
  End If
End For
Output: $\mathbf{P}_k^{\mathrm{NE}} = \mathbf{P}_k^{(t)}$ for $\forall k$.

---

5 Numeric Results

The goal of this part is to show the performance of the proposed algorithms. Notice if $N_t = N_r$, (10) degenerates to (8) which conserves the optimality of best response. For this situation, we choose $N_t = N_r = 2$ with $K = 2$ users. A sufficient large power budget is chosen such that the BR is included in the feasible action set $\overline{P}_k = 10\,\mathrm{mW}$ for $\forall k \in \{1, 2\}$ and the circuit power is $P_c = 1\,\mathrm{mW}$. The error tolerated for Algorithm 2 is $\epsilon_1 = \epsilon_2 = 0.001$. In Fig. 1, the achievable utility region, the average performance under NE found by Algorithm 2 and the averaged performance achieved by uniform power allocation (UPA) are depicted. All results are averaged over 1000 randomly generated channel samples. It is observed that the performance achieved by deploying uniform power allocation (UPA) is Pareto-dominated by NE which can be found by Algorithm 2. Furthermore, the NE found by Algorithm 2 is close to the Pareto frontier achieved by some centralized algorithms which suggest the efficiency using Algorithm 2 is higher than UPA.

Moreover, define the social welfare for a given action profile as $w(\boldsymbol{p}) = \sum_{k\in\mathcal{K}} u_k(\boldsymbol{p}_k, \boldsymbol{p}_{-k})$. Then the average social welfare as function of the power budget of user in Fig. 2 respectively. There are two different regions for social

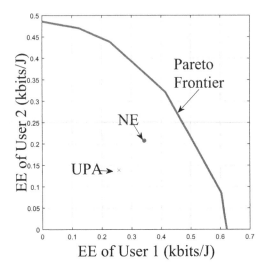

Fig. 1. Energy Efficiency under NE and uniform power allocation with $N_t = N_r = 2$ for 2-user situation. Policy found by our algorithms outperforms than UPA policy.

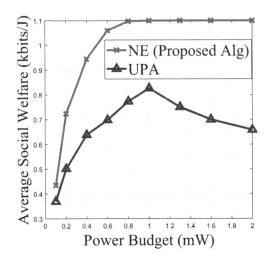

Fig. 2. Performance under NE and UPA as function of the power budget of user with $N_t = N_r = 2$ for 2-user situation. There are two different regions: one corresponds to Proposition 2. In the region uncovered by Proposition 2, proposed algorithm still dominates UPA.

welfare. In the first region where the power budget is sufficiently large, the NE found by our proposed algorithm is independent of the power budget while the performance of UPA is decreasing w.r.t. the increase of the power budget. In the second region where Proposition 2 is no more valid in this region. Nevertheless, the performance achieved by our algorithm is still better than UPA. Then a more probable situation is considered where $N_t < N_r$ meaning that the number of antennas in user terminal is less than the one in base station. The discussion in Sect. 4 shows that the proposed suboptimal algorithm is actually suboptimal due to the usage of ε-approximate best response. For numeric demonstration, we choose $N_t = 2 < N_r = 4$. The performance of Algorithm 2 is illustrated in Fig. 3. The sub-optimality is clearly demonstrated in this figure. However, the resulted policy actually Pareto-dominates the exact NE found by Algorithm 1 and the dispersion is relatively small in terms of average performance. This remark entails that even the policy found by Algorithm 2 is not the NE of the game in its sub-optimal region however its performance does slightly outperforms the exact NE.

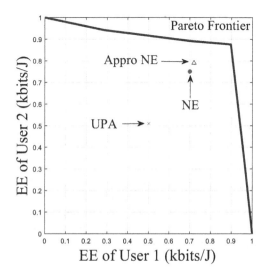

Fig. 3. Performance achieved by Algorithm 1 (NE) and Algorithm 2 (Approximate NE) and UPA with $N_t = 2$ and $N_r = 4$ for 2-user situation. Approximated solution found by Algorithm 2 is very near to the exact NE and Pareto-dominates it. Moreover, two policies found by proposed algorithms both outperform UPA.

6   Conclusions

In this paper, a game where the individual utility function is the energy efficiency in a MIMO multiple access channel system is considered. The existence and the uniqueness of Nash Equilibrium is proved and an exact algorithm and a sub-optimal algorithm is proposed to find the NE of this game. Simulation results show that performance under NE found by proposed algorithms is always better than uniform power allocation policy for both inside or outside the range covered by the main proposition of the paper. Besides, our sub-optimal algorithm by deploying an ε-approximate best response yields surprisingly a policy Pareto-dominates the exact NE of the game. Other techniques such as pricing might be useful to improve the efficiency of the overall system. The situation where each user is allowed to freely choose its covariance matrix merely constrained to the maximum power is the natural extension of this paper. Moreover, the discussion over the effect of successive interference cancellation and multiple carrier seems to be complicated and serve as the challenge of the future works.

References

1. Zou, H., Zhang, C., Lasaulce, S., Saludjian, L., Panciatici, P.: Energy-efficient MIMO multiuser systems: nash equilibrium analysis. In: Habachi, O., Meghdadi, V., Sabir, E., Cances, J.-P. (eds.) UNet 2019. LNCS, vol. 12293, pp. 68–81. Springer, Cham (2020). https://doi.org/10.1007/978-3-030-58008-7_6
2. Zappone, A., Björnson, E., Sanguinetti, L., Jorswieck, E.: Globally optimal energy-efficient power control and receiver design in wireless networks. IEEE Trans. Signal Process. **65**, 2844–2859 (2017)
3. Zappone, A., Jorswieck, E.: Energy efficiency in wireless networks via fractional programming theory. Found. Trends® Commun. Inform. Theory **11**(3–4), 185–396 (2015)
4. Zappone, A., Chong, Z., Jorswieck, E.: Energy-aware competitive power control in relay-assisted interference wireless networks. IEEE Trans. Wirel. Commun. **12**, 1860–1871 (2013)
5. Zhang, C., Lasaulce, S., Agrawal, A., Visoz, R.: Distributed power control with partial channel state information: performance characterization and design. IEEE Trans. Veh. Technol. **68**(9), 8982–8994 (2019)
6. Belmega, E.V., Lasaulce, S.: Energy-efficient precoding for multiple-antenna terminals. IEEE Trans. Signal Process. **59**(1), 329–340 (2011)
7. Debreu, G.: A social equilibrium existence theorem. Natl. Acad. Sci. **38**, 886–893 (1952)
8. Dinkelbach, W.: On nonlinear fractional programming. Manage. Sci. **13**(7), 492–498 (1967)
9. Varma, V., Lasaulce, S., Debbah, M., Elayoubi, S.: An energy efficient framework for the analysis of MIMO slow fading channels. IEEE Trans. Signal Process. **61**(10), 2647–2659 (2013)
10. Raghavendra, P.S., Daneshrad, B.: An energy-efficient water-filling algorithm for OFDM systems. In: 2010 IEEE International Conference on Communications (ICC), pp. 1–5, May 2010
11. Li, Y., Jiang, T.: Fixed-point algorithms for energy-efficient power allocation in spectrum-sharing wireless networks. In: IEEE Global Communications Conference (GLOBECOM), pp. 1–6 (2016)
12. Du, B., Pan, C., Zhang, W., Chen, M.: Distributed energy-efficient power optimization for CoMP systems with max-min fairness. IEEE Commun. Lett. **18**(6), 999–1002 (2014)
13. Nash, J.: Equilibrium points in n-person game. Proc. Natl. Acad. Sci. **36**(1), 48–49 (1950)
14. Belmega, E.V., Lasaulce, S., Debbah, M.: Power allocation games for MIMO multiple access channels with coordination. IEEE Trans. Wirel. Commun. **8**(6), 3182–3192 (2009)

Derivative-Free Optimization over Multi-user MIMO Networks

Olivier Bilenne$^{1(\boxtimes)}$, Panayotis Mertikopoulos^1, and E. Veronica Belmega^2

^{1} Univ. Grenoble Alpes, CNRS, Inria, Grenoble INP, LIG, 38000 Grenoble, France
olivier.bilenne@inria.fr, panayotis.mertikopoulos@imag.fr
^{2} ETIS UMR 8051, CY University, ENSEA, CNRS, 95000 Cergy, France
belmega@ensea.fr

Abstract. In wireless communication, the full potential of multiple-input multiple-output (MIMO) arrays can only be realized through optimization of their transmission parameters. Distributed solutions dedicated to that end include iterative optimization algorithms involving the computation of the gradient of a given objective function, and its dissemination among the network users. In the context of large-scale MIMO, however, computing and conveying large arrays of function derivatives across a network has a prohibitive cost to communication standards. In this paper we show that multi-user MIMO networks can be optimized without using any derivative information. With focus on the throughput maximization problem in a MIMO multiple access channel, we propose a "derivative-free" optimization methodology relying on very little feedback information: a single function query at each iteration. Our approach integrates two complementary ingredients: exponential learning (a derivative-based expression of the mirror descent algorithm with entropic regularization), and a single-function-query gradient estimation technique derived from a classic approach to derivative-free optimization.

Keywords: Derivative-free optimization · Zeroth-order optimization · Exponential learning · MIMO systems · Throughput maximization · SPSA

1 Introduction

The appeal of multiple-input and multiple-output (MIMO) technologies in wireless communication is their ability to increase throughputs significantly and to improve the systems' robustness to ambient noise and channel fluctuations [1,7]. On this account, large-scale deployment of *multiple-input and multiple-output*

The authors are grateful for financial support from the French National Research Agency (ANR) projects ORACLESS (ANR–16–CE33–0004–1) and ELIOT (ANR–18–CE40–0030 and FAPESP 2018/12579–7). The research of P. Mertikopoulos has also received financial support from the COST Action CA 16228 'European Network for Game Theory' (GAMENET).

© Springer Nature Switzerland AG 2021
S. Lasaulce et al. (Eds.): NetGCooP 2021, CCIS 1354, pp. 17–24, 2021.
https://doi.org/10.1007/978-3-030-87473-5_3

(MIMO) terminals is perceived as one of the key enabling technologies for next-generation wireless networks.

Releasing the full potential of large MIMO arrays requires, however, a principled approach to optimization, with the aim of minimizing computational overhead and related expenditures.

An essential aspect of the emblematic *throughput maximization* problem resides in the optimization of MIMO transmission parameters (such as the users' signal covariance matrices) [2,5,15,16,18]. In multi-user networks, conventional optimization methods involve the use of water-filling (WF) techniques [13,14,18], which invariably rely on the availability of perfect channel state information at the transmitter (CSIT), and are vulnerable to observation noise, asynchronicities, and other operational impediments that arise in real-world networks.

More recently proposed in [12] as an alternative to water-filling, the *matrix exponential learning* (MXL) algorithm proceeds incrementally by combining (stochastic) gradient steps with a matrix exponential mapping that ensures feasibility of the users' signal covariance variables. In so doing, MXL guarantees fast convergence in cases where WF methods demonstrably fail. On the negative side, an important implementation bottleneck of MXL is the requirement to (*i*) invert a relatively large matrix at the receiver; and (*ii*) broadcast the resulting matrix to all connected users[1]. In consequence, the computation and communication overhead of MXL quickly becomes prohibitive in larger MIMO systems.

In this paper, we focus on the problem (stated in Sect. 2) of throughput maximization in a MIMO multiple access channel (MAC), with the objective to overcome the above limitations of the MXL by means of zeroth-order optimization, i.e., by making no gradient computations whatsoever. Following a classic approach from the *simultaneous perturbation stochastic approximation* (SPSA) framework [6,17], we devise in Sect. 3 a "gradient-free" optimization algorithm by plugging into the chassis of the original MXL method a gradient estimator based no longer on first-order feedback but on function queries (a single one at each iteration). Our developments are followed by a discussion on the performances and potential of gradient-free matrix exponential learning (Sect. 4).

Notation. We use bold capital letters for matrices, saving the letters k, l for user assignments and t, s for time indices, so that e.g., matrix $\mathbf{Q}_k$ relates to user k, $\mathbf{Q}_t$ to time t, and $\mathbf{Q}_{k,t}$ to user k at time t.

2 Problem Statement

Consider a MIMO network where K users are transmitting simultaneously to a wireless receiver equipped with N antennas over a shared Gaussian vector MAC, modeled by

$$\mathbf{y} = \sum_{k=1}^{K} \mathbf{H}_k \mathbf{x}_k + \mathbf{z},$$

---

[1] In a MIMO array with $N = 128$ receive antennas, this would correspond to transmitting approximately 65 kB of data per frame, thus exceeding typical frame size limitations by a factor of $50\times$ to $500\times$ depending on the specific standard [9].

where $\mathbf{y} \in \mathbb{C}^N$ is the signal at the reception, M_k, $\mathbf{x}_k \in \mathbb{C}^{M_k}$ and $\mathbf{H}_k \in \mathbb{C}^{N \times M_k}$ respectively denote the number of antennas, the transmitted messages and the channel matrix of user k $(k = 1, \ldots, K)$, and $\mathbf{z} \in \mathbb{C}$ models additive zero-mean Gaussian noise with unit covariance. Without loss of generality, we assume every user to possess at least two antennas $(M_k \geq 2)$. Let P_k be the maximum mean power consumption of user k due to transmissions, and let

$$\mathbf{Q}_k = \frac{1}{P_k} \mathbb{E}[\mathbf{x}_k \mathbf{x}_k^\dagger]$$

denote the normalized covariance matrix of $\mathbf{x}_k$. By definition, the matrix $\mathbf{Q}_k$ is Hermitian—we write $\mathbf{Q}_k \in \mathrm{Herm}\,(M_k)$—and positive semidefinite.

Our goal is to maximize, under the maximum available transmit power constraint $\mathrm{tr}(\mathbf{Q}_k) \leq 1$ for $k = 1, \ldots, K$, the achievable sum rate under successive interference cancellation (SIC),

$$R(\mathbf{Q}) = \log \det \left(\mathbf{I} + \sum_{k=1}^K P_k \, \mathbf{H}_k \mathbf{Q}_k \mathbf{H}_k^\dagger \right) \tag{1}$$

where the aggregate form $\mathbf{Q} = (\mathbf{Q}_k, \ldots, \mathbf{Q}_K)$ contains all the unknowns of the problem. Since the maximum sum rate is achieved at a boundary point $\mathbf{Q}$ where $\mathrm{tr}(\mathbf{Q}_1) = \cdots = \mathrm{tr}(\mathbf{Q}_K) = 1$, the search domain of the problem is confined to the Cartesian product set $\mathcal{Q} = \mathcal{Q}_1 \times \cdots \times \mathcal{Q}_K$, where

$$\mathcal{Q}_k = \{\mathbf{Q}_k \in \mathrm{Herm}\,(M_k) : \mathrm{tr}(\mathbf{Q}_k) = 1, \mathbf{Q}_k \succeq 0\}$$

is a compact subset of a d_k-dimensional real subspace, with $d_k = M_k^2 - 1 > 0$ for every user k.

The throughput maximization problem can be stated as the convex program:

$$\begin{aligned} &\text{maximize } R(\mathbf{Q}) \\ &\text{subject to } \mathbf{Q} \in \mathcal{Q}. \end{aligned} \tag{RM}$$

The structure of the feasible set $\mathcal{Q}$ makes the problem amenable to parallel optimization settings where (RM) is regarded as a collection of K sub-problems

$$\begin{aligned} &\text{maximize } R(\mathbf{Q}_k; \mathbf{Q}_{-k}) \\ &\text{subject to } \mathbf{Q}_k \in \mathcal{Q}_k \end{aligned} \tag{RM$_k$}$$

to be solved in parallel by the users. Equivalently, (RM$_k$) can be interpreted as maximizing the achievable transmission rate of user k when single-user decoding (SUD) is performed at the receiver,

$$R_k(\mathbf{Q}_k; \mathbf{Q}_{-k}) := R(\mathbf{Q}) - R(\mathbf{Q}_1, \ldots, \mathbf{Q}_{k-1}, 0, \mathbf{Q}_{k+1}, \ldots, \mathbf{Q}_K), \tag{2}$$

given the covariance matrices of the remaining users, thus regarding the interference due to the signals sent by other users as colored noise. Since the achievable sum rate (1) is a concave *potential function* for the game defined by (2), the solutions of (RM) are the solutions of the Nash equilibrium problem defined by (2), i.e. any solution $\mathbf{Q}^\star$ of (RM) satisfies, for $k = 1, \ldots, K$,

$$R_k(\mathbf{Q}_k^\star; \mathbf{Q}_{-k}^\star) \geq R_k(\mathbf{Q}_k; \mathbf{Q}_{-k}^\star) \quad \forall \mathbf{Q}_k \in \mathcal{Q}_k, \tag{NE}$$

and conversely. In other words, maximizing the achievable sum rate under SIC is equivalent to equilibrating the individual transmission rates (2) under SUD.

Many optimization methods rely on derivative information. Differentiation of the achievable sum rate (1) gives us the gradient $\nabla R = (\nabla_1 R, \ldots, \nabla_K R)$ where, for $k = 1, \ldots, K$,

$$\nabla_k R(\mathbf{Q}) = P_k \mathbf{H}_k^\dagger \left[\mathbf{I} + \sum_{l=1}^{K} P_l \, \mathbf{H}_l \mathbf{Q}_l \mathbf{H}_l^\dagger \right]^{-1} \mathbf{H}_k. \tag{3}$$

Making the derivatives $\nabla_k R(\mathbf{Q})$ available to the users implies the inversion of the $N \times N$ Hermitian matrix $\mathbf{I} + \sum_{l=1}^{K} P_l \, \mathbf{H}_l \mathbf{Q}_l \mathbf{H}_l^\dagger$ at the receiver, followed by the broadcast of the result towards the users, which then are able to compute (3) locally. On account that the communication overhead induced by the dissemination of the gradient may be prohibitive, we proceed under the assumption that the gradient is not accessible to the users, which instead are required to compute their own estimates of $\nabla_k R$, based no longer on derivative information but on mere measurements of $R(\mathbf{Q})$.

3   Derivative-Free Matrix Exponential Learning

3.1   The MXL Algorithm

Among the existing (derivative-based) methods of solution for (RM) is the *matrix exponential learning* (MXL) [10], which in our developments will serve both as reference and as a starting point. We refer to [3,11] for a characterization of the MXL algorithm as an instance of the mirror descent algorithm implemented with the *von Neumann relative entropy* for Bregman divergence. Given an initial point $\mathbf{Y}_0 = \mathbf{Y}_1 = (0, \ldots, 0)$ in the space of the gradients $\mathcal{Q}^* = \mathcal{Q}_1^* \times \cdots \times \mathcal{Q}_K^*$, where $\mathcal{Q}_k^* = \{ \mathbf{Y}_k \in \mathrm{Herm}\,(M_k) : \mathrm{tr}(\mathbf{Y}_k) = 0 \}$, the t-th step of the algorithm is defined for $t \geq 1$ by

$$\begin{aligned} \mathbf{Q}_t &= \mathbf{\Lambda}(\mathbf{Y}_t), \\ \mathbf{Y}_{t+1} &= \mathbf{Y}_t + \gamma_t \hat{\mathbf{V}}_t, \end{aligned} \tag{MXL}$$

where $\{\mathbf{Q}_t\}$ denotes the issued sequence of estimates for the optimal configuration, $\{\mathbf{Y}_t\}$ is a sequence generated in the space of the gradients, $\{\gamma_t\}$ is a sequence of positive step-sizes, $\hat{\mathbf{V}}_t = (\hat{\mathbf{V}}_{1,t}, \ldots, \hat{\mathbf{V}}_{K,t}) \in \mathcal{Q}^*$ is an estimate of the gradient $\nabla R(\mathbf{Q}_t)$, and we set $\mathbf{\Lambda} = (\mathbf{\Lambda}_1, \ldots, \mathbf{\Lambda}_K)$, where the exponential learning mapping $\mathbf{\Lambda}_k$ is defined by

$$\mathbf{\Lambda}_k(\mathbf{Y}_k) = \frac{\exp(\mathbf{Y}_k)}{\mathrm{tr}(\exp(\mathbf{Y}_k))},$$

in which exp denotes the (matrix) complex exponential function.

In contrast to the available implementations of MXL, which rely on full/noisy [10] or partial [8] gradient feedback for the computation of $\hat{\mathbf{V}}_t$, the gradient estimates $\hat{\mathbf{V}}_t$ in this work are derived without gradient information, as explained in Sect. 3.2.

3.2 Derivative-Free MXL

Description of the the Gradient-Free MXL. Our developments build on an early approach to derivative-free optimization [6,17] which, in time, has been seen as the cornerstone to the field of simultaneous perturbation stochastic approximation (SPSA). After translation into our distributed, Hermitian setting, the SPSA approach can be described as follows.

In the absence of any gradient feedback, each user k infers an estimate $\hat{\mathbf{V}}_k \approx \nabla_k R$ of their individual gradient, derived from randomized queries of the sum rate R in the close neighborhood of the current iterate. For $k = 1, \ldots, K$, let $\rho_k > 0$ and $\mathbf{C}_k \in \mathcal{Q}_k$ such that the ball $\mathbf{C}_k + \rho_k \mathbb{B}_{d_k}$ is entirely contained by $\mathcal{Q}_k$. Concretely, each user k draws randomly, uniformly, and independently, a matrix $\mathbf{Z}_k$ on the sphere $\mathbb{S}_{d_k-1} = \{\mathbf{Z}_k \in \mathcal{Q}_k^* : \|\mathbf{Z}_k\|_2 = 1\}$ living in the d_k-dimensional space $\mathcal{Q}_k^*$, and we let $\mathbf{Z} = (\mathbf{Z}_1, \ldots, \mathbf{Z}_K)$ aggregate the random matrices of all users. The gradient estimator for user $k = 1, \ldots, K$ is then defined as

$$\hat{\mathbf{V}}_k(\mathbf{Q}) = \frac{d_k}{\delta} R(\hat{\mathbf{Q}}) \, \mathbf{Z}_k, \qquad \text{(SPSA)}$$

where $\hat{\mathbf{Q}} = (\hat{\mathbf{Q}}_1, \ldots, \hat{\mathbf{Q}}_K)$, and each test matrix

$$\hat{\mathbf{Q}}_k = \mathbf{Q}_k + \frac{\delta}{\rho_k}(\mathbf{C}_k - \mathbf{Q}_k) + \delta \mathbf{Z}_k \qquad (4)$$

is derived from $\mathbf{Q}_k$ after deviation by random quantity $\delta \mathbf{Z}_k$, and prior shrinking of $\mathcal{Q}_k$ so as to keep the test configuration $\hat{\mathbf{Q}}$ inside the feasible set. The presence in (SPSA) of the factor $d_k = M_k^2 - 1$ can be explained as the ratio between the volumes of the sphere $\mathbb{S}_{d_k-1}$ (where $\mathbf{Z}_k$ is picked) and the containing ball $\mathbb{B}_{d_k} = \{\mathbf{Z}_k \in \mathcal{Q}_k^* : \|\mathbf{Z}_k\|_2 \leq 1\}$.

The distinguishing property of (SPSA) lies in that the *bias* of the gradient estimator can be controlled by the parameter δ as this bias is uniformly bounded over $\mathcal{Q}$:

$$\|\mathbb{E}[\hat{\mathbf{V}}_k(\mathbf{Q}, \mathbf{Z}; \rho) - \nabla_k R(\mathbf{Q})]\|_* = O(\delta) \qquad (5)$$

Besides, the *norm* of the gradient estimator (SPSA) satisfies

$$\|\hat{\mathbf{V}}_k(\mathbf{Q}, \mathbf{Z}; \rho)\|_* = O\left(\frac{1}{\delta}\right) \qquad (6)$$

uniformly on $\mathcal{Q}$. Equations (5) and (6) thus unveil a tradeoff between the $O(\delta)$ bias of the estimator and its $O(\frac{1}{\delta})$ deviation from the true derivative. This bias–variance tradeoff induces in the present context strict restrictions on the choice of the query radius δ and of the step-size policy of the MXL algorithm, with

---

Algorithm 1: Gradient-free MXL

    Parameters : $\{\gamma_t\}_{t=1}^{\infty}$, $\{\delta_t\}_{t=1}^{\infty}$
    Init.: $t \leftarrow 1$, $\mathbf{Y} \leftarrow 0$, $\forall k$: **transmit with** $\mathbf{Q}_k \leftarrow \frac{P_k}{M_k}\mathbf{I}_k$

1: **Repeat until stopping criterion is reached**
2:     **For** $k \in \{1,\ldots,K\}$ **do in parallel**
        Sample $\mathbf{Z}_k$ **uniformly in** $\mathbb{S}_{d_k-1}$
        Transmit with $\hat{\mathbf{Q}}_k \leftarrow \mathbf{Q}_k + \frac{\delta_t}{\rho_k}(\mathbf{C}_k - \mathbf{Q}_k) + \delta_t\mathbf{Z}_k$
        Receive feedback $r \leftarrow R(\hat{\mathbf{Q}})$
        $\hat{\mathbf{V}}_k \leftarrow \left(\frac{M_k^2-1}{\delta_t}\right) r\,\mathbf{Z}_k$
        $\mathbf{Y}_k \leftarrow \mathbf{Y}_k + \gamma_t\hat{\mathbf{V}}_k$
        $\mathbf{Q}_k \leftarrow \mathbf{\Lambda}_k(\mathbf{Y}_k)$
3:     $t \leftarrow t+1$

---

consequences on the performance of the algorithm, as discussed towards the end of the section.

See Algorithm 1 for a pseudocode description of the gradient-free optimization algorithm obtained after combining MXL with (SPSA). Given a (typically non-increasing) query radius sequence $\{\delta_t\}$ and a step-size sequence $\{\gamma_t\}$, the task of user k at time step t consists of (i) sampling a random direction $\mathbf{Z}_{k,t} \in \mathcal{Q}_k^*$, (ii) implementing the test covariance matrix $\hat{\mathbf{Q}}_{k,t}$ obtained as in (4) by variation of the current covariance estimate $\mathbf{Q}_{k,t}$, (iii) receiving the value of the achievable total transmission rate $R(\hat{\mathbf{Q}}_t)$, (iv) inferring an estimate $\hat{\mathbf{V}}_{k,t}$ of the gradient along user direction k, and (v) updating $\mathbf{Y}_{k,t}$ and $\mathbf{Q}_{k,t}$ in accordance with (MXL).

Convergence of the Gradient-Free MXL. The convergence of the gradient-free version of MXL is guaranteed with probability 1 on condition that the implementation parameters are chosen with care [3,4]. Indeed, if Algorithm 1 is implemented with non-increasing step-size and query-radius policies satisfying the conditions

$$\text{(a) } \sum_{t=1}^{\infty}\gamma_t = \infty, \quad \text{(b) } \sum_{t=1}^{\infty}\frac{\gamma_t^2}{\delta_t^2} < \infty, \quad \text{(c) } \delta_t \downarrow 0, \quad \text{(d) } \delta_t < \min_k \frac{1}{\sqrt{M_k(M_k-1)}} \quad (\forall t), \quad (7)$$

then the sequences of estimates $\{\mathbf{Q}_t\}$ and of test configurations $\{\hat{\mathbf{Q}}_t\}$ converge almost surely towards the optimum $\mathbf{Q}^*$.

Numerical Simulations. Figure 1 reports experimental results for a network with 16 antennas at the receiver and 20 homogeneous users equipped with, on average, 3 antennas. A comparison is made between the transmission rates iteratively realized by Algorithm 1 and those of the reference MXL algorithm with perfect gradient feedback. The gradient-free algorithm is run with decreasing step-size and query-radius sequences chosen in accordance with (7).

$(y - R(\mathbf{Q}_1))/(R^\star - R(\mathbf{Q}_1))$

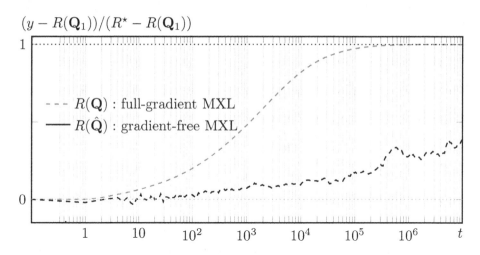

Fig. 1. Convergence of the gradient-free algorithm ($N = 16$, $K = 20$, $\mathbb{E}[M_k] = 3$): Algorithm 1 is run with policies $(\gamma_t, \delta_t) = (0.01\,t^{-3/4}, 0.1\,t^{-1/4})$, while MXL with full gradient feedback run with decreasing step size policy $\gamma_t = 0.01\,t^{-1/2}$.

It can be seen on Fig. 1 that the gradient-based algorithm finds optimal configuration within a handful of iterations. If the gradient-free algorithm also progresses towards the optimum, its convergence is less straightforward and much slower than with full gradient feedback. This tendency to slowness, which can be explained by the bias–variance tradeoff induced by the gradient estimator (SPSA), is only exacerbated in networks of larger sizes, where high problem dimensionality creates a bottleneck implying prohibitively slow convergence. In [3] it is shown that the convergence rate of Algorithm 1 is at best $O(1/\sqrt[4]{T})$ after T iterations, in contast to the considerably faster $O(1/\sqrt{T})$ rates that can be expected from the first-order methods.

4   Discussion and Perspectives

Besides the very light nature of the feedback information it requires (a single query of the objective function per iteration), the distributed, zeroth-order (derivative-free) optimization methodology presented in this paper owes to the MXL algorithm the desirable feature that it is both easy to implement, and flexible in the sense that it can be run asynchronously for the users (cf. [3]). As seen in the previous section, its major drawback is slow convergence compared to gradient-based methods. The slowness issue is addressed in detail in our more recent work [3], where the formulation of the gradient estimator (SPSA) is revisited thoroughly in order to meet the $O(1/\sqrt{T})$ convergence rate of the first-order methods. The interested reader is referred to the developments and discussions of [3] for an extensive analysis of the performances, possibilities, and guarantees of single-query zeroth-order optimization methods in the vein of Algorithm 1.

References

1. Andrews, J.G., et al.: What will 5G be? IEEE J. Sel. Areas Commun. **32**(6), 1065–1082 (2014)
2. Belmega, E.V., Lasaulce, S., Debbah, M., Jungers, M., Dumont, J.: Power allocation games in wireless networks of multi-antenna terminals. Telecommun. Syst. **47**(1–2), 109–122 (2011)
3. Bilenne, O., Mertikopoulos, P., Belmega, E.V.: Fast gradient-free optimization in distributed multi-user MIMO systems, March 2020. Submitted. https://hal. archives-ouvertes.fr/hal-02861460
4. Bravo, M., Leslie, D.S., Mertikopoulos, P.: Bandit learning in concave N-person games. In: NIPS 2018: Proceedings of the 32nd International Conference on Neural Information Processing Systems (2018)
5. Cheng, R.S., Verdú, S.: Gaussian multiaccess channels with ISI: capacity region and multiuser water-filling. IEEE Trans. Inf. Theory **39**(3), 773–785 (1993)
6. Flaxman, A.D., Kalai, A.T., McMahan, H.B.: Online convex optimization in the bandit setting: gradient descent without a gradient. In: SODA 2005: Proceedings of the 16th annual ACM-SIAM Symposium on Discrete Algorithms, pp. 385–394 (2005)
7. Larsson, E.G., Edfors, O., Tufvesson, F., Marzetta, T.L.: Massive MIMO for next generation wireless systems. IEEE Commun. Mag. **52**(2), 186–195 (2014)
8. Li, W., Assaad, M.: Matrix exponential learning schemes with low informational exchange. IEEE Trans. Signal Process. **67**(12), 3140–3153 (2019)
9. Liao, R., Bellalta, B., Oliver, M., Niu, Z.: MU-MIMO MAC protocols for wireless local area networks: a survey. IEEE Commun. Surv. Tuts. **18**(1), 162–183 (2016)
10. Mertikopoulos, P., Belmega, E.V., Moustakas, A.L.: Matrix exponential learning: distributed optimization in mimo systems. In: 2012 IEEE International Symposium on Information Theory Proceedings, pp. 3028–3032, July 2012. https://doi.org/10. 1109/ISIT.2012.6284117
11. Mertikopoulos, P., Belmega, E.V., Negrel, R., Sanguinetti, L.: Distributed stochastic optimization via matrix exponential learning. IEEE Trans. Signal Process. **65**(9), 2277–2290 (2017)
12. Mertikopoulos, P., Moustakas, A.L.: Learning in an uncertain world: MIMO covariance matrix optimization with imperfect feedback. IEEE Trans. Signal Process. **64**(1), 5–18 (2016)
13. Scutari, G., Palomar, D.P., Barbarossa, S.: Simultaneous iterative water-filling for Gaussian frequency-selective interference channels. In: ISIT 2006: Proceedings of the 2006 International Symposium on Information Theory (2006)
14. Scutari, G., Palomar, D.P., Barbarossa, S.: Asynchronous iterative waterfilling for Gaussian frequency-selective interference channels. IEEE Trans. Inf. Theory **54**(7), 2868–2878 (2008)
15. Scutari, G., Palomar, D.P., Barbarossa, S.: Optimal linear precoding strategies for wideband non-cooperative systems based on game theory – part I: nash equilibria. IEEE Trans. Signal Process. **56**(3), 1230–1249 (2008)
16. Scutari, G., Palomar, D.P., Barbarossa, S.: Optimal linear precoding strategies for wideband non-cooperative systems based on game theory – part II: algorithms. IEEE Trans. Signal Process. **56**(3), 1250–1267 (2008)
17. Spall, J.C.: A one-measurement form of simultaneous perturbation stochastic approximation. Automatica **33**(1), 109–112 (1997)
18. Yu, W., Rhee, W., Boyd, S., Cioffi, J.M.: Iterative water-filling for Gaussian vector multiple-access channels. IEEE Trans. Inf. Theory **50**(1), 145–152 (2004)

A Friendly Interference Game in Wireless Secret Communication Networks

Zhifan Xu[1] and Melike Baykal-Gürsoy[2(✉)]

[1] Department of Industrial and Systems Engineering, Rutgers University,
Piscataway, NJ 08854, USA
zhifan.xu@rutgers.edu
[2] Department of Industrial and Systems Engineering, RUTCOR and CAIT,
Rutgers University, Piscataway, USA
gursoy@soe.rutgers.edu

Abstract. This paper considers a parallel wireless network in which multiple individuals exchange confidential information through independent sender-receiver links. An eavesdropper can intercept encrypted information through a degraded channel of each sender-receiver link. A friendly jammer, by applying interference to the eavesdropping channels, can increase the level of secrecy of the network. The optimal power allocation strategy of the friendly jammer under a power constraint is derived. A convex optimization model is used when all channels are under the threat of an eavesdropping attack and a non-zero sum game model is analyzed when the eavesdropper can only attack a limited quantity of channels.

Keywords: Friendly jammer · Eavesdropping · Non-zero sum game

1 Introduction and Problem Formulation

Eavesdropping attacks are major threats for wireless communication networks due to their multi-cast nature. Instead of depending only on encryption and randomness in coding schemes [8,10,13], various efforts have been made to investigate possibilities to facilitate the security of wireless communication networks. Recent investigations reveal that intentionally generated interference signals can decrease the eavesdropping capacity of communication channels [6], which leads to the practice of employing a friendly jammer to counter eavesdropping attacks [9,11,12].

Due to the limitations of battery and power technology in current state, algorithms for efficient power control are crucial in wireless networks, and game theory has been widely adopted when an intelligent adversary exists. Altman *et al.* [1] obtained the base station's optimal power allocation strategy in jamming games. Garnaev and Trappe [4,5] investigated the optimal transmission power

This material is based upon work supported by the National Science Foundation (Grant No. 1901721).

S. Lasaulce et al. (Eds.): NetGCooP 2021, CCIS 1354, pp. 25–37, 2021.
https://doi.org/10.1007/978-3-030-87473-5_4

allocation problem against an eavesdropper using zero-sum games. Garnaev *et al.* [3] described the interaction between a friendly jammer and an eavesdropper using a zero-sum game. It is shown that Nash Equilibria of such games exhibits a water-filling scheme [2,7].

This paper considers a wireless communication network consisting of n parallel legitimate sender-receiver links where a friendly jammer can assign J_i amount of power to interfere a potential eavesdropper at channel i. The total amount of power that can be utilized by the friendly jammer is bounded by J. Moreover, the fact that it is almost inevitable for the interference signals to degrade legitimate sender-receiver channels is taken into consideration.

For each legitimate sender-receiver link $i \in \{1, ..., n\}$, the communication capacity that can be used to transmit messages under friendly jamming is

$$C_{L_i}(J_i) = \ln\left(1 + \frac{g_i^L T_i}{\sigma + h_i^L J_i}\right),$$

where T_i is pre-decided transmission power applied to channel i, σ is the Gaussian noises, g_i^L is the channel gain of transmission signals on channel i, and h_i^L is the channel gain of interference signals on channel i. At the same time, the eavesdropper can intercept information transmitted through channel i using an eavesdropping channel with capacity

$$C_{E_i}(J_i) = \ln\left(1 + \frac{g_i^E T_i}{\sigma + h_i^E J_i}\right),$$

where g_i^E is the channel gain of transmission signals and h_i^E is the channel gain of interference signals on eavesdropping channel i. We assume that $g_i^L > g_i^E$, $\forall i = 1, ..., n$ to represent the fact that every eavesdropping channel is a degraded version of the corresponding communication channel. We also assume that $h_i^L < h_i^E$, $\forall i = 1, ..., n$ to represent the fact that interference signals are more effective on eavesdropping channels than communication channels. So a power allocation policy for the friendly jammer is a vector $\boldsymbol{J} = (J_1, ..., J_n)$ such that $\sum_{i=1}^n J_i \leq J$.

Without the threat of an eavesdropping attack, legitimate users can utilize channel i's full communication capacity $C_{L_i}(J_i)$ to transmit messages securely. Meanwhile, channel i's capacity that can be used to transmit secret messages under an eavesdropping attack is defined as its secrecy capacity $C_{S_i}(J_i)$ (see [8,9,13]), which is

$$C_{S_i}(J_i) = \left(C_{L_i}(J_i) - C_{E_i}(J_i)\right)^+ = \left(\ln\left(1 + \frac{g_i^L T_i}{\sigma + h_i^L J_i}\right) - \ln\left(1 + \frac{g_i^E T_i}{\sigma + h_i^E J_i}\right)\right)^+.$$

Note that $C_{L_i}(0) > C_{E_i}(0)$, $\forall i = 1, ..., n$, under the assumption $g_i^L > g_i^E$, $\forall i = 1, ..., n$. Thus, $C_{S_i}(0) > 0$, $\forall i = 1, ..., n$, which means all channels have positive secrecy capacity without friendly interference. Also note that it is always true that $\ln\left(1 + \frac{g_i^L T_i}{\sigma + h_i^L J_i}\right) > \ln\left(1 + \frac{g_i^E T_i}{\sigma + h_i^E J_i}\right)$, $\forall J_i \geq 0, i = 1, ..., n$, since $g_i^L > g_i^E$

and $h_i^L < h_i^E$, $\forall i = 1, ..., n$. Hence, the expression of a secrecy capacity $C_{L_i}(J_i)$ w.r.t. $J_i \geq 0$ can be simplified as

$$C_{S_i}(J_i) = \ln \left(1 + \frac{g_i^L T_i}{\sigma + h_i^L J_i}\right) - \ln \left(1 + \frac{g_i^E T_i}{\sigma + h_i^E J_i}\right).$$

In addition, assume $\frac{d}{dJ_i} C_{S_i}(J_i = 0) > 0$, so that the friendly jammer has an incentive to increase channel i's secrecy capacity under an eavesdropping attack.

The structure of the paper is as follows: Sect. 2 discusses the key properties of the secrecy capacity functions $C_{S_i}(J_i)$'s and considers a single player power allocation problem where all communication channels are under the threat of an eavesdropping attack. Section 3 is the main part of the paper, which reveals the water-filling structure of the Nash Equilibrium in a friendly interference game where the eavesdropper can only attack a limited number of channels. Based on these theoretical results, Sect. 3.2 presents a computational algorithm to determine the optimal power allocation strategy. Section 4 demonstrates numerical examples. Section 5 summarizes the conclusions and discusses possible future research.

2 Basic Optimization Model

This section considers the scenario in which communication channels are under the threat of eavesdropping attacks all at the same time, so the friendly jammer is the only decision maker. The friendly jammer aims to maximize the overall secrecy capacity of this network.

2.1 Properties of $C_{S_i}(J_i)$'s

The following lemmas present the key properties of the secrecy capacity functions $C_{S_i}(J_i)$, $\forall i = 1, ..., n$.

Lemma 1. $C_{S_i}(J_i)$ is unimodal w.r.t. $J_i \geq 0$ and has a unique maximum at $J_i = \bar{J}_i$ such that $\frac{d}{dJ_i} C_{S_i}(\bar{J}_i) = 0$.

Proof. Let $c_i(J_i) := \frac{d}{dJ_i} C_{S_i}(J_i)$, then,

$$c_i(J_i) = \frac{g_i^E h_i^E T_i}{(g_i^E T_i + \sigma + h_i^E J_i)(\sigma + h_i^E J_i)} - \frac{g_i^L h_i^L T_i}{(g_i^L T_i + \sigma + h_i^L J_i)(\sigma + h_i^L J_i)} = \frac{A_i(J_i)}{B_i(J_i)},$$

$$where \begin{cases} A_i(J_i) = (g_i^E h_i^L - g_i^L h_i^E) h_i^E h_i^L T_i J_i^2 + (g_i^E - g_i^L) \cdot 2\sigma h_i^E h_i^L T_i J_i + C_i, \\ B_i(J_i) = (g_i^E T_i + \sigma + h_i^E J_i)(\sigma + h_i^E J_i)(g_i^L T_i + \sigma + h_i^L J_i)(\sigma + h_i^L J_i), \\ C_i = \left[g_i^E h_i^E (g_i^L T_i + \sigma) - g_i^L h_i^L (g_i^E T_i + \sigma)\right] \sigma T_i. \end{cases}$$

Note that $A_i(0) > 0$ since $c_i(0) > 0$ and $A_i(J_i)$ is a concave quadratic function since $g_i^E h_i^L - g_i^L h_i^E < 0$, then there exists a unique value $\bar{J}_i > 0$ such that $A_i(\bar{J}_i) = 0$. Also note that $B_i(J_i) > 0$, $\forall J_i \geq 0$. Thus, $J_i = \bar{J}_i$ is the unique solution to $c_i(J_i) = 0$ w.r.t. $J_i \geq 0$.

Moreover, note that $A_i(J_i) > 0$, $\forall 0 \leq J_i < \bar{J}_i$ and $A_i(J_i) < 0$, $\forall J_i > \bar{J}_i$, so $c_i(J_i) > 0$, $\forall 0 \leq J_i < \bar{J}_i$ and $c_i(J_i) < 0$, $\forall J_i > \bar{J}_i$, since $c_i(J_i) = \frac{A_i(J_i)}{B_i(J_i)}$ and $B_i(J_i) > 0$, $\forall J_i \geq 0$. Furthermore, $c_i(J_i)$ is also continuous w.r.t. $J_i \geq 0$, so $c_i(J_i)$ crosses the horizontal axis exactly once in $[0, \infty)$ at $J_i = \bar{J}_i$. Thus, $C_{S_i}(J_i)$ is unimodal w.r.t. $J_i \geq 0$ and it has a unique maximum at $J_i = \bar{J}_i$. □

Since $C_{S_i}(J_i)$ is unimodal w.r.t. $J_i \geq 0$ and it has a unique maximum at $J_i = \bar{J}_i$, then $C_{S_i}(J_i) < C_{S_i}(\bar{J}_i)$, $\forall J_i > \bar{J}_i$. Hence, the friendly jammer will put at most $\bar{J}_i$ power to channel i if she aims to maximize channel i's secrecy capacity. To solve the friendly jammer's power allocation problem, it is enough to consider $C_{S_i}(J_i)$'s properties w.r.t. the refined feasible region $J_i \in \left[0, \bar{J}_i\right]$, $\forall i = 1, ..., n$.

Lemma 2. $C_{S_i}(J_i)$ *is concave and strictly increasing w.r.t.* $J_i \in \left[0, \bar{J}_i\right]$ *for all* $i = 1, ..., n$.

Proof. Let $c_i(J_i)$, $A_i(J_i)$ and $B_i(J_i)$ be defined as in lemma 1. Note that:

(a) $A_i(J_i)$ is a concave quadratic function and is strictly decreasing w.r.t. $J_i \geq 0$ since $g_i^E h_i^L - g_i^L h_i^E < 0$ and $g_i^E - g_i^L < 0$,
(b) $B_i(J_i)$ is strictly increasing w.r.t. $J_i \geq 0$,
(c) $A_i(J_i) \geq 0$ and $B_i(J_i) > 0$ for all $J_i \in \left[0, \bar{J}_i\right]$.

Thus, $c_i(J_i) = \frac{A_i(J_i)}{B_i(J_i)}$ is strictly decreasing w.r.t. $J_i \in \left[0, \bar{J}_i\right]$, which implies $C_{S_i}(J_i)$ is a concave function on $J_i \in \left[0, \bar{J}_i\right]$.

Also, $C_{S_i}(J_i)$ is strictly increasing w.r.t. $J_i \in \left[0, \bar{J}_i\right]$, since $c_i(J_i) > 0$, $\forall J_i \in \left[0, \bar{J}_i\right)$ as proved in Lemma 1. □

Although it is impossible for the friendly jammer's optimal power allocation policy $\boldsymbol{J}^*$ to have $J_i^* \geq \bar{J}_i$, we would like to present a full description of the shape of $C_{S_i}(J_i)$ w.r.t. $J_i \in [0, +\infty)$ without going into tedious proofs of all details.

First, consider $c_i(J_i)$ w.r.t. $J_i \in \left[\bar{J}_i, +\infty\right)$. Note that: (a) $A_i(J_i)$ is concave, quadratic, strictly decreasing and negative w.r.t. $J_i > \bar{J}_i$, (b) $B_i(J_i)$ is quartic, strictly increasing and positive w.r.t. $J_i > \bar{J}_i$. Thus, it can be seen that

$$\lim_{J_i \to +\infty} c_i(J_i) = \frac{A_i(J_i)}{B_i(J_i)} = -0,$$

and it can be proved by the mean value theorem that there exists a point $\tilde{J}_i \in (\bar{J}, +\infty)$ such that $\frac{d}{dJ_i} c_i(\tilde{J}_i) = 0$. Moreover, $\tilde{J}_i$ is actually the unique solution to $\frac{d}{dJ_i} c_i(J_i) = 0$ w.r.t. $J_i \in [0, +\infty)$ given the properties of $A_i(J_i)$ and $B_i(J_i)$. Since $c_i(J_i)$ is decreasing at $J_i = \bar{J}_i$, it can also be verified that $c_i(J_i)$ is strictly decreasing in $[\bar{J}_i, \tilde{J}_i]$, and then strictly increasing in $(\tilde{J}_i, +\infty)$. In summary, for $J_i \in [0, +\infty)$, $c_i(J_i)$ has the following properties as shown in Fig. 1b:

– $c_i(J_i) \geq 0$, $\forall J_i \in \left[0, \bar{J}_i\right]$, and $c_i(J_i) < 0$, $\forall J_i \in (\bar{J}_i, +\infty)$.
– $c_i(J_i)$ is strictly decreasing w.r.t. $J_i \in [0, \tilde{J}_i]$, and is strictly increasing w.r.t $J_i \in (\tilde{J}_i, +\infty)$.

Recall that $c_i(J_i) = \frac{d}{dJ_i} C_{S_i}(J_i)$ by definition, so $C_{S_i}(J_i)$ is concave w.r.t. $J_i \in [0, \tilde{J}_i]$, since $c_i(J_i)$ is strictly decreasing w.r.t $J_i \in [0, \tilde{J}_i]$. In summary, for $J_i \in [0, +\infty)$, $C_{S_i}(J_i)$ has the following properties as shown in Fig. 1a:

- $C_{S_i}(J_i) > 0, \ \forall J_i \in [0, +\infty)$.
- $C_{S_i}(J_i)$ is concave w.r.t $J_i \in [0, \tilde{J}_i]$ and reaches its maximum at $J_i = \bar{J}_i < \tilde{J}_i$.
- $C_{S_i}(J_i)$ is convex and decreasing w.r.t $J_i \in (\tilde{J}_i, +\infty)$.

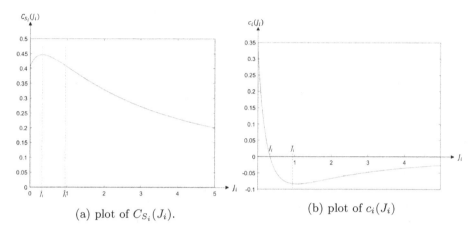

(a) plot of $C_{S_i}(J_i)$.

(b) plot of $c_i(J_i)$

Fig. 1. Plots of $C_{S_i}(J_i)$ and $c_i(J_i)$, with $g_i^L = 2, g_i^E = 1, h_i^L = 1, h_i^E = 2, T_i = 1, \sigma = 1$.

2.2 The Optimal Power Allocation Policy

To find the optimal power allocation policy for the network, the friendly jammer needs to solve the following problem:

$$\max_{\boldsymbol{J}} \ v_J(\boldsymbol{J}) = \sum_{i=1}^{n} C_{S_i}(J_i) = \sum_{i=1}^{n} \left[\ln \left(1 + \frac{g_i^L T_i}{\sigma + h_i^L J_i} \right) - \ln \left(1 + \frac{g_i^E T_i}{\sigma + h_i^E J_i} \right) \right]$$

$$\text{s.t.} \ \sum_{i=1}^{n} J_i \leq J, \tag{1}$$

$$0 \leq J_i \leq \bar{J}_i, \ \forall i = 1, ..., n.$$

Note that $v_J(\boldsymbol{J})$ is a concave function on $\boldsymbol{J} \in \mathcal{J}$ where $\mathcal{J}$ is the feasible region of optimization problem 1. To prove it, simply let $v_i(\boldsymbol{J}) := C_{S_i}(J_i)$, then $v_i(\boldsymbol{J})$ is concave on $\boldsymbol{J} \in \mathcal{J}$ since $C_{S_i}(J_i)$ is concave on $J_i \in [0, \bar{J}_i]$. Thus, $v_J(\boldsymbol{J})$ is concave on $\boldsymbol{J} \in \mathcal{J}$ since it is a sum of $v_i(\boldsymbol{J})$'s. So optimization problem 1 is a convex optimization problem.

Theorem 1. *The considered convex optimization problem has a unique optimal solution $\boldsymbol{J}^* = (J_1^*, ..., J_n^*)$ that is subject to water-filling scheme. Let the sequence of sender-receiver links be ordered according to $c_i(0)$ such that $c_1(0) > ... > c_n(0)$ assuming $c_i(0) \neq c_j(0), \ \forall i \neq j$ for the sake of simplicity.*

a) If $\sum_{i=1}^{n} \bar{J}_i \geq J$, then there exists $w \geq 0$ and a threshold integer $k > 0$ such that

$$\begin{cases} c_1(J_1^*) = ... = c_k(J_k^*) = w, \\ \sum_{i=1}^{k} J_i^* = J, \\ c_i(0) \leq w, J_i^* = 0, \quad \forall k < i \leq n. \end{cases} \quad (2)$$

and J^* is the solution of equations system 2.
b) If $\sum_{i=1}^{n} \bar{J}_i < J$, then $k = n$, $w = 0$, and J^* is the solution to

$$c_1(J_1^*) = ... = c_n(J_n^*) = w = 0. \quad (3)$$

and $\sum_{i=1}^{n} J_i^* < J$.

Proof. We provide a proof in the appendix.   □

Based on Theorem 1, the optimal value of w in the system of Eq. 2 can be determined using numerical methods such as bisection search.

3   A Friendly Interference Game

This section considers another scenario where the attacker can eavesdrop on only one of n channels due to resource constraints. To intercept as much information as possible, the attacker tries to maximize the expected eavesdropping capacity. Let the attack strategy of the eavesdropper be $y = (y_1, ..., y_n)$ where y_i represents the probability that the eavesdropper picks channel i as target. Naturally, we have $y_i \geq 0$, $\forall i = 1, ..., n$ and $\sum_{i=1}^{n} y_i = 1$. Thus, the payoff for the attacker under interference signals is

$$v_E(J, y) = \sum_{i=1}^{n} y_i C_{E_i}(J_i) = \sum_{i=1}^{n} y_i \ln\left(1 + \frac{g_i^E T_i}{\sigma + h_i^E J_i}\right). \quad (4)$$

Meanwhile, the friendly jammer still tries to maximize the total capacity that can be used to transmit messages securely, so the friendly jammer's payoff is

$$v_J(J, y) = \sum_{i=1}^{n} [C_{L_i}(J_i) - y_i C_{E_i}(J_i)]$$
$$= \sum_{i=1}^{n} \left[\ln\left(1 + \frac{g_i^L T_i}{\sigma + h_i^L J_i}\right) - y_i \ln\left(1 + \frac{g_i^E T_i}{\sigma + h_i^E J_i}\right)\right], \quad (5)$$

and the friendly jammer's power allocation startegy J is still subject to the total power constraint.

Now, we have a non-zero sum game with two players, namely the friendly jammer and the eavesdropper. We shall look for the Nash Equilibrium, that is, we want to find a strategy pair (J^*, y^*) such that

$$v_J(J, y^*) \leq v_J(J^*, y^*), \quad \forall J \in \mathcal{J},$$
$$v_E(J^*, y) \leq v_E(J^*, y^*), \quad \forall y \in \mathcal{Y},$$

where $\mathcal{J}$ is the region containing all possible power allocation strategies J and $\mathcal{Y}$ is the region containing all probabilistic attack strategies y of this game.

3.1   Theoretical Analysis

Consider the friendly jammer's problem given a fixed attack strategy y^* of the adversary. It can be seen that $\frac{\partial v_J(J,y^*)}{\partial J_i}$ share similar properties with $c_i(J_i)$, and interfering any channel i where $\frac{\partial v_J(J,y^*)}{\partial J_i}|_{J_i=0} \leq 0$ will be a dominated strategy for the friendly jammer. Let I be the set of channels where $\frac{\partial v_J(J,y^*)}{\partial J_i}|_{J_i=0} > 0$, let $\bar{J}_i(y^*) > 0$ be a real value such that $\frac{\partial v_J(J,y^*)}{\partial J_i}|_{J_i=\bar{J}_i(y^*)} = 0$, $\forall i \in I$, then the friendly jammer needs to solve the following optimization problem:

$$\max_{J_i, \forall i \in I} \sum_{i \in I} \left[\ln \left(1 + \frac{g_i^L T_i}{\sigma + h_i^L J_i} \right) - y_i^* \ln \left(1 + \frac{g_i^E T_i}{\sigma + h_i^E J_i} \right) \right]$$

$$\text{s.t.} \quad \sum_{i \in I} J_i \leq J, \tag{6}$$

$$0 \leq J_i \leq \bar{J}_i(y^*), \quad \forall i \in I.$$

Optimization problem 6 is similar to optimization problem 1 but with a subset of channels as targets and with smaller coefficients $y_i^* \leq 1$ in the objective function. So the objective function of optimization problem 6 is concave w.r.t. its feasible region. Thus, an optimal power allocation strategy $\{J_i^*, i \in I\}$ should still have the properties implied by KKT conditions. That is,

$$\frac{\partial v_J(J^*, y^*)}{\partial J_i} = \frac{y_i^* g_i^E h_i^E T_i}{(g_i^E T_i + \sigma + h_i^E J_i^*)(\sigma + h_i^E J_i^*)} - \frac{g_i^L h_i^L T_i}{(g_i^L T_i + \sigma + h_i^L J_i^*)(\sigma + h_i^L J_i^*)}$$

$$\begin{cases} = w_D, & for \ J_i^* > 0, \quad \forall i \in I, \\ \leq w_D, & for \ J_i^* = 0, \quad \forall i \in I, \end{cases} \tag{7}$$

where $w_D \geq 0$, and $w_D(\sum_{i \in I} J_i^* - J) = 0$.

Similarly, given the friendly jammer's strategy J^*, the eavesdropper can find his optimal strategy by solving a convex optimization problem where an optimal solution y^* should satisfy

$$\frac{\partial v_E(J^*, y^*)}{\partial y_i} = \ln \left(1 + \frac{g_i^E T_i}{\sigma + h_i^E J_i^*} \right) \begin{cases} = w_A, & for \ y_i^* > 0, \quad \forall i = 1, ..., n, \\ \leq w_A, & for \ y_i^* = 0, \quad \forall i = 1, ..., n, \end{cases} \tag{8}$$

where $w_A \geq 0$ and $\sum_{i=1}^n y_i^* = 1$.

Theorem 2. *Define $\Theta_i(J_i) := \frac{\partial v_E(J,y)}{\partial y_i}$. Let the sequence of sender-receiver links be ordered according to $\Theta_i(0)$ such that $\Theta_1(0) > ... > \Theta_n(0)$ assuming $\Theta_i(0) \neq \Theta_j(0)$, $\forall i \neq j$ for the sake of simplicity. Let $k > 0$ be the largest integer such that*

$$\begin{cases} \Theta_1(J_1) = ... = \Theta_k(J_k) = w_A \geq 0, \\ \sum_{i=1}^k J_i = J, \\ \Theta_i(0) \leq w_A, \quad J_i = 0, \quad \forall k < i \leq n. \end{cases} \tag{9}$$

Let $J' = (J_1', ..., J_n')$ be the solution of system of Eqs. 9.

Define $\Delta_{J_i'}(y_i) := \frac{\partial v_J(J,y)}{\partial J_i}\big|_{J_i=J_i'}$. *Let* $m > 0$ *be the largest integer such that*

$$\begin{cases} \Delta_{J_1'}(y_1) = ... = \Delta_{J_k'}(y_m) = w_D \geq 0, \\ \sum_{i=1}^{m} y_i \leq 1. \end{cases} \tag{10}$$

a) If $k \leq m$, *a Nash Equilibrium strategy pair* (J^*, y^*) *can be found by solving*

$$\begin{cases} J^* = J', \\ \Delta_{J_1^*}(y_1^*) = ... = \Delta_{J_k^*}(y_k^*) = w_D \geq 0, \\ \sum_{i=1}^{k} y_i^* = 1. \end{cases} \tag{11}$$

b) If $k > m$, *there exists a positive integer* $h \leq k$ *such that a Nash Equilibrium strategy pair* (J^*, y^*) *can be found where* J^* *is the solution of*

$$\begin{cases} \Theta_1(J_1^*) = ... = \Theta_h(J_h^*) = w_A' > w_A, \\ J_i^* = 0, \quad \forall h < i \leq n, \\ \Theta_h(0) > w_A' \geq \Theta_{h+1}(0), \end{cases} \tag{12}$$

and y^* *is the solution of*

$$\begin{cases} \Delta_{J_1^*}(y_1^*) = ... = \Delta_{J_h^*}(y_h^*) = 0, \\ y_{h+1}^* = 1 - \sum_{i=1}^{h} y_i^*, \quad \Delta_{J_{h+1}^*}(y_{h+1}^*) \leq 0, \\ y_{h+1}^* = 0, \quad if \ w_A' \neq \Theta_{h+1}(0), \\ y_i^* = 0, \quad \forall h+1 < i \leq n. \end{cases} \tag{13}$$

Proof. We provide a proof in the appendix. □

3.2 Algorithm to Find (J^*, y^*)

This sections presents an algorithm based on the bisection methods and Theorem 2 to approximate a pair of Nash Equilibrium strategies (J^*, y^*) within a given tolerance factor, δ. An explicit value for δ is used to make the algorithm terminate within reasonable CPU time.

Algorithm
Inputs. Parameters of the communication network: $T_i, g_i^L, g_i^E, h_i^L, h_i^E, \forall i = 1, ..., n$. The background noise σ. And the explicit tolerance $\delta \leq 0.01$.
Step 1. Let k, w_A and J' be the solution of system of Eq. 9.
Step 2. Let m, w_D and y^* be the solution of system of Eq. 10.
Step 3. If $k \leq m$. Let $J^* \leftarrow J'$. Let y^* be the solution of system of Eq. 11. (J^*, y^*) is a pair of NE strategies and the algorithm is terminated. Otherwise, go to step 4.

Step 4. If $k > m$. Let $h \leftarrow k$. Let $w_A^{LB} \leftarrow w_A$.

    Step 4a. Let $h \leftarrow h - 1$ and then $w_A' \leftarrow \Theta_{h+1}(0)$.

    Step 4b. Let J' be the solution of system of Eq. 12. Let $w_D \leftarrow 0$. Let m and y^* be the solution of system of Eq. 10 with the value of J'. Go to step 5.

Step 5. If $h = m$. Let $J^* \leftarrow J'$, then (J^*, y^*) is a pair of NE strategies and the algorithm is terminated. Otherwise, go to step 6.

Step 6. If $h > m$. Let $w_A^{LB} \leftarrow \Theta_{h+1}(0)$. Go to step 4a. Otherwise, go to step 7.

Step 7. If $h < m$. Let $w_A^{UB} \leftarrow \Theta_{h+1}(0)$ and then $h \leftarrow h + 1$.

    Step 7a. Let $w_A' \leftarrow \frac{1}{2}(w_A^{UB} + w_A^{LB})$. Let J' be the solution of system of Eq. 12. Let $w_D \leftarrow 0$. Let m and y^* be the solution of system of Eq. 10 with the value of J'.

    Step 7b. If $h = m$ and $y_{h+1}^* \leq \delta$. Let $J^* \leftarrow J'$ and $y_{h+1}^* \leftarrow 0$. Then (J^*, y^*) is NE strategies and the algorithm is terminated. Otherwise, go to step 7c.

    Step 7c. If $h = m$ and $y_{h+1}^* > \delta$, or if $h < m$. Let $w_A^{UB} \leftarrow w_A'$. Go to step 7a. Otherwise, go to step 7d.

    Step 7d. If $h > m$. Let $w_A^{LB} \leftarrow w_A'$. Go to step 7a.

4 Numerical Illustrations

This section presents a few numerical examples. First, consider a 5 parallel channel communication network with $g_i^L = p^{i-1}$ for $i \in [1, 5]$ where $p \in (0, 1)$, which corresponds to Rayleigh fading in orthogonal frequency-division multiplexing (OFDM) systems. Similarly, let $g_i^E = q^{i-1}$ for $i \in [1, 5]$ where $q \in (0, 1)$ for the eavesdropper. Let $p = 0.65$ and $q = 0.5$ such that the assumption $g_i^L > g_i^E$, $\forall i = 1, ..., 5$ is satisfied. Also set $\delta = 0.1$ and $T_i = 1$, $\forall i = 1, ..., 5$. Finally, we set $h_i^E = 0.45$, $h_i^L = 0.05$ for all $i = 1, ..., 5$ and $J = 1$ for the sake of simplicity. The assumption $c_i(0) > 0$, $\forall i = 1, ..., 5$ is satisfied.

Following the algorithm in Sect. 3.2, we get $k = 3$ and $m = 4$. So the threshold index is $h = 3$. The approximated NE strategies are found in 0.0313 CPU time, with $J^* = (0.731, 0.254, 0.016, 0, 0)$, $y^* = (0.535, 0.305, 0.16, 0, 0)$, $w_A = 1.204$ and $w_D = 0.074$. It can be clearly seen that (J^*, y^*) is subject to water-filling scheme where both players focus on the channels with higher initial eavesdropping capacities.

Now increase h_i^L's by 50%, meaning that the legitimate users suffer more from the interference signals. Using the algorithm in Sect. 3.2, we get $k = 3$ and $m = 2$ in 0.0469 CPU time. And the final threshold index is $h = 2$, with $J^* = (0.667, 0.222, 0, 0, 0)$, $y^* = (0.541, 0.339, 0.12, 0, 0)$, $w_A = 1.253$ and $w_D = 0$. Compared to the previous example, the friendly jammer protects fewer channels and leaves channel 3 with no protection under attack even though there is unused jamming power (Fig. 2).

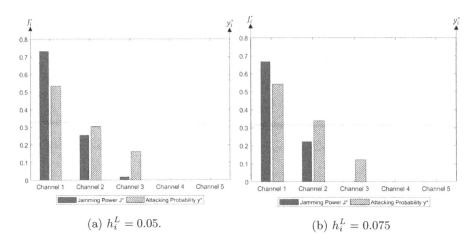

(a) $h_i^L = 0.05$.　　　　　　　　　　(b) $h_i^L = 0.075$

Fig. 2. Plots of $\boldsymbol{J^*}$ and $\boldsymbol{y^*}$.

5　Conclusions and Future Research

In this paper, we consider a friendly interference game where a friendly jammer is
employed to interfere eavesdroppers in a wireless network. We prove the existence
of the optimal power allocation strategy of the friendly jammer as part of a pair
of Nash Equilibrium strategies in a non-zero sum game. It turns out that the
optimal power allocation strategy will be subject to a water-filling scheme. An
algorithm to approximate the optimal power allocation strategy to within a given
tolerance is presented. We also show that the effect of interference signals on
legitimate users is a key parameter that affect the performance of this approach.
The interference signals should be carefully tuned such that no interference power
will be wasted.

　　Of interest for future research is an extension of this model to include the
legitimate users as decision makers. For instance, a base station controlling transmission power among multiple channels may cooperate with a friendly jammer.

Appendix

Proof of Theorem 1. Consider the KKT conditions of convex optimization
problem 1. A vector $\boldsymbol{J^*} = (J_1^*, ..., J_n^*)$ is the optimal solution if there exists a
group of non-negative numbers $w, \lambda_1, ..., \lambda_n, \mu_1, ..., \mu_n$ such that

$$\begin{cases} \sum_{i=1}^n J_i^* - J \le 0, \\ w(J - \sum_{i=1}^n J_i^*) = 0, \\ \lambda_i J_i^* = 0, \quad \mu_i(\bar{J}_i - J_i^*) = 0, \quad \forall i = 1, ..., n, \\ w = c_i(J_i^*) + \lambda_i - \mu_i, \quad \forall i = 1, ..., n. \end{cases} \quad (14)$$

It should be noted that actually we must have $\mu_i = 0$ for all $i = 1, ..., n$. To show that, suppose $\mu_i > 0$ for some i, then we have $J_i^* = \bar{J}_i$, which leads to $\lambda_i = 0$. Thus, we must have $w = c_i(\bar{J}_i) - \mu_i = -\mu_i < 0$, which is impossible. Thus, the KKT condition can be simplified to

$$
\begin{cases}
\sum_{i=1}^{n} J_i^* - J \leq 0, \\
w(J - \sum_{i=1}^{n} J_i^*) = 0, \\
\lambda_i J_i^* = 0, \quad \forall i = 1, ..., n, \\
w = c_i(J_i^*) + \lambda_i, \quad \forall i = 1, ..., n.
\end{cases}
\tag{15}
$$

a) Given $\sum_{i=1}^{n} \bar{J}_i \geq J$, it must be true that $\sum_{i=1}^{n} J_i^* - J = 0$. To show that, suppose $\sum_{i=1}^{n} J_i^* - J < 0$, then $w = 0$. Thus, $\lambda_i = 0, c_i(J_i^*) = 0, \forall i = 1, ..., n$. Then $J_i^* = \bar{J}_i, \forall i = 1, ..., n$ follows. Hence, $\sum_{i=1}^{n} J_i^* = \sum_{i=1}^{n} \bar{J}_i \geq J$, which is against the assumption $\sum_{i=1}^{n} J_i^* - J < 0$. So $\sum_{i=1}^{n} J_i^* - J = 0$ must be true given $\sum_{i=1}^{n} \bar{J}_i \geq J$.

The equality condition $\sum_{i=1}^{n} J_i^* - J = 0$ implies that there must exist some i's such that $J_i^* > 0$. Let $k > 0$ be the largest index i such that $J_i^* > 0$, then we have $\lambda_k = 0$ and

$$0 \leq w = c_k(J_k^*) < c_k(0) < c_j(0), \quad \forall 1 \leq j < k.$$

Now, to satisfy $w = c_j(J_j^*) + \lambda_j, \forall j = 1, ..., k - 1$ knowing $\lambda_j \geq 0$, we must have $J_j^* > 0, \forall j = 1, ..., k - 1$, which leads to $\lambda_j = 0, \forall j = 1, ..., k - 1$ and $w = c_k(J_k^*), j = 1, ..., k - 1$. So $\boldsymbol{J}^*$ can be an optimal solution when system of Eq. 2 has a solution with $w \geq 0$. It is easy to verify that the solution will be unique if it exists.

b) Given $\sum_{i=1}^{n} \bar{J}_i < J$, it must be true that $\sum_{i=1}^{n} J_i^* - J < 0$ since $J_i^* \leq \bar{J}_i, \forall i = 1, ..., n$. Then $w = 0$ follows, and $\boldsymbol{J}^*$ must satisfy $0 = c_i(J_i^*) + \lambda_i, \forall i = 1, ..., n$, which leads to $J_i^* > 0, \forall i = 1, ..., n$ and $\lambda_i = 0, \forall i = 1, ..., n$. So $\boldsymbol{J}^*$ can be an optimal solution when system of Eq. 3 has a solution and it is easy to verify that you can't have a solution of system of Eq. 2 at the same time.

Proof of Theorem 2. We will first show that a NE strategy pair $(\boldsymbol{J}^*, \boldsymbol{y}^*)$ of this non-zero sum game will be subject to a water-filling scheme.

Let $(\boldsymbol{J}^*, \boldsymbol{y}^*)$ be a NE strategy pair and let h be the largest integer such that $J_h^* > 0$. According to condition 7, it is true that $h \in I$ and $y_h^* > 0$. Then, by condition 8,

$$w_A = \Theta_h(J_h^*) < \Theta_h(0) < \Theta_i(0), \quad \forall i = 1, ..., h - 1,$$

which implies $J_i^* > 0, \forall i = 1, ..., h - 1$, in order for condition 8 to be satisfied. Then, by condition 7 again, one has $y_i^* > 0, \forall i = 1, ..., h - 1$.

In summary, a NE strategy pair $(\boldsymbol{J}^*, \boldsymbol{y}^*)$ should be subject to a water-filling scheme with a threshold index h such that

$$
\begin{cases}
J_i^* > 0, \ y_i^* > 0, \ \forall i = 1, ..., h, \\
J_i^* = 0, \ \forall i = h + 1, ..., N.
\end{cases}
\tag{16}
$$

a) If $k \leq m$, the solution to system of Eqs. 9 and 11 satisfy conditions 7, 8 and 16, which defines the NE strategy pair $(\boldsymbol{J}^*, \boldsymbol{y}^*)$ of this game. Thus, $I = \{1, ..., k\}$ in this case.

b) If $k > m$, there is no feasible solution to system of Eqs. 9 and 11 since it is impossible to have $\Delta_{J_k^*}(y_k^*) = w_D \geq 0$. Besides, by the definition of k, it is impossible to have $h > k$.

Notice that solving system of Eq. 9 instead with $w_A' > w_A$ and $\sum_{i=1}^k J_i \leq J$ will provide solutions containing smaller J_i''s and threshold index $h \leq k$, and smaller J_i''s lead to larger m as defined by system of Eq. 10. Thus, one can find a NE strategy pair $(\boldsymbol{J}^*, \boldsymbol{y}^*)$ by increasing the value of w_A until conditions 7, 8 and 16 are satisfied.

Now, one can search for the threshold index $h \leq k$ with an increased $w_A' > w_A$ as shown in system of Eq. 12. Clearly, $\sum_{i=1}^n J_i^* < J$ in this case, which leads to $w_D = 0$ by constraint $w_D(\sum_{i=1}^n J_i^* - J) = 0$. Also, under condition 8, it is possible to have $y_{h+1}^* = 1 - \sum_{i=1}^h y_i^* > 0$ if and only if $w_A' = \Theta_{h+1}(0)$, but the constraint $\Delta_{J_{h+1}^*}(y_{h+1}^*) \leq w_D = 0$ must be satisfied at the same time as implied by condition 7 and $J_{h+1}^* = 0$.

In summary, the solution of system of Eqs. 12 and 13 form a Nash Equilibrium strategy pair $(\boldsymbol{J}^*, \boldsymbol{y}^*)$ of this game in this case.

References

1. Altman, E., Avrachenkov, K., Garnaev, A.: A jamming game in wireless networks with transmission cost. In: Chahed, T., Tuffin, B. (eds.) NET-COOP 2007. LNCS, vol. 4465, pp. 1–12. Springer, Heidelberg (2007). https://doi.org/10.1007/978-3-540-72709-5_1
2. Altman, E., Avrachenkov, K., Garnaev, A.: Closed form solutions for water-filling problems in optimization and game frameworks. Telecommun. Syst. 47(1–2), 153–164 (2011)
3. Garnaev, A., Baykal-Gürsoy, M., Poor, H.V.: Incorporating attack-type uncertainty into network protection. IEEE Trans. Inf. Forensics Secur. 9(8), 1278–1287 (2014)
4. Garnaev, A., Trappe, W.: An eavesdropping game with SINR as an objective function. In: Chen, Y., Dimitriou, T.D., Zhou, J. (eds.) SecureComm 2009. LNICST, vol. 19, pp. 142–162. Springer, Heidelberg (2009). https://doi.org/10.1007/978-3-642-05284-2_9
5. Garnaev, A., Trappe, W.: Secret communication when the eavesdropper might be an active adversary. In: Jonsson, M., Vinel, A., Bellalta, B., Belyaev, E. (eds.) MACOM 2014. LNCS, vol. 8715, pp. 121–136. Springer, Cham (2014). https://doi.org/10.1007/978-3-319-10262-7_12
6. Goel, S., Negi, R.: Guaranteeing secrecy using artificial noise. IEEE Trans. Wirel. Commun. 7(6), 2180–2189 (2008)
7. Lai, L., El Gamal, H.: The water-filling game in fading multiple-access channels. IEEE Trans. Inf. Theory 54(5), 2110–2122 (2008)
8. Leung-Yan-Cheong, S., Hellman, M.: The Gaussian wire-tap channel. IEEE Trans. Inf. Theory 24(4), 451–456 (1978)

9. Rabbachin, A., Conti, A., Win, M.Z.: Intentional network interference for denial of wireless eavesdropping. In: 2011 IEEE Global Telecommunications Conference - GLOBECOM 2011, pp. 1–6, December 2011
10. Shannon, C.E.: Communication theory of secrecy systems. Bell Syst. Tech. J. **28**(4), 656–715 (1949)
11. Tang, X., Liu, R., Spasojevic, P., Poor, H.V.: Interference assisted secret communication. IEEE Trans. Inf. Theory **57**(5), 3153–3167 (2011)
12. Tang, X., Liu, R., Spasojevic, P., Poor, H.V.: The Gaussian wiretap channel with a helping interferer. In: 2008 IEEE International Symposium on Information Theory, pp. 389–393. IEEE (2008)
13. Wyner, A.D.: The wire-tap channel. Bell Syst. Tech. J. **54**(8), 1355–1387 (1975)

Slow-Link Adaptation Algorithm for Multi-source Multi-relay Wireless Networks Using Best-Response Dynamics

Ali Al Khansa[1,2(✉)], Stefan Cerovic[1], Raphael Visoz[1], Yezekael Hayel[2], and Samson Lasaulce[3]

[1] Orange Labs, Chatillon, France
{ali.alkhansa,stefan.cerovic,raphael.visoz}@orange.com
[2] LIA, Avignon University, Avignon, France
yezekael.hayel@univ-avignon.fr
[3] CRAN, CNRS, Vandoeuvre-lès-Nancy, France
samson.lasaulce@centralesupelec.fr

Abstract. In this paper, we propose a link adaptation algorithm for slow-fading half-duplex orthogonal multiple access multiple relay channel under a centralized node scheduling approach. During the first phase, the sources transmit in turn. During the second phase, a scheduled node (relay or source) transmits incremental redundancies on its correctly decoded source messages. The proposed algorithm aims at maximizing the average spectral efficiency under individual QoS targets for a given modulation and coding scheme family. The main principle of the algorithm is to reduce the complexity, and this is achieved using Best-Response Dynamics (BRD) tools. The rates are first initialized and then an iterative rate correction is applied. The resulting scheduling offers a tractable complexity under practical knowledge of channel states and yields performance close to the corresponding exhaustive search approaches as demonstrated by Monte-Carlo simulations.

Keywords: Slow-link adaptation · Best-response dynamics · Multi-source multi-relay wireless network · QoS target · Spectral efficiency

1 Introduction

1.1 Research Context

Spectral efficiency of wireless networks is one of the main concerns of different researches nowadays. Cooperative communication is one promising concept, which aims at increasing the efficiency in order to answer the increasingly challenging demands from users. Its principle is to allow users to share their resources, so that they can improve their transmission and reception. Fundamental principles and general problems of cooperative communications are

ⓒ Springer Nature Switzerland AG 2021
S. Lasaulce et al. (Eds.): NetGCooP 2021, CCIS 1354, pp. 38–47, 2021.
https://doi.org/10.1007/978-3-030-87473-5_5

introduced in [1], where the three-terminal Relay Channel (RC) is studied. Inner and outer bounds for the capacity of different relay channels are derived in [2], many of those still not being surpassed today. The corresponding bounds for Multiple-Access Relay Channel (MARC) are derived in [3].

In this paper, we study a Multiple Access Multiple Relay Channel (MAMRC), which can be viewed as a generalization of previously mentioned models. In particular, a centralized link adaptation (rate allocation) algorithm is proposed for Orthogonal Multiple Access Multiple Relay Channel (OMAMRC), where a time-slotted communication is adopted. The algorithm here is based on two steps: step one, choosing the initial source rates; and step two, using Best-Response Dynamics (BRD) method to correct the previous selected initial rates. Briefly speaking, best-response dynamics work in a way, where each user will try to choose the best rate choice, which optimizes the same profit function defined to all users. When all rates reach a value, and tend not to change it, we reach our convergence point where no more modifications are needed to the allocated rates.

In OMAMRC, at least two sources communicate with a single destination with the help of at least two relaying nodes. A relaying node can be either a dedicated relay, which does not have its own message to transmit, or a source itself, which does have its own message and that can relay the messages of the other sources in some cases. The latter case is often referred to as user cooperation and, thus, falls into the scope of the paper. Relaying nodes are half-duplex, meaning that they cannot listen to source messages and transmit at the same time.

Channel State Information (CSI) is available at the receiver for each direct link. This means that the centralized scheduler located at the destination only has the CSI of source-to-destination and relay-to-destination links. On the other hand, CSI of source-to-relay, relay-to-relay and source-to-source links is not available. Our main focus is a design of a link adaptation algorithm under the scenario where the Channel Distribution Information (CDI) of all links (e.g., average SNR of all links) is reported to the destination on a long-term basis in order to derive the (slow) rate allocation of the sources. This kind of scenario typically corresponds to fast varying radio conditions, e.g., for high mobility conditions.

1.2 Scope of the Paper

Our main contribution is the proposal of a slow-link adaptation algorithm, which aims at maximizing the average spectral efficiency after a fixed number of retransmissions subject to Quality of Service (QoS) constraints on individual Block Error Rate (BLER) per source. The decisions of the algorithm are based on CDI of all the links in the network, while the utility metric is obviously conditional on the node scheduling (selection) strategy used.

To solve the multi-dimensional rate allocation optimization problem, we followed the best-response dynamics tools [4]. Such game theoretic based algorithms are widely used in several wireless problems like power control in cellular networks[5] and wireless systems in general [6,7]. First, each source rate is derived independently in a certain arbitrary strategy. Then, the choice of each

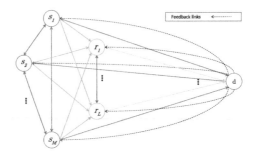

Fig. 1. Cooperative Orthogonal Multiple Access Multiple Relay Channel (OMAMRC) with feedback.

source rate depends on the previous choices of all other sources rates, in the aim to achieve the optimal total average spectral efficiency. Accordingly, we modify these choices dynamically, and iteratively, until we reach a convergence point where any change in the set of source rates will make the efficiency worse.

2   System Model

A slow-fading OMAMRC is considered, where sources belonging to the set $\mathcal{S} = \{1, \ldots, M\}$ communicate with a common destination with the help of the other sources, which perform user cooperation, and a set of relays $\mathcal{R} = \{M+1, \ldots, M+L\}$ (see Fig. 1). A message $\mathbf{u}_s \in \mathbb{F}_2^{K_s}$ of a source s has a length of K_s information bits. That length depends on the selected Modulation and Coding Schemes (MCS) for that source. The messages of all sources are mutually independent. Relays operate in half-duplex mode and they do not have their own messages to transmit. We define the set of all source and relay nodes as $\mathcal{N} = \mathcal{S} \cup \mathcal{R} = \{1, \ldots, M+L\}$. In other words, a source s_i will be the node i in set $\mathcal{N}$, and a relay r_i will be the node $i + M$ in set $\mathcal{N}$.

CSI is available at the receiver of each link and is assumed perfect. That means that a priori, only the CSI $\mathbf{h}_{\text{dir}} = [\mathbf{h}_{\text{s,d}}, \mathbf{h}_{\text{r,d}}] = [h_{1,d}, \ldots, h_{M+L,d}]$ of source-to-destination (S-D), and relay-to-destination (R-D) links are perfectly known by the destination, while the CSI of source-to-source (S-S), source-to-relay (S-R) and relay-to-relay (R-R) links are unknown.

We assume that the radio-links between the different nodes are fixed within a frame transmission (slow fading assumption). It is assumed that all nodes transmit with the same power, where each node is equipped with one antenna only. In the rest of the paper, the following notations are used:

- $x_{a,k} \in \mathbb{C}$ is the coded modulated symbol for channel use k, sent from node $a \in \mathcal{S} \cup \mathcal{R}$, whose power is normalized to unity.
- $y_{a,b,k}$ is a received signal at node $b \in \mathcal{S} \cup \mathcal{R} \cup \{d\} \setminus \{a\}$, originating from node a described previously.
- $\gamma_{a,b}$ is the average signal-to-noise ratio (SNR) that captures both path-loss and shadowing effects.

- $h_{a,b}$ are the channel fading gains, which are independent and follow a zero-mean circularly symmetric complex Gaussian distribution with variance $\gamma_{a,b}$.
- $n_{a,b,k}$ are independent and identically distributed AWGN samples, which follow a zero-mean circularly-symmetric complex Gaussian distribution with unit variance.

Using the previous notation, the received signal at node $b \in \mathcal{S} \cup \mathcal{R} \cup \{d\} \setminus \{a\}$, originating from node $a \in \mathcal{S} \cup \mathcal{R}$ can be represented as:

$$y_{a,b,k} = h_{a,b} x_{a,k} + n_{a,b,k},\tag{1}$$

where k denotes a given channel use. During the first phase, $k \in \{1, \ldots, N_1\}$, while during the second phase, $k \in \{1, \ldots, N_2\}$, where N_1 and N_2 represent the number of channel uses in each phase respectively.

3   Problem Formulation

Let us define the initial transmission rate of source s as $R_s = K_s/N_1$ in bits per complex dimension or bits per channel use [b.c.u]. In the slow-link adaptation scenario, the long-term transmission rate $\bar{R}_s$ per source is defined as the number of transmitted bits over the total number of channel uses spent:

$$\begin{aligned}
\bar{R}_s^{\text{sla}} &= \frac{K_s}{MN_1 + N_2 \mathbb{E}(T_{\text{used}})} \\
&= \frac{R_s}{M + \alpha \mathbb{E}(T_{\text{used}})},
\end{aligned}\tag{2}$$

where we take into account the average number of retransmission rounds used in the second phase: $\mathbb{E}(T_{\text{used}}) = \sum_{t=1}^{T} t \Pr\{T_{\text{used}} = t\}$ (for a number of frames that tends to infinity), and $\alpha = N_2/N_1$. Note that the denominator of Eq. 2 represents the number of channel uses corresponding to both transmission and retransmission phases. The transmission phase corresponds to M time slots, each with N_1 channel uses, and the retransmission phase corresponds to T_{used} time slots, each with N_2 channel uses.

In the same scenario, the average spectral efficiency, which is used as a performance metric in this work, can be defined as:

$$\eta^{\text{sla}} = \sum_{i=1}^{M} \bar{R}_i^{\text{sla}} (1 - \Pr\{\mathcal{O}_{i,T}\}),\tag{3}$$

where $\mathcal{O}_{s,T}$ is the event that source s is not decoded correctly at the destination after round T, called the "individual outage event of source s after round T" in the following. This metric is chosen since it describes how well the spectral assets are utilized in the MAMRC model adapted. It is composed of the sum over all sources participating in each frame transmission. The sum will include the transmission rate per source multiplied by the probability of transmission success of this source message.

The individual outage event $\mathcal{O}_{s,t}(a_t, \mathcal{S}_{a_t,t-1}|\mathbf{h}_{\mathrm{dir}}, \mathcal{P}_{t-1})$, of source s after round t, depends on selected node $a_t \in \mathcal{N}$ and associated decoding set $\mathcal{S}_{a_t,t-1}$. It is conditional on the knowledge of $\mathbf{h}_{\mathrm{dir}}$ and $\mathcal{P}_{t-1}$, where $\mathcal{P}_{t-1}$ denotes the set collecting the nodes $\hat{a}_l$ which were selected in rounds $l \in \{1,\ldots,t-1\}$ prior to round t together with their associated decoding sets $\mathcal{S}_{\hat{a}_l,l-1}$, and the decoding set of the destination $\mathcal{S}_{d,t-1}$ ($\mathcal{S}_{d,0}$ is the destination's decoding set after the first phase). Similarly, we define the "common outage event after round t" $\mathcal{E}_t(a_t, \mathcal{S}_{a_t,t-1}|\mathbf{h}_{\mathrm{dir}}, \mathcal{P}_{t-1})$ as the event that at least one source is not decoded correctly at the destination at the end of round t. In the rest of the paper, in order to simplify the notation, the dependency on $\mathbf{h}_{\mathrm{dir}}$ and $\mathcal{P}_{t-1}$ is omitted.

Analytically, the common outage event of a given subset of sources is declared if the vector of their rates lies outside of the corresponding MAC capacity region. For some subset of sources $\mathcal{B} \subseteq \bar{\mathcal{S}}_{d,t-1}$, where $\bar{\mathcal{S}}_{d,t-1} = \mathcal{S} \backslash \mathcal{S}_{d,t-1}$ is the set of non-successfully decoded sources at the destination after round $t-1$, for candidate node a_t this event can be expressed as:

$$\mathcal{E}_{t,\mathcal{B}}(a_t, \mathcal{S}_{a_t,t-1}) = \bigcup_{\mathcal{U} \subseteq \mathcal{B}} \left[\sum_{i \in \mathcal{U}} R_i > \sum_{i \in \mathcal{U}} I_{i,d} + \sum_{l=1}^{t-1} \alpha I_{\hat{a}_l,d}[\mathcal{C}_{\hat{a}_l}] + \alpha I_{a_t,d}[\mathcal{C}_{a_t}] \right], \quad (4)$$

where $I_{a,b}$ denotes the mutual information between the nodes a and b, $[P]$ representing the Iverson bracket which gives 1 if the event P is satisfied, and 0 if not, and where $\mathcal{C}_{\hat{a}_l}$ and $\mathcal{C}_{a_t}$ have the following definitions:

$$\begin{aligned}
\mathcal{C}_{\hat{a}_l} &= \left[[\mathcal{S}_{\hat{a}_l,l-1} \cap \mathcal{U} \neq \emptyset] \wedge [\mathcal{S}_{\hat{a}_l,l-1} \cap \mathcal{I} = \emptyset] \right], \\
\mathcal{C}_{a_t} &= \left[[\mathcal{S}_{a_t,t-1} \cap \mathcal{U} \neq \emptyset] \wedge [\mathcal{S}_{a_t,t-1} \cap \mathcal{I} = \emptyset] \right].
\end{aligned} \quad (5)$$

In (5), the sources that belong to $\mathcal{I} = \bar{\mathcal{S}}_{d,t-1} \backslash \mathcal{B}$ are considered as interference, with $\wedge$ standing for the logical and. In (4), for each subset $\mathcal{U}$ of set $\mathcal{B}$, we check if the sum-rate of sources contained in $\mathcal{U}$ is higher than the accumulated mutual information at the destination (since IR-type of HARQ is used). The accumulated mutual information is split into three summations, which originate from the direct transmissions from sources contained in $\mathcal{U}$, the transmission of previously activated nodes during the second phase: $\sum_{l=1}^{t-1} \alpha I_{\hat{a}_l,d}[\mathcal{C}_{\hat{a}_l}]$, and the transmission of the candidate node a_t during the second phase: $\alpha I_{a_t,d}[\mathcal{C}_{a_t}]$.

The individual outage event of source s after round t for candidate node a_t can be defined as (check [8] prop. 1):

$$\begin{aligned}
\mathcal{O}_{s,t}(a_t, \mathcal{S}_{a_t,t-1}) &= \bigcap_{\mathcal{I} \subset \bar{\mathcal{S}}_{d,t-1}, \mathcal{B}=\bar{\mathcal{I}}, s \in \mathcal{B}} \mathcal{E}_{t,\mathcal{B}}(a_t, \mathcal{S}_{a_t,t-1}), \\
&= \bigcap_{\mathcal{I} \subset \bar{\mathcal{S}}_{d,t-1}} \bigcup_{\mathcal{U} \subseteq \bar{\mathcal{I}}: s \in \bar{\mathcal{I}}} \left[\sum_{i \in \mathcal{U}} R_i > \sum_{i \in \mathcal{U}} I_{i,d} + \sum_{l=1}^{t-1} \alpha I_{\hat{a}_l,d}[\mathcal{C}_{\hat{a}_l}] + \alpha I_{a_t,d}[\mathcal{C}_{a_t}] \right],
\end{aligned} \quad (6)$$

where $\bar{\mathcal{I}} = \bar{\mathcal{S}}_{d,t-1} \backslash \mathcal{I}$.

The individual outage probability $\Pr\{\mathcal{O}_{s,t}\} = \Pr\{\mathcal{O}_{s,t}(\hat{a}_t, \mathcal{S}_{\hat{a}_t,t-1})\}$ stands in practice for the average BLER of source s after t retransmissions. In the following, we denote it either $\Pr\{\mathcal{O}_{s,t}\}$ or $BLER_{s,t}$.

4 Link Adaptation Best-Response Algorithm

Let the source s transmit with the rate R_s, and let us denote the average BLER after T rounds with $BLER_{s,T}(R_s)$. In the point-to-point link scenario, the individual throughput of the source s is given by $\eta_s = \bar{R}_s \times (1 - BLER_{s,T}(R_s))$. In order to maximize it, the usual practice would be to find the optimal pair $(R_s, BLER_{s,T}(R_s))$. In the MAMRC setup however, $BLER_{s,T}$ depends on the vector of rates $(R_1, \ldots, R_M)$. The reason for the dependence of $BLER_{s,T}$ on all sources' rates lies in the fact that the decoding set of the selected node in the given round depends on all the rates, which influences the probability of non-successful decoding of the message of the source s. Hence, in theory, all $(R_1, \ldots, R_M)$ need to be optimized jointly in order to reach the optimal solution.

In order to be precise with the notation, we distinguish hereafter $\hat{R}_s$, the rate of source s after the optimization, and R_i, one possible value of $\hat{R}_s$ taken from the set of possible rates $\{\tilde{R}_1, \ldots, \tilde{R}_{n_{\mathrm{MCS}}}\}$, where n_{MCS} is the number of different modulation and coding schemes.

In general, the optimization problem that is needed to be solved under the given individual QoS constraints for slow-link adaptation is given by (7). It should be noted that $\Pr\{\mathcal{O}_{s,T}\}$ is conditional on the node activation sequence.

$$(\hat{R}_1, \ldots, \hat{R}_M) = \underset{(R_1, \ldots, R_M) \in \{\tilde{R}_1, \ldots, \tilde{R}_{n_{\mathrm{MCS}}}\}^M}{\mathrm{argmax}} \sum_{i=1}^{M} \frac{R_i}{M + \alpha \mathbb{E}(T_{used})} \Big(1 - \Pr\{\mathcal{O}_{i,T}\}\Big),$$

$$\text{subject to } \Pr\{\mathcal{O}_{i,T}\} \leq BLER_{\mathrm{QoS},i}, \forall i \in \{1, \ldots, M\}.$$

$$(7)$$

Solution for the given problem is analytically intractable to the knowledge of the authors. Namely, it is hard to predict what is the decoding set of each node in a given round, as it depends on the rates of sources, which we are trying to find in the first place. The same conclusion holds even in the case where the node activation sequence is pre-determined and known in advance.

Concerning the starting point of our BRD algorithm, we can resort to an approach that is based on a "Genie-Aided" (GA) assumption, where all the sources $s \in \mathcal{S}\setminus i = \{1, 2, \ldots, i-1, i+1, \ldots, M\}$ except the one for which we want to allocate the rate, i, are assumed to be decoded correctly at the destination and relaying nodes. Such an assumption can reduce the complexity of the problem, and helps us start with a clever starting point. In this paper, we mainly focus on the BRD algorithm, rather than the GA approach. For this reason, and for the sake of brevity, the detailed explanation, as well as the step by step algorithm of the GA approach will be omitted.

After setting the starting point of the rates, our BRD algorithm starts. In a given iteration t, all the sources' rates are updated in a cyclic fashion. The rate

---

Algorithm 1. Best-Response algorithm under the QoS constraints on individual BLER targets $BLER_{\text{QoS},i}, \forall i \in \{1, \ldots, M\}$.

1: $t \leftarrow 0.$ ▷ Counter of iterations.
2: Set the candidate rates: $\{\tilde{R}_1, \ldots, \tilde{R}_{n_{MCS}}\}$.
3: Rate initialization under GA assumption with a random node selection: $[\hat{R}_1(0), \ldots, \hat{R}_M(0)] \leftarrow [R_1^{\text{GA}}, \ldots, R_M^{\text{GA}}]$.
4: $\hat{R}_i(-1) \leftarrow 0$ for all $i \in \{1, \ldots, M\}$ ▷ To force loop to start
5: **while** $(|\hat{R}_i(t) - \hat{R}_i(t-1)| > 0)$, for some $i \in \{1, \ldots, M\}$ **do**
6:     $t \leftarrow t + 1.$
7:     **for** $i \leftarrow 1$ **to** M **do**   ▷ for all sources, choose:
8:        $\hat{R}_i(t) \leftarrow \text{argmax}_{R_i \in \{\tilde{R}_1, \ldots, \tilde{R}_{n_{MCS}}\}} \eta^{sla}\big(\hat{R}_1(t), \ldots \hat{R}_{i-1}(t), R_i, R_{i+1}(t-1), \ldots, \hat{R}_M(t-1)\big)$ ▷ the rate which maximizes η
       such that $\Pr\{\mathcal{O}_{i,T}\} \leq BLER_{\text{QoS},i}$ ▷ while satisfying the constraint
9:     **end for**
10: **end while**

---

of source i is a function of the sources' rates updated in the same iteration prior to source i (sources with index $i' < i$), and the rates updated for the last time in the previous iteration, $t-1$ (sources with index $i'' > i$). Since our goal is to maximize the total average spectral efficiency, we recall that our profit function for each user i is the performance metric defined in Eq. (3).

Comparing with Eq. (7) where the multivariate optimization problem is considered, here, the problem is relaxed such that every source will only work with the source rate allocated to it particularly, where other sources' rates are fixed at a current iteration. Following the BRD tools, and in an iterative algorithm and a circular manner, each user will choose the most suitable rate, which optimizes the associated profit function. We recall that each user is subjected to a QoS constraint $\Pr\{\mathcal{O}_{i,T}\} \leq BLER_{\text{QoS},i}$ for $i \in \{1, \ldots, M\}$, and thus has no right to break any of these constraints while choosing its rate. In other words, a source has no right to change its source rate to one which increases the spectral efficiency but breaks any of the QoS constraints. Following this description, each user will have to choose the optimal rate that meets the constraints. Note that this procedure is repeated until no user tends to change its source rate, where at this point, we reach convergence because any change will either decrease our profit function or break the QoS constraints.

The principle of the update function for source i is to check whether the average spectral efficiency η_{current} increases if we choose another source rate from the available source rates set. Additionally, we check if the condition of the individual BLER is below the corresponding target, $BLER_{\text{check}}$, is true. If both conditions are met, we choose the source rate which maximizes the efficiency and meets the BLER target. Once we reach our "optimal" source rate for a given user, we proceed to the next user, following the same steps.

The complexity of the proposed iterative rate correction algorithm (see algo. 1) is much smaller than in the case of the exhaustive search approach algorithm. In the latter, the calculation of each individual BLER is performed $n_{\text{MCS}}{}^M$

times, while in the proposed algorithm, in one iteration the same calculation is performed $n_{\mathrm{MCS}} \times M$ times. The complexity of an iteration corresponds to the $argmax$ expression mentioned in algorithm 1, in step 8. Since each source node will pass through all possible values of n_{MCS}, we reach the result of $n_{\mathrm{MCS}} \times M$. Finally, we mention that due to the nature of the rate-efficiency relationship, and as the simulation results showed, the number of iterations used in the BRD algorithm will be on average between one or two iterations.

5   Numerical Results

In this section, we validate the proposed slow-link algorithm by performing Monte-Carlo simulations. We consider (3, 3, 1)-OMAMRC, with $T = 4$ and $\alpha = 0.5$. The allocated rates are chosen from a discrete MCS family whose rates belong to the set $\{0.5,1,1.5,2,2.5,3,3.5\}$ [b.c.u]. We assume independent Gaussian distributed channel inputs (with zero mean and unit variance), with $I_{a,b} = \log_2(1 + |h_{a,b}|^2)$. Note that some other formulas could be also used for calculating $I_{a,b}$ but they would not have any impact on the basic concepts of this work [9].

Asymmetric link configuration is assumed, where the average SNR of each link is set as follows. First, the average SNR of each link is set to γ. Second, the average SNR of each link that includes source 2 is reduced by 4 dB and which includes source 3 is reduced by 7 dB. Finally, the average SNR of the link between the sources 2 and 3 is set to $\gamma - 5$ dB. In that way, we have set on purpose the source 1 to be in the best propagation conditions, while the source 3 is in the worst ones. Here, we validate the performance of the slow-link adaptation algorithm under the constraint on the individual BLER target after T retransmissions for each source set to $BLER_{\mathrm{QoS}}^{(1)} = 1$. This kind of choice for the individual BLER targets corresponds to a pure maximization of the average spectral efficiency. Figure 2 shows the average spectral efficiency that can be obtained using the allocated rates that result from different algorithms, as a function of γ. Five different algorithms are considered: the proposed best-response dynamics algorithm, the algorithm based solely on GA assumption with random node selection (referred to as "Genie-Aided approach"), two trivial algorithms where the rates are predetermined and set to the minimum ("Min rates: $R_1 = 0.5, R_2 = 0.5, R_3 = 0.5$ [b.c.u.]") and the maximum ("Max rates: $R_1 = 3.5, R_2 = 3.5, R_3 = 3.5$ [b.c.u.]") possible ones, and the exhaustive search approach algorithm. We observe that the average spectral efficiency obtained by the proposed slow-link adaptation algorithm coincides with the one obtained by the exhaustive search approach algorithm, for each γ. This indicates that our proposed algorithm is converging to the optimal solution. In addition we see that our first step GA approach lacks this convergence. Also, our algorithm provides the gain of approx. 2 dB compared with the strategy of predetermined maximum possible allocated rates, the difference becoming lower as γ grows (it converges to zero for $\gamma = 15$ dB). The performance of the strategy of predetermined minimum possible allocated rates quickly becomes highly non-optimal starting already from $\gamma = -4$ dB, as γ grows.

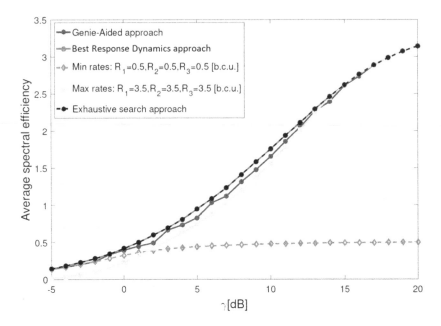

Fig. 2. Average spectral efficiency that corresponds to different slow-link adaptation algorithms s.t. $BLER_{QoS}^{(1)}$ target.

6 Conclusion

In this paper, a slow-link adaptation algorithm is proposed for slow-fading orthogonal multiple access multiple relay channel. The relays are half-duplex and the orthogonal multiple access is performed in time under the scheduling control of the destination. During the first phase, the sources transmit in turn. The proposed slow-link adaptation algorithm aims at maximizing the average spectral efficiency under individual QoS targets for a given modulation and coding scheme family. To reduce the complexity of proposed algorithms found in the literature, a best-response dynamics algorithm is performed to correct iteratively the previously chosen rates. The resulting scheduling and link adaptation algorithm offers a tractable complexity under practical knowledge of the channel qualities and yield performance close to corresponding exhaustive search approaches as demonstrated by Monte-Carlo simulations.

References

1. Van Der Meulen, E.C.: Three-terminal communication channels. Adv. Appl. Probab. **3**(1), 120–154 (1971)
2. Cover, T., Gamal, A.E.: Capacity theorems for the relay channel. IEEE Trans. Inf. Theory **25**(5), 572–584 (1979)
3. Kramer, G., van Wijngaarden, A.J.: On the white Gaussian multiple-access relay channel. In: Proceedings of IEEE ISIT 2000, Sorrento, Italy, June 2000

4. Bistritz, I., Leshem, A.: Convergence of approximate best-response dynamics in interference games. In: Proceedings of IEEE CDC (2016)

5. Douros, V., Toumpis, S., Polyzos, G.C.: Power control under best response dynamics for interference mitigation in a two-tier femtocell network. In: Proceedings of WiOpt (2012)

6. Han, Z.: Game Theory in Wireless and Communication Networks: Theory, Models and Applications. C. U. Press (2012)

7. Lasaulce, S., Tembine, H.: Game Theory and Learning for Wireless Networks: Fundamentals and Applications. Academic Press (2011)

8. Mohamad, A., Visoz, R., Berthet, A.O.: Outage analysis of various cooperative strategies for the multiple access multiple relay channel. In: 2013 IEEE 24th Annual International Symposium on Personal, Indoor, and Mobile Radio Communications (PIMRC), pp. 1321–1326. IEEE (2013)

9. Polyanskiy, Y., Poor, H.V., Verdu, S.: Channel coding rate in the finite blocklength regime. IEEE Trans. Inf. Theory **56**(5), 2307–2359 (2010)

Stochastic Models for Network
Performance Analysis

Vaccination in a Large Population: Mean Field Equilibrium Versus Social Optimum

Josu Doncel[1]([✉]), Nicolas Gast[2], and Bruno Gaujal[2]

[1] University of the Basque Country, 48940 Leioa, Spain
josu.doncel@ehu.eus
[2] Univ. Grenoble Alpes, Inria, CNRS, LIG, 38000 Grenoble, France

Abstract. We analyze a virus propagation dynamics in a large population of agents (or nodes) with three possible states (Susceptible, Infected, Recovered) where agents may choose to vaccinate. We show that this system admits a unique symmetric equilibrium when the number of agents goes to infinity. We also show that the vaccination strategy that minimizes the social cost has the same threshold structure as the mean field equilibrium, but with a shorter threshold. This implies that, to encourage optimal vaccination behavior, vaccination should always be subsidized.

1 SIRV Dynamics

A large number N of *agents* (that can be nodes in a communication network, persons in a crowd, or abstract players in a mathematical game) are subject to interactions: they meet (or communicate with) each other according to a uniform process, described in the following. Each agent (or player[1]) has 3 possible states: Susceptible, Infected, Recovered (S, I, R).

When an agent in state S meets an agent in state I, it gets infected. An agent in state I will eventually recover and go to state R. An agent in state R stays in R forever. This is a classical SIR model of virus propagation among the agents.

This model can be seen as over-simplistic but actually, it has a good predictive power for human epidemics and other diffusion processes. This model was introduced as early as in 1927 by Kermack and McKendrick in a series of papers [1] and *"because of their seminal importance to the field of theoretical epidemiology, these articles were republished in the Bulletin of Mathematical Biology in 1991"* (from wikipedia). This so-called SIR model has been studied ever since, being the subject of papers in mathematics, computer science, health studies and bio-mathematics.

Here, we consider a stochastic evolution of the population, whose state changes are driven by Poisson events and players can decide to vaccinate (hence the V in the name of our model).

– A player encounters other players with rate γ (activity of the player). A slight generalization is possible without any impact on the following analysis. The

---

[1] Both terms will be used in the following, interchangeably.

© Springer Nature Switzerland AG 2021
S. Lasaulce et al. (Eds.): NetGCooP 2021, CCIS 1354, pp. 51–59, 2021.
https://doi.org/10.1007/978-3-030-87473-5_6

rate of encounters with susceptible players is equal to γ_1 and the rate of encounters with infected players is equal to $\gamma_2(< \gamma_1)$ to take into account a lesser activity of infected players. If the first player is Susceptible and the second is Infected in an encounter, then the first one becomes Infected.
- An Infected agent Recovers at rate ρ.
- A Susceptible agent can *decide* to get vaccinated. It chooses its vaccination rate $\pi(t) \in [0, \theta]$.
- Once an agent is vaccinated or recovered, its state becomes R in both cases and does not change after this point.

Similar models have been studied in [2,3], although the vaccination policies in these papers do not depend explicitly on the susceptible population.

Let $(m_S(t), m_I(t), m_R(t))$ be the proportion of the agents in states S, I, R respectively at time t. The Markovian evolution of one player is given by a non-homogeneous continuous time Markov chain displayed in Fig. 1.

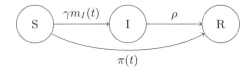

Fig. 1. Markov chain driving the evolution of one agent in the population

2 Cost Functions and Objectives

In addition to the state evolution, the system is endowed with cost functions.

The cost of being infected is c_I per time unit.

The vaccination cost is linear in its rate: For one player, the cost to vaccinate at rate π is $c_V \pi$.

We are now ready to state the problem to be solved: *under full information and common knowledge (the state of all players at time t and their vaccination strategy is known to all), each player wants to choose a vaccination strategy that minimizes its cost up to a time horizon T.*

This is not a well defined problem because the optimal strategy of any player depends on the strategy of any other player, who in turn is trying to optimize its vaccination strategy that depends on the first player's strategy. One classical way to get around this difficulty is to consider "stable points", that is Nash equilibria and social optimum.

Definition 1 (Symmetric Nash Equilibrium (SNE)). *A Symmetric Nash Equilibrium is a vaccination strategy π_{NE} such that if all the players use π_{NE}, then any player's optimal strategy is to use π_{NE}.*

Definition 2 (Social Optimal). *A social optimal is a vaccination strategy π_{SO} imposed upon all players that minimizes the sum of the costs of all the players.*

SNE always exist in our SIRV model (the proof is based on *Kakutani* fixed point theorem, not detailed here). SO always exist in our SIRV model (The proof uses the compacity of the strategy space for the weak topology and is not detailed here). For more details on these two proofs of existence, see [4].

Unfortunately both strategies are very hard to compute when N is large because of the combinatorial explosion of the state space.

3   Mean Field Limit

Since the players are indistinguishable, the state of the system is given by the *population distribution*, $(m_S(t), m_I(t), m_R(t))$.

When $N \to \infty$, $(m_S(t), m_I(t), m_R(t))$ behaves as a fluid, whose evolution follows the Kolmogorov equations of the individual Markov chain (see for example [4,5]). Under a given vaccination strategy π.

$$\begin{cases} \dot{m}_S(t) = -\gamma m_S(t) m_I(t) - \pi(t) m_S(t) \\ \dot{m}_I(t) = \gamma m_S(t) m_I(t) - \rho m_I(t) \\ \dot{m}_R(t) = \rho m_I(t) + \pi(t) m_S(t). \end{cases} \tag{1}$$

There are some technicalities here, because $\pi(t)$ may not be continuous. In this case, one can still use the Carathéodory Existence Theorem to show that the solution of these equation is well defined and unique once $(m_S(0), m_I(0), m_R(0))$ are given (see [4] for a rigorous proof).

When $\pi(\cdot) \equiv 0$, this yields the classical SIR dynamics of *Kermack and McKendrick (1927)*. Even in this simple case, completely closed form solutions have not been found (see the recent results in [6]). However, the mean field framework is still easier to analyze qualitatively than the case with a finite number of players because, at the limit, the strategy chosen by a single player will not alter the behavior of the whole population. To make this more precise, we introduce the best response correspondence in the mean field case.

3.1   Best Response

Let us pick one player among the infinite population (it is called player 0 in the following). Let us define the *state probabilities* (p_S^0, p_I^0, p_R^0) of Player 0 as follows: For $X \in \{S, I, R\}$,

$$p_X^0(t) = \mathbb{P}(\text{ Player 0 is in state } X \text{ at time } t).$$

If player 0 uses strategy π_0 while the population uses strategy π, then the state probabilities (p_S^0, p_I^0, p_R^0) of Player 0 has an evolution given by its local Kolmogorov equation:

$$\begin{cases} \dot{p}_S^0(t) = -\gamma p_S^0(t) m_I(t) - \pi_0(t) p_S^0(t) \\ \dot{p}_I^0(t) = \gamma p_S^0(t) m_I(t) - \rho p_I^0(t) \\ \dot{p}_R^0(t) = \rho p_I^0(t) + \pi_0(t) p_S^0(t). \end{cases}$$

The population distribution $(m_S(t), m_I(t), m_R(t))$ used in these equations are the solutions of the mean field limit Eq. (1) and do not depend on the strategy chosen by Player 0.

Using the foregoing notations, the expected individual cost for Player 0 is:

$$W(\pi_0, \pi) = \int_0^T \left(c_V \pi_0(t) p_S^0(t) + c_I p_I^0(t) \right) dt,$$

where c_V is the vaccination cost and c_I is the unit time cost of being infected, as defined earlier.

A *best response* of Player 0 to a population using strategy π is a strategy π_0^* that minimizes its cost.

This best response can be computed using a Bellman optimality equation. If we denote by $W_X^*(t)$ the optimal total cost from t to T of Player 0 when it is in state X at time t, they satisfy

$$W_R^*(T) = W_S^*(T) = W_I^*(T) = 0.$$

and

$$- \dot{W}_S^*(t) = \min_{\pi_0(t)} \left[\pi_0(t) \left(c_V - W_S^*(t) \right) + \gamma m_I(t)(W_I^*(t) - W_S^*(t)) \right] \qquad (2)$$

$$-\dot{W}_I^*(t) = c_I - \rho W_I^*(t). \qquad (3)$$

The best response strategy when in state S at time t is given by

$$\pi_0^*(t) = \arg \min_{\pi_0(t)} \left[\pi_0(t) \left(c_V - W_S^*(t) \right) + \gamma m_I(t)(W_I^*(t) - W_S^*(t)) \right]. \qquad (4)$$

Let us denote by $BR(\pi) = \pi_0^*$, the *best responses* of Player 0 to π.

Definition 3 (Mean Field Equilibrium (MFE)). *If $\pi \in BR(\pi)$ then π is a mean field equilibrium.*

3.2 Threshold Policy

By analyzing the Bellman Equations (2)–(3), we can show that the best response strategy is unique and has a specific structure.

Lemma 1. *For any population strategy π, the best-response $\pi_0^* \in BR(\pi)$ is a threshold strategy: There exists a critical time $t_c^0(\pi)$ s.t.*

$$\pi_0^*(t) = \begin{cases} \theta & \text{if } t < t_c^0(\pi), \\ 0 & \text{if } t > t_c^0(\pi). \end{cases}$$

The full proof is available in [7], is essentially based on the analysis of

$$\pi_0^*(t) = \arg \min_{\pi_0(t)} \left[\pi_0(t) \left(c_V - W_S^*(t) \right) + \gamma m_I(t)(W_I^*(t) - W_S^*(t)) \right],$$

which implies that $\pi_0(t)$ must be 0 whenever $c_V > W_S^*(t)$ and θ whenever $c_V < W_S^*(t)$.

This induces the following theorem.

Theorem 1. *SIRV has a unique mean-field equilibrium. This equilibrium is deterministic (players do not randomize their decisions) and is a threshold strategy. The threshold will be denoted by τ_c^{MFE}.*

In 2004, Francis ([2] showed that the stable point for each player in the SIRV equations has a threshold. This result precedes the introduction of mean field games and mean field equilibria, only defined in 2007 in [8]. It is quite remarkable (and also natural) that the two notions should coincide.

Theorem 1 is proved in the long version of this paper, available as a Research Report [7].

We believe that t_c^{MFE} cannot be computed in closed form as a function of $T, c_I, c_V, \theta, \rho$ and γ. However, the complexity reduction from finding a stable strategy (whose domain has an infinite dimension) to determining one real number makes numerical computations feasible (see Sect. 5).

4   Social Optimal Strategy

We denote by $C(\pi)$ the average cost incurred by the population under strategy π, i.e.,

$$C(\pi) = \int_0^T \left(c_I m_I(t) + c_V \pi(t) m_S(t) \right) dt.$$

A *social optimal* strategy minimizes the total cost:

$$\pi^{opt} \in \arg\min_\pi C(\pi).$$

It is known that the solution of this problem is a bang-bang strategy [9] (only vaccination rates 0 and *theta* will be used). A further analysis of the dynamics shows the following result:

Proposition 1. *The strategy that minimizes the total cost is a threshold strategy: There exists a critical time τ^{opt} s.t.*

$$\pi^{opt} = \begin{cases} \theta & \text{if } t < \tau^{opt}, \\ 0 & \text{if } t > \tau^{opt}. \end{cases}$$

Proof. (Sketch) The proof is based on the *Pontryagin* maximum principle: If π^{opt} is an optimal strategy, then there exist two Lagrange multipliers $\lambda_S(t)$ and $\lambda_I(t)$ such that $\lambda_S(T) = 0$, $\lambda_I(T) = 0$ and for any $t < T$,

$$-\dot{\lambda}_S = c_V \pi^{opt}(t) + (-\gamma m_I^{opt}(t) - \pi^{opt}(t))\lambda_S + \gamma m_I^{opt}(t)\lambda_I$$
$$-\dot{\lambda}_I = c_I - \gamma m_S^{opt}(t)\lambda_S + (\gamma m_S^{opt}(t) - \rho)\lambda_I$$
$$\pi^{opt}(t) = \arg\min[c_V \pi(t) m_S^{opt}(t) + c_I m_I^{opt}(t)$$
$$+ (\gamma m_S^{opt}(t) m_I^{opt}(t) - \pi(t) m_S^{opt}(t))\lambda_S$$
$$+ (\gamma m_S^{opt}(t) m_I^{opt}(t) - \rho m_I^{opt}(t))\lambda_I],$$

where $m_I^{opt}(t), m_S^{opt}(t)$ are the proportions of the population in states I and S respectively, at time t, under the optimal strategy. By straightforward simplifications, one gets

$$- \dot{\lambda}_S = \inf_{\pi} \left(\pi(t)(c_V - \lambda_S) + \gamma m_I^{opt}(t)(\lambda_I - \lambda_S) \right) \tag{5}$$

$$-\dot{\lambda}_I = c_I - \rho\lambda_I + \gamma m_S^{opt}(t)(\lambda_I - \lambda_S) \tag{6}$$

$$\pi^{opt}(t) = \arg \min \left(\pi(t) m_S^{opt}(t)(c_V - \lambda_S) \right). \tag{7}$$

One can notice a similarity between Eqs. (5)–(6) for the Lagrange multipliers and Eqs. (2)–(3) for the costs of the best response.

This induces a similar shape for the strategy that optimizes the social cost: There exists a critical time τ^{opt} s.t.

$$\pi^{opt}(t) = \theta \text{ if } t < \tau^{opt} \text{ and } 0 \text{ if } t > \tau^{opt}.$$

Proposition 2. *The threshold of the optimal social cost is larger than the threshold of the mean field equilibrium: $\tau^{opt} \geq \tau^{MFE}$.*

Proof. Again, the proof is based on the fact that Eqs. (5)–(6) and Eqs. (2)–(3) are similar up to the additional term $\gamma m_S^{opt}(t)(\lambda_I - \lambda_S)$ for λ_I. Using this, the comparison between the optimal strategy and the mean field equilibrium bowls down to the comparisons of the Lagrange Multipliers λ_S, λ_I and the costs J_S, J_I.

One easy case is when c_V is larger than c_I/ρ. In this case, for all t, $\lambda_S \leq \lambda_I \leq c_I/\rho \leq c_V$ so that the jump time of the mean field equilibrium is $\tau^{eq} = 0$. Therefore, the socially optimal jump time τ^{opt} can only be larger than τ^{eq}.

Let us now consider the case when $c_V < c_I/\rho$. In this case, τ^{opt} is the time when λ_S gets below c_V. By examining the Lagrange multipliers λ_S and λ_I between τ^{opt} and T, one can show that they must satisfy the following properties:

- $\lambda_S(T) = 0,\ \lambda_I(T) = 0,$
- $\forall t \in [\tau^{opt}, T], \lambda_S(t) \leq \lambda_I(t).$

Indeed, if there is a time t such that $\lambda_S(t) = \lambda_I(t)$, then their derivatives become comparable $(\dot{\lambda}_S(t) \leq \dot{\lambda}_I(t))$. Therefore, the additional term $\gamma m_S^{opt}(t)(\lambda_I - \lambda_S)$ in (5) remains positive so that $\lambda_I(t) \leq J_I(t), \forall \tau^{opt} \leq t \leq T$. In turn this implies that $\lambda_S(t) \geq J_S(t), \forall \tau^{opt} \leq t \leq T$.

This implies that $J_S(\tau^{opt}) \leq \lambda_S(\tau^{opt}) = c_V$. Finally, this implies that τ^{MFE} (the time when J_S crosses level c_V) is smaller that τ^{opt}, with equality only when $\tau^{opt} = 0$.

This result should be intuitive: Each member of the population has a personal incentive to stops vaccinating when the individual risk becomes low enough, while the social planner has a higher standard for the virus disappearance before deciding to stop the vaccination.

5 Numerical Comparisons

5.1 Comparison of MFE and Social Optimum

In this section, we report a numerical evaluation of the threshold of the two strategies[2]. We consider the same system parameters as in [3], which is based on the epidemiological study of the H1N1 epidemic of 2009–2010 in France : $\rho = 36.5$, $\gamma = 73$, $\theta = 10$, $c_I = 36.5$ and $c_V = 0.5$. Besides, we consider that the proportion of susceptible and infected population at time 0 are both equal to 0.4.

For these parameters, we compute the optimal strategy and the mean field equilibrium over a time horizon of a year, that is, $T = 1$. The results are reported in Fig. 2a where the population state space is divided into three regions that represent the decisions taken by both strategies at time 0, as a function of the initial state. In the white region, both strategies vaccinate at maximum rate. In the dark gray region, the strategy of the social optimum is to vaccinate at maximum rate and the strategy of the equilibrium is to not vaccinate. In the light gray region, both strategies are to not vaccinate.

We also plot the trajectories corresponding to both strategies when the proportion of infected population and of susceptible population at time 0 are both equal to 0.4. In Fig. 2a (see Fig. 2b for a zoomed figure), we plot with a solid line the behavior of the equilibrium vaccination strategy, and with a dashed line, the behavior of the social optimum. The obtained cost for the equilibrium vaccination strategy for the parameters under consideration is 0.6824, whereas for the social optimum vaccination strategy is 0.6818.

5.2 Mechanism Design

For any vaccination cost c_V, while the other parameters remain fixed, we denote by $\tau^{opt}(c_V)$ (resp. $\tau^{MFE}(c_V)$) the jump time of the socially optimal strategy (resp. equilibrium strategy). It can be shown that in both cases, the jump times are decreasing in c_V: the more costly is the vaccination, the less people vaccinate (for the socially optimal situation as well as for the mean field equilibrium). Figure 3 confirms that the jump times decrease with c_V and also shows that the jump times are never equal for this range of parameters.

Therefore, if the vaccination decisions are let to individuals, then vaccination should be subsidized, by offering a subsidy g off the vaccination cost so that both jump times coincide, i.e.,

$$\tau^{MFE}(c_V - g) = \tau^{opt}(c_V).$$

For example, with the same parameters as in the simulation of Fig. 2, and for $c_V = 0.8$, the jump time of the social optimum is 0.0106, while the jump time of the equilibrium is 0. When $c_V = 0.65$, the jump time of the equilibrium is 0.034. This simulation shows that the subsidy required to encourage selfish individuals to vaccinate optimally consists of a reduction of the vaccination cost of $g = 0.15$.

---

[2] The codes to reproduce these experiments are available at https://github.com/josudoncel/MeanFieldGameAnalysisSIRModelVaccinations.

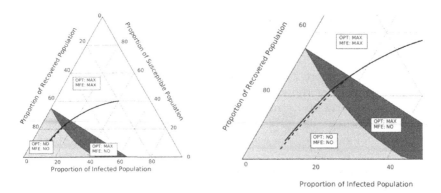

(a) Population dynamics.   (b) Pop. dynamics (zoomed).

Fig. 2. Population dynamics under the equilibrium strategy (dashed line) and the socially optimal strategy (solid line). Three zones are displayed: *(i)* in the white region, the social optimum and the equilibrium vaccinate with maximum rate; *(ii)* in the dark gray region, the social optimum vaccinates with maximum rate, while the equilibrium does not vaccinate; and *(iii)* in the light gray region, neither the social optimum nor the equilibrium vaccinates. $m_I(0) = m_S(0) = 0.4$.

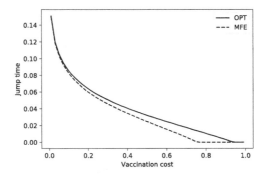

Fig. 3. Jump times comparison when c_V varies from 0.01 to 1. The jump time of the mean field equilibrium (MFE) is represented with a solid line and the jump time of the social optimum (OPT) with a dotted line. The horizontal distance is the subsidy to be granted to incentive players to use the optimal vaccination strategy.

Acknowledgements. The work of Nicolas Gast was supported by the French National Research Agency (ANR) through REFINO Project under Grant ANR-19-CE23-0015. The work of Josu Doncel has been supported by the Department of Education of the Basque Government through the Consolidated Research Group MATH-MODE (IT1294-19), by the Marie Sklodowska-Curie grant agreement No 777778 and by the Spanish Ministry of Science and Innovation with reference PID2019-108111RB-I00 (FEDER/AEI).

References

1. Kermack, W.O., McKendrick, A.G.: A contribution to the mathematical theory of epidemics. Proc. R. Soc. Lond. A Math. Phys. Eng. Sci. **115**(772), 700–721 (1927)
2. Francis, P.J.: Optimal tax/subsidy combinations for the flu season. J. Econ. Dyn. Control **28**(10), 2037–2054 (2004)
3. Laguzet, L., Turinici, G.: Individual vaccination as Nash equilibrium in a SIR model: the interplay between individual optimization and societal policies (2015). hal-01100579, version 1. https://hal.archives-ouvertes.fr/hal-01100579v1
4. Doncel, J., Gast, N., Gaujal, B.: Discrete mean field games: existence of equilibria and convergence. J. Dyn. Games **6**(3), 1–19 (2019). https://hal.inria.fr/hal-01277098
5. Gomes, D.A., Mohr, J., Souza, R.R.: Discrete time, finite state space mean field games. J. Mathématiques Pures et Appliquées **93**(3), 308–328 (2010). http://www.sciencedirect.com/science/article/pii/S002178240900138X
6. Harko, T., Lobo, F.S., Mak, M.K.: Exact analytical solutions of the susceptible-infected-recovered (SIR) epidemic model and of the sir model with equal death and birth rate. Appl. Math. Comput. **236**, 184–194 (2014)
7. Doncel, J., Gast, N., Gaujal, B.: A mean-field game analysis of SIR dynamics with vaccination (2017). Working Paper. https://hal.inria.fr/hal-01496885
8. Lasry, J.-M., Lions, P.-L.: Mean field games. Jpn. J. Math. **2**(1), 229–260 (2007)
9. Behncke, H.: Optimal control of deterministic epidemics. Optim. Control Appl. Methods **21**(6), 269–285 (2000)

Controlling Packet Drops to Improve Freshness of Information

Veeraruna Kavitha[1(✉)] and Eitan Altman[2]

[1] IEOR, Indian Institute of Technology Bombay, Mumbai, India
vkavitha@iitb.ac.in
[2] INRIA Sophia Antipolis, LINCS Paris, and CERI/LIA, Avignon University,
Avignon, France

Abstract. Many systems require frequent and regular updates of certain information. These updates have to be transferred regularly from the source(s) to a common destination. We consider scenarios in which an old packet (entire information unit) becomes completely obsolete, in the presence of a new packet. We consider transmission channels with unit storage capacity; upon arrival of a new packet, if another packet is being transmitted then one of the packets is lost. We consider the control problem that consists of deciding which packet to discard so as to maximise the average age of information (AAoI). We derive drop policies that optimize the AAoI. We show that the state independent (static) policies like dropping always the old packets or dropping always the new packets are optimal in many scenarios, among an appropriate set of stationary Markov policies.

Keywords: Age of information · Freshness of information · Lossy systems · Renewal processes · Dynamic and static policies

1 Introduction

The performance measures that have been studied traditionally in queueing systems have been related to delays and losses. Recently, with the advent of applications demanding frequent and regular updates of a certain information, there is also significant focus towards the freshness of information. Timely updates of the information is an important aspect of such systems, e.g., sensor networks, news feed (social network) over mobile networks or remote control/monitoring of autonomous vehicles etc. Many more such applications are mentioned in [3,7,8]. Most of the times the regular updates are transferred from the source of information to the destination using wireless communication systems.

To measure the freshness of information, the concept of age of information (AoI), has been introduced. AoI is defined as the difference between the current time and the generation time of the latest available information [3]. Peak age

E. Altman—This work was financed by the ANR "Maestro 5G".

S. Lasaulce et al. (Eds.): NetGCooP 2021, CCIS 1354, pp. 60–77, 2021.
https://doi.org/10.1007/978-3-030-87473-5_7

of information (PAoI) and Average age of Information (AAoI) are the relevant performance measures, introduced recently in [3,7]. The study of AAoI/PAoI differs significantly from the conventional performance metrics, such as expected transmission delay, expected number of losses etc.

There has been considerable work in this direction since its recent introduction, we discuss a relevant few of them. In [3] authors discuss the optimal rate of information generation that minimizes the AAoI for various queuing systems. They showed that the smallest age under FCFS can be achieved if a new packet is available exactly when the packet in service finishes service. In [1] the authors consider AoI only for the packets waiting to be transferred/processed. When the queue is empty their AoI is zero, their definition accounts for the oldness of the information waiting at the head of the line. In [2] authors study PAoI and generalize the previously available results to the systems with heterogeneous service time distributions. The authors consider update rates that minimize the maximum PAoI among all the sources. In [8] authors discuss attempt probabilities for slotted aloha system that optimize AAoI.

Most of the work, discussed above, considers lossless systems, where all the packets are transferred (possibly after some delays). However often in systems that require regular updates of the same information, the old packet[1] becomes obsolete once a new one is available. Then the old packet may be dropped. Thus it is appropriate to consider lossy systems, for such scenarios. As an example, in sensor networks the information is consolidated (to generate data packets) from random sets of nodes at random instances of times. Further the transmission of information to the final destination can be over wireless links, which is again random. Further more in some sensor-based applications, the update rates could be significantly high leading to possible availability of a new packet(s) before the old one is completely transferred and then the later becomes obsolete.

If a new packet arrives at source while an old packet is being transferred, it appears upfront that the transfer of the old packet (entire information unit) has to be abandoned. But if the transfer of the old packet is on the verge of getting completed, and since the new packets may require considerable time for transmission, it might be better to discard the new packet and continue the transmission of old packet. Further, the packet transfer times have large fluctuations when the packets are transferred through wireless medium. Thus it is not clear as to which packet is to be discarded. In this work we study the way in which the choice of the packet to be dropped influences the freshness of the information.

We showed that dropping the old packets (always) is optimal for AAoI, when the packet transfer times are distributed according to exponential or hyper exponential distribution. This is a static policy as the drop decision does not

---

[1] Throughout we refer an entire information unit as a packet, that could stand for message or a post or a frame, which needs to be updated frequently. Here are few examples: messages describing weather forecast, or cricket score, or the sensed events related to the entire area in a sensor network, or the information from stock exchanges etc.

depend upon the state of the system, but is optimal among all the stationary
Markov and randomized (SMR) policies. The SMR dynamic policies depend
upon the age of information at an appropriate decision epoch. We also establish certain conditions under which dropping the new packets (always) is optimal among SMR policies. For transfer time distributions like uniform, Weibull,
Poisson, log-normal etc., one of the two static policies is optimal (even among
dynamic/Markov policies) based on the parameters. The AAoI cost is the *ratio
of two time-average costs resulting in non standard Markov decision processes
(MDPs)*, and the above conclusions were obtained by solving these. With the
aid of numerical computations we showed for almost all cases that, one of the
two static policies is near optimal.

2   System with Losses and Fresh Updates

Consider source(s) sending regular updates of a certain information to a destination. The information update packets (entire information units) arrive at any
source according to a Poisson process with rate λ. The packets are of constant
length or of random lengths, and the transfer times depend upon the (random)
medium. In all, the source requires IID (independent and identically distributed)
times $\{T_i\}$ to deliver the packets to the destination, which are equivalently the
job times in the queue. Our focus is on measures related to the freshness of
information available at the destination.

Age/Freshness of Information. The age of information (AoI), from the given
source and at the given destination, at time t is defined

$$G(t) := t - r_t,$$

where r_t is the time at which the last successfully received packet (at destination) before time t, is generated. Our aim is to study the (time) average age of
information (AAoI), defined as below[2]:

$$\bar{a} := \lim_{T \to \infty} \frac{\int_0^T G(t)dt}{T}. \tag{1}$$

We consider freshness of information in a lossy system, and our focus is on
the packet to be dropped when there are two simultaneous packets. We begin
with analysis of the system that drops new packets, when busy.

2.1   Drop the New Packets (DNP)

The source does not interrupt transmission of any packet. If a new update packet
arrives, during transmission, it is dropped. Once the transfer is complete (after

---

[2] Limit exists almost surely in all our scenarios, as will be shown in respective proofs.

random time T), the source waits for new packet, and starts transmission of the new packet immediately after. And this continues (see Fig. 1).

The age of the information $G(t)$ grows linearly with time at unit rate, at all time instances, except for the one at which a packet is just received at the destination. At that time epoch the age drops to T_k, because: a) T_k is the time taken to transfer the (new) packet from source to destination, after its arrival at the source queue; and b) this represents the age of the new packet at destination.

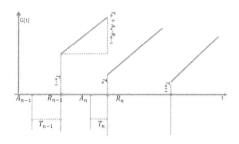   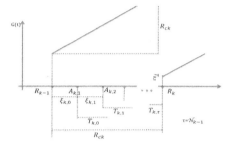

Fig. 1. DNP scheme, renewal cycles   **Fig. 2.** DOP scheme, a renewal cycle

Thus we have a process (resembling a renewal process) as in Fig. 1. Here $\{R_k\}$ are the epochs at which a message is transferred successfully (these would become the renewal instances in two concatenated processes, see footnote 3), while $\{A_k\}$ are the arrival instances of the packets (at the source and of those transferred) governed by Poisson point process (PPP). Let $\{\xi_k\}$ represent these (residual) inter-arrival times, note $\xi_k = A_k - R_{k-1}$ for each k. As seen from the figure, the age of the information is given by sawtooth waveform. Further more, clearly, the alternate renewal cycles are independent of one another, thus by RRT[3] the long run time average of the age of the information (1) equals:

---

[3] The adjacent cycles (e.g., intervals between R_k, R_{k+1} and R_{k+1}, R_{k+2}) are not independent, because of $G_k = T_k$, however the alternate ones are. Concatenate odd and even cycles to obtain two separate renewal process, observe that $R_k \to \infty$ as $k \to \infty$ and apply (Renewal Reward Theroem) RRT to both the processes to obtain:

$$\bar{a} = \lim_{k \to \infty} \frac{\sum_{l \le k} \int_{R_{l-1}}^{R_l} G(t)dt}{R_k}$$

$$= \lim_{k \to \infty} \left(\frac{\sum_{2l \le k} \int_{R_{2l-1}}^{R_{2l}} G(t)dt}{\sum_{2l \le k} R_{2l} - R_{2l-1}} \frac{\sum_{2l \le k} R_{2l} - R_{2l-1}}{R_k} + \frac{\sum_{2l+1 \le k} \int_{R_{2l}}^{R_{2l+1}} G(t)dt}{\sum_{2l \le k} R_{2l+1} - R_{2l}} \frac{\sum_{2l \le k} R_{2l+1} - R_{2l}}{R_k} \right)$$

$$= \frac{E\left[\int_{R_1}^{R_2} G(s)ds \right]}{E[R_2 - R_1]} \frac{1}{2} + \frac{E\left[\int_{R_2}^{R_3} G(s)ds \right]}{E[R_3 - R_1]} \frac{1}{2} = \frac{E\left[\int_{R_1}^{R_2} G(s)ds \right]}{E[R_2 - R_1]} \quad \text{a.s., as the two processes are identical.}$$

For renewal process with even cycles, the time intervals between two successful packet receptions $\{(R_{2k} - R_{2k-1})\}_k$ form the renewal periods and the time integral of the costs in (1) for each of even renewal periods, $\left\{ \int_{R_{2k-1}}^{R_{2k}} G(s)ds \right\}_k$, form the rewards.

$$\bar{a} = \frac{E\left[\int_{R_{k-1}}^{R_k} G(s)ds\right]}{E[R_k - R_{k-1}]}, \text{ almost surely (a.s.), and with } G_k := G(R_k).$$

$$= \frac{E[G_{k-1}(R_k - R_{k-1})] + 0.5E[(R_k - R_{k-1})^2]}{E[R_k - R_{k-1}]} = E[G_{k-1}] + \frac{E[(R_k - R_{k-1})^2]}{2E[R_k - R_{k-1}]} a.s. \tag{2}$$

The last line follows by independence (Fig. 1) and memoryless property of PPP. For DNP, $G_{k-1} = T_{k-1}$, and

$$\bar{a}_{DNP} = E[T_{k-1}] + \frac{1}{2}\frac{E[(T_k + \xi_k)^2]}{E[T_k + \xi_k]} \text{ a.s.,}$$

where ξ_k, the inter-arrival time, is exponentially distributed with parameter λ and is independent of the transfer times T_k, T_{k-1}. Simplifying

$$\bar{a}_{DNP} = E[T] + \frac{1}{\lambda} + \frac{E[T^2]}{2E[T]}\frac{\rho}{1+\rho} \text{ with } \rho := \lambda E[T]. \tag{3}$$

2.2   Drop the Old Packets (DOP)

When source receives a new message, the ongoing transfer (if any) of the old packet is stopped and the old packet is dropped. The source immediately starts transfer of the new packet. The new message would imply a more fresh information, but might also imply longer time (because we now require the transfer of the entire message) before the information at the destination is updated. However the variability in transfer times $\{T_k\}$ might imply interruption is better for average freshness under certain conditions, and we are studying this aspect.

The renewal points (as in footnote 3) will again be the instances at which a message is successfully received. But note that only when a message transfer is not interrupted by a new arrival, we have a successful message reception. Thus the renewal cycles in Figure (1) get prolonged appropriately (see Fig. 2). Let $A_{k,1}$ be the first arrival instance after the $(k-1)$-th renewal epoch R_{k-1}. Let $\xi_{k,0}$ be the corresponding inter-arrival time (which is exponentially distributed). Its service (i.e., message transfer) starts immediately and let $T_{k,0}$ be the job size, or the (random) time required to transfer this message. In case a second arrival occurs (after inter-arrival time $\xi_{k,1}$) within this service, we start the service of the new packet by discarding the old one. This happens with probability $1 - \gamma$ where $\gamma := P(T_{k,0} \leq \xi_{k,1})$. The renewal cycle is completed after second transfer, in case the second message transfer is not interrupted. The second can also get interrupted, independent of previous interruptions and once again with the same probability $1 - \gamma$, because of IID nature of the transfer times and the inter arrival times. If second is also interrupted the transfer of the third one starts immediately and this continues till a job is not interrupted (i.e., with probability γ). And then the renewal cycle is completed.

Once again the alternate cycles are IID, RRT can be applied to AAoI given by (1) and AAoI is given by equation (2). However the renewal cycles $\{R_k - R_{k-1}\}_k$

are more complex now, and we proceed with deriving their moments. The k-th renewal cycle can be written precisely as below, using the arrival sequence $\{\xi_{k,i}\}_{i\geq 0}$ and transfer times sequence $\{T_{k,i}\}_{i\geq 0}$ belonging to k-th renewal cycle:

$$R_{ck} := R_k - R_{k-1} = \xi_{k,0} + \sum_{i=1}^{\mathcal{N}_k-1} \xi_{k,i} + T_{k,\mathcal{N}_k-1} \quad = \xi_{k,0} + \boldsymbol{\Gamma}_k, \tag{4}$$

$$\boldsymbol{\Gamma}_k := \sum_{i=1}^{\mathcal{N}_k-1} \xi_{k,i} + T_{k,\mathcal{N}_k-1}, \text{ and} \qquad \mathcal{N}_k := \inf\{i \geq 1 : \xi_{k,i} > T_{k,i-1}\}. \tag{5}$$

In the above $\mathcal{N}$ is the number of interruptions before successful transfer, and it is geometrically distributed with parameter $1 - \gamma$ and $\boldsymbol{\Gamma}$ (given by (5)) is the time taken to complete one packet transfer, in the midst of interruptions by new arrivals. The above random variables are specific to a given renewal cycle, but are also IID across different cycles. Further, $G_k = G(R_k)$ is now a 'special' transfer time (represented by $\underline{T}$): one which is not interrupted. Thus

$$G_k = \underline{T}_k := T_{k,\mathcal{N}_k-1}, \text{ and, } E[G_k] = E[\underline{T}_k] = E[T|T \leq \xi] \text{ for any } k. \tag{6}$$

Hence the AAoI of DOP scheme (again by independence of alternate cycles, exactly as in footnote 3) equals (see (2)):

$$\bar{a}_{DOP} = E[T|T \leq \xi] + \frac{E[(\xi + \boldsymbol{\Gamma})^2]}{2E[\xi + \boldsymbol{\Gamma}]} \text{ almost surely (a.s.).} \tag{7}$$

To complete the analysis we require the first two moments[4] of $\boldsymbol{\Gamma}$ (see equation (4)) and finally the AAoI for DOP scheme is obtained in the following (Proof in Appendix A):

Lemma 1. *The first two moments of the renewal cycle are (with $\gamma = E[T < \xi]$):*

$$E[R_c] = \frac{1}{\lambda\gamma} \text{ and } E[R_c^2] = \frac{2}{\lambda^2\gamma^2} - \frac{2E[Te^{-\lambda T}]}{\lambda\gamma^2}. \tag{8}$$

Further, the AAoI for DOP scheme equals:

$$\bar{a}_{DOP} = \frac{1}{\lambda\gamma} = E[R_c]. \tag{9}$$

∎

Thus the AAoI with DOP scheme exactly equals the expected renewal cycle, while that with DNP scheme is strictly bigger than the expected renewal cycle (from (3), $\bar{a}_{DNP} = E[R_k - R_{k-1}] + \frac{E[T^2]}{2E[T]}\frac{\rho}{1+\rho}$). It is not guaranteed that the expected renewal cycle with DOP scheme is smaller than that with DNP scheme. Thus it is not clear upfront as to which scheme is better. But it is equally (or more) important to understand if any scheme with controlled drops can perform better than these two schemes.

---
[4] At first glance $\boldsymbol{\Gamma}$ may appear like busy period of $M/G/\infty$ queue, but it is not true.

3   Controlled Drops

In the previous section two 'extreme' and static schemes are considered: in one all
the old packets are dropped while in the other all the new packets are dropped.
Now we investigate if there exists a better scheme with partial/controlled drops.
We also study the conditions under which DOP is better than DNP. With message successful transfer epochs $\{R_k\}_k$ as the decision epochs, we consider a
dynamic decision about the (DOP/DNP) scheme to be used. The dynamic decision depends upon the state[5], the age of information G_k, at decision epoch R_k.

Threshold Policies: We initially restrict ourselves to special type of *dynamic
policies, called threshold policies:* DNP scheme is selected if age (G_k) is above
a threshold (say $\theta \geq 0$) and DOP is selected other wise. With DNP scheme,
new packets are dropped (other than the first one in that renewal cycle) till the
message transfer is complete. With DOP decision, old packets are dropped and
transmission of new packet starts immediately, whenever the former is interrupted. This continues till a message is transferred completely. Further dropping
of (old/new) packets depends upon the decision at the next decision epoch.
    In contrast to the previous subsections, the length of renewal cycles $\{R_{ck}\}_k$
are no more identically distributed. The distribution of R_{ck} depends upon the
scheme chosen at the decision epoch R_k. It is easy to observe that the length of
the renewal cycle R_{ck} does not depend upon the absolute value of state G_k, but
only upon the state dependent binary (DOP/DNP) decision. Thus the distribution of R_{ck} can be one among two types and precisely equals (see (4)):

$$R_{ck+1} = \begin{cases} \xi_{k+1,0} + T_{k+1,0}, & \text{with DNP (i.e., with } G_k > \theta) \\ \xi_{k+1,0} + \Gamma_{k+1} & \text{else, and using (6),} \end{cases} \quad (10)$$

$$G_{k+1} = \begin{cases} T_{k+1,0}, & \text{with DNP } (G_k > \theta) \\ \underline{T}_{k+1}, & \text{other wise.} \end{cases} \quad (11)$$

Observe that $\theta = 0$ implies DNP, while DOP is obtained by considering
$\theta \to \infty$. *For ease of notation we say* $\theta = \infty$ *when DOP is selected for all G_k.*
    For every θ, the random variables $\{G_k\}_k$ and $\{R_{ck}\}_k$ constitute a Markov
chain. Using these, one can rewrite AAoI (1) as:

$$\bar{a}(\theta) = \lim_{k \to \infty} \frac{\int_0^{R_k} G(t)dt}{R_k} \quad \text{a.s.,}$$

because $R_k \to \infty$ a.s., as $k \to \infty$, and this is because

$$R_k = \sum_{l \leq k} R_{cl} \geq \sum_{l \leq k} \xi_{l,0} \text{ for all } k \text{ and } \sum_{l \leq k} \xi_{l,0} \stackrel{k \to \infty}{\to} \infty \text{ a.s.}$$

---

[5] The source can easily have access to $\{G_k\}$, as it can easily keep track of success-
ful/unsuccessful prior transmissions.

Thus,

$$\bar{a}(\theta) = \lim_{k \to \infty} \frac{\sum_{l \leq k} \left(G_{l-1} R_{cl} + 0.5 R_{cl}^2 \right)}{k} \frac{k}{\sum_{l \leq k} R_{cl}}. \tag{12}$$

As already discussed, the distribution of R_{ck} (for any k) can be of two types depending only upon the event $\{G_k < \theta\}$ (see (10). Let G_* and R_{c*} represent the random quantities corresponding to stationary distributions of G_k and R_{ck} respectively. As before, the stationary distribution R_{c*} depends only upon the stationary event $\{G_* < \theta\}$. Thus it suffices to obtain the stationary distribution of $\{G_k\}_k$. In fact the transitions of $\{G_k\}$ given by (11) also depend only upon the events $\{G_{k-1} < \theta\}$. Thus it further suffices to study the two state Markov chain $X_k := 1_{\{G_k < \theta\}}$ (1_A is the indicator of the event A) and the rest of the random quantities can be studied using this two state chain. The Markov chain has the following evolution

$$X_{k+1} = \begin{cases} 1_{\{T_{k+1,0} < \theta\}} & \text{if } X_k = 0, \\ 1_{\{\underline{T}_{k+1} < \theta\}} & \text{else.} \end{cases} \tag{13}$$

When $\theta = \infty$, $X_k \equiv 1$ for all k. The transition probabilities (with $\theta \neq \infty$) are:

$$P(X_{k+1} = x' | X_k = x) = \begin{cases} p_\theta & \text{if } x = 0, \ x' = 1 \\ q_\theta & \text{if } x = 1, \ x' = 0 \end{cases} \quad \text{where} \tag{14}$$

$$p_\theta := P(T < \theta) \text{ and } q_\theta := P(\underline{T} > \theta) = P(T > \theta | T \leq \xi).$$

This chain has unique stationary distribution given by

$$\pi_\theta(0) = \frac{q_\theta}{q_\theta + p_\theta} 1_{\{\theta \neq \infty\}} = 1 - \pi_\theta(1), \text{ and } P(X_* = 0) = \pi_\theta(0), \tag{15}$$

where X_* is the random quantity corresponding to stationary distribution of $\{X_k\}$ (see [12]). The stationary distribution of the remaining quantities is dictated by that of $\{X_k\}$: for example the stationary distribution of G_* is the same as that of T, a typical transfer time when $X_* = 0$ and equals that of $\underline{T} = T | T \leq \xi$ (conditional distribution) when $X_* = 1$.

The Markov chain $\{X_k\}$ is clearly ergodic, the rest of the stationary random quantities R_{c*}, G_* depend just upon X_*, hence strong law of large numbers (SLLN) (e.g., [9]) can be applied[6] separately to the numerator and denominator of (12) to obtain:

$$\bar{a}(\theta) = \frac{E_{\pi_\theta}[G_{l-1} R_{cl}] + 0.5 E_{\pi_\theta}[R_{cl}^2]}{E_{\pi_\theta}[R_{cl}]} \quad \text{a.s. ,}$$

---

[6] One can not apply the usual renewal theory based analysis, as the process is (the odd/even cycles are also) Markovian and can not be modelled as a Renewal process, with IID renewal cycles.

where $E_{\pi_\theta}[\cdot]$ is the stationary expectation. It is easy to verify (see (10)) by appropriate conditioning that:

$$
\begin{aligned}
E_{\pi_\theta}[G_{l-1}R_{cl}] &= E_{\pi_\theta}[G_{l-1}R_{cl}; X_{l-1} = 1] + E_{\pi_\theta}[G_{l-1}R_{cl}; X_{l-1} = 0] \\
&= \left(\frac{1}{\lambda} + E[T]\right) E[G_*; G_* > \theta] + E[G_* \; ; \; G_* < \theta]\left(\frac{1}{\lambda} + E[\Gamma]\right) \text{ and}
\end{aligned}
$$
$$
E[G_*; G_* > \theta] = E[T; T > \theta]\pi_\theta(0) + E[T; T > \theta | T \le \xi]\pi_\theta(1).
$$

Using similar logic,

$$
E_{\pi_\theta}[G_{l-1}R_{cl}] + \frac{1}{2}E_{\pi_\theta}[R_{cl}^2] = d_n E[G_*; G_* > \theta] + d_o E[G_* \; ; \; G_* < \theta] + 0.5E[R_{c*}^2]
$$
$$
= \beta_\theta(0)\pi_\theta(0) + \beta_\theta(1)\pi_\theta(1), \tag{16}
$$
$$
E_{\pi_\theta}[R_{cl}] = d_n\pi_\theta(0) + d_o\pi_\theta(1),
$$

with the following definitions:

$$
\begin{aligned}
\beta_\theta(0) &:= d_n E[T; T > \theta] + d_o E[T; T \le \theta] + 0.5c_n, \\
\beta_\theta(1) &:= d_n E[T; T > \theta | T \le \xi] + d_o E[T; T \le \theta | T \le \xi] + 0.5c_o, \\
c_n &:= E[(\xi + T)^2], \qquad c_o := E\left[(\xi + \Gamma)^2\right] \text{ and} \\
d_n &:= E[T + \xi], \qquad d_o := E[\xi + \Gamma].
\end{aligned}
$$

Thus the AAoI equals

$$
\bar{a}(\theta) = \frac{\beta_\theta(0)\pi_\theta(0) + \beta_\theta(1)\pi_\theta(1)}{d_n\pi_\theta(0) + d_o\pi_\theta(1)} \quad \text{a.s.} \tag{17}
$$

Optimal Threshold θ: We are interested in optimal threshold, θ^* and hence consider:

$$
\min_{\theta \ge 0} \bar{a}(\theta). \tag{18}
$$

The objective function depends upon θ in a complicated manner, further the dependence is influenced by the distribution of the transfer times. However one can derive the optimal policies by using an appropriate lower bound function. We first consider the case: $E[T] > E[\Gamma], or d_n > d_o$. From (16) and the definitions following (16) and because of positivity of the terms:

$\bar{a}(\theta) \ge f_o(\theta)$ for any $\theta \ge 0$, with function,

$$
\begin{aligned}
f_o(\theta) &:= \frac{d_o\big(b_n\pi_\theta(0) + b_o\pi_\theta(1)\big) + 0.5(c_n\pi_\theta(0) + c_o\pi_\theta(1))}{d_n\pi_\theta(0) + d_o\pi_\theta(1)} \\
&= \frac{d_o\big((b_n - b_o)\pi_\theta(0) + b_o\big) + 0.5(c_n - c_o)\pi_\theta(0) + 0.5c_o}{(d_n - d_o)\pi_\theta(0) + d_o}, \text{ with}
\end{aligned}
$$
$$
b_n := E(T), \quad b_o := E(T | T < \xi) = \frac{E[Te^{-\lambda T}]}{E[e^{-\lambda T}]}. \tag{19}
$$

Further using (7) we have[7] (e.g., $\pi_\theta(0) \to 0$ as $\theta \to \infty$):

$$\lim_{\theta \to \infty} f_o(\theta) = \lim_{\theta \to \infty} \bar{a}(\theta) = \bar{a}_{DOP}. \tag{20}$$

If the DOP scheme is optimal for the lower bound function $f_o(\theta)$, i.e., if

$$\min_\theta f_o(\theta) = \lim_{\theta \to \infty} f_o(\theta), \tag{21}$$

then DOP would be optimal for AAoI, because then using (20):

$$\bar{a}_{DOP} \geq \min_\theta \bar{a}(\theta) \geq \min_\theta f_o(\theta) = \lim_{\theta \to \infty} f_o(\theta) = \bar{a}_{DOP}.$$

We prove that (21) is true when DOP renewal cycle is smaller, and hence show the optimality of DOP (proof in Appendix A and in [12]):

Theorem 1. *If $d_n \geq d_o$ then DOP is optimal, i.e.,*

$$\min_{\theta \geq 0} \bar{a}(\theta) = \lim_{\theta \to \infty} \bar{a}(\theta) = \bar{a}_{DOP}. \qquad \blacksquare$$

It is clear from (3) and (7) that the DOP scheme is better than the DNP scheme when its expected renewal cycle is smaller, i.e., when $d_n \geq d_o$. Theorem 1 proves much more under the same condition, the DOP scheme is better than any other threshold scheme.

We now study the reverse case, i.e., when $d_n < d_o$ or equivalently when $E[T] < E[\Gamma]$. In this case $\bar{a}(\theta) > f_n(\theta)$ where

$$f_n(\theta) := \frac{d_n(b_n \pi(0) + b_o \pi(1)) + 0.5 c_n \pi(0) + 0.5 c_o \pi(1)}{d_n \pi(0) + d_o \pi(1)}.$$

As in the previous case, if DNP is proved optimal for this lower bound function, then DNP is optimal for controlled AAoI, and this is proved in the following (proof in Appendix A and in [12]):

Theorem 2. *If $d_n < d_o$ and (note that[8] $1 - \lambda E\left[Te^{-\lambda T}\right] > 0$)*

$$\rho \frac{E[T^2]}{2E[T]} - (1 + \rho)(d_o - d_n)\left(1 - \lambda E\left[Te^{-\lambda T}\right]\right) \leq 0, \tag{22}$$

then DNP is optimal,

$$\min_{\theta \geq 0} \bar{a}(\theta) = \bar{a}(0) = \bar{a}_{DNP}. \qquad \blacksquare$$

---

[7] It is not difficult to establish the continuity of the relevant functions as $\theta \to \infty$ and it is not difficult to show that the limit equals that with DOP scheme.

[8] because ξ is exponential,

$$1 - \lambda E\left[Te^{-\lambda T}\right] = \lambda(E[\xi] - E[T; T \leq \xi]) = \lambda(E[\xi; T > \xi] + E[\xi - T; T \leq \xi]) > 0.$$

.

Stationary Markov Randomized Policies: We now generalize the results to Stationary Markov Randomized (SMR) policies. As seen from (12) the *objective function AAoI is the ratio of two average costs and hence the usual techniques of Markov decision processes may not be applicable. Nevertheless we could use exactly the same techniques as in previous subsection to show the optimality of DNP/DOP policy even under SMR policies.* This is true under the assumptions of Theorems 1–2.

Let α^∞ be any Stationary Markov Randomized policy: $\alpha(G)$ represents the probability with which DNP scheme is selected when the state $G_k = G$, and this is true for all decision epochs k (∞ implies same state-dependent decision for all decision epochs). Re define $X_k = 1$ if DOP scheme is selected (i.e., if old packet is dropped), else $X_k = 0$. Like before, the random variables X_k, R_{ck} and G_k depend mainly upon X_{k-1}, and same is the case with their stationary distributions. Let π_α represent the stationary probability that $\{X^* = 0\}$, when policy α^∞ is used and note that:

$$\pi_\alpha := \pi_\alpha(0) = \frac{q_\alpha}{q_\alpha + p_\alpha} \text{ with} \tag{23}$$
$$q_\alpha := E[\alpha(\underline{T})] = E[\alpha(T)|T \le \xi] \text{ and } p_\alpha = E[1 - \alpha(T)].$$

As before, the stationary expectation (see (12))

$$
\begin{aligned}
E_{\pi_\alpha}[G_{l-1}R_{cl}] &= E_{\pi_\alpha}[G_{l-1}R_{cl}; X_{l-1} = 1] + E_{\pi_\alpha}[G_{l-1}R_{cl}; X_{l-1} = 0] \\
&= d_n E[G_* E[X^* = 1|G*]] + d_o E[G_* E[X^* = 0|G_*]] \\
&= d_n E[G_* \alpha(G*)] + d_o E[G_*(1 - \alpha(G*))] \\
&= d_n \Big(E[T\alpha(T)]\pi_\alpha(1) + E[T\alpha(T)|T \le \xi]\pi_\alpha(0) \Big) \\
&\quad + d_o \Big(E[T(1 - \alpha(T))]\pi_\alpha(1) + E[T(1 - \alpha(T))|T \le \xi]\pi_\alpha(0) \Big).
\end{aligned}
$$

Similarly

$$E_{\pi_\alpha}[R_{ck}^2] = E_{\pi_\alpha}[R_{ck}^2; X_{k-1} = 0] + E_{\pi_\alpha}[R_{ck}^2; X_{k-1} = 1] = c_o \pi_\alpha(1) + c_n \pi_\alpha(0),$$

Proceeding exactly as in the case of threshold policies:

$$\bar{a}(\alpha) = \frac{\beta_\alpha(0)\pi_\alpha(0) + \beta_\alpha(1)\pi_\alpha(1)}{d_n \pi_\alpha(0) + d_o \pi_\alpha(1)} \text{ with}$$
$$\beta_\alpha(0) := d_n E[T\alpha(T)] + d_o E[T(1 - \alpha(T))] + 0.5c_n \text{ and}$$
$$\beta_\alpha(1) := d_n E[T\alpha(T)|T \le \xi] + d_o E[T(1 - \alpha(T))|T \le \xi] + 0.5c_o.$$

Using the lower bound functions, $f_o(\cdot)$ and $f_n(\cdot)$, and following exactly the same logic one can extend Theorems 1–2:

Theorem 3. *a)* If $d_n \ge d_o$ then DOP is optimal among SMR policies, i.e.,

$$\min_{\alpha^\infty \in SMR} \bar{a}(\alpha) = \bar{a}_{DOP}.$$

b) If $d_n < d_o$ and (22) of Theorem 2 is true then DNP is optimal,

$$\min_{\alpha^\infty \in SMR} \bar{a}(\alpha) = \bar{a}_{DNP}. \qquad \blacksquare$$

We thus have that the static policies DOP/DNP are optimal among stationary Markov (dynamic) policies for all the conditions, except when $d_n < d_o$ and (22) of Theorem 2 is not true. Using numerically aided study of the next section, we will show that these 'exception conditions' are 'rare'.

3.1 Numerically Aided Study

DOP Optimal Among SMR Policies. We considered several distributions for transfer times and tested the conditions required for DOP/DNP optimality. The results are summarized in Table 1. By direct substitution one can show that $d_n = d_o$ for exponential and $d_n > d_o$ for hyper exponential distribution. Thus by Theorem 3, DOP is optimal for these transfer times.

DNP/DOP Is Almost Optimal. When $d_n < d_o$, but (22) is not satisfied, we do not have theoretical understanding of the optimal policy. We study such test cases by numerically optimizing (17) over threshold policies. One such example is plotted in Fig. 3, which considers Erlang distributed transfer times. The AAoI is plotted as a function of θ, it decreases as $\theta \to \infty$, hence confirming that the AAoI is minimized by DOP scheme.

A second example is considered in Fig. 4 with uniformly distributed transfer times, distributed between $(0, \phi)$. Here again AAoI $\bar{a}(\theta)$ is plotted as a function of θ for two different parameters. An intermediate $\theta^* \in (0, \infty)$ is optimal in both the examples of this figure, however DOP and DNP perform almost similar. Further AAoI at θ^* is close to that at DNP/DOP (Fig. 4). We considered many more such case studies and observed similar pattern: DOP/DNP scheme is (almost) optimal. These examples include truncated exponential, Log normal, Poisson distributed and Erlang transfer times etc.

Best Among DNP/DOP. Thus either DNP or DOP scheme is (almost) optimal among the threshold policies. Hence it is important to derive the conditions that suggest the best among the two. One can find the best among DNP/DOP schemes by directly using (3) and (9), i.e., DNP is better than DOP iff (recall $\rho = \lambda E[T]$)

$$E[T] - \frac{1-\gamma}{\lambda\gamma} + \frac{E[T^2]}{2E[T]}\frac{\rho}{1+\rho} < 0 \text{ or iff } 1 > \left(\frac{E[T^2]}{2(E[T])^2}\frac{\rho^2}{1+\rho} + 1 + \rho \right)\gamma. \qquad (24)$$

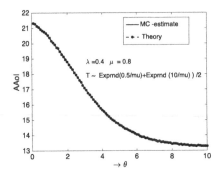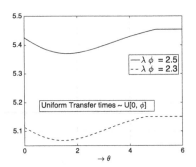

Fig. 3. When $d_o = 13.19 > d_n = 12.97$ and condition (22) negated: optimizer is DOP

Fig. 4. AAoI versus θ for uniform transfer times: Intermediate θ optimal but DNP/DOP almost optimal

Table 1. Criterion for $\bar{a}_{DOP} \leq \bar{a}_{DNP}$, for different types of T

Distribution	CDF ($P(T \leqslant x)$)	$\bar{a}_{DOP} \leq \bar{a}_{DNP}$ when
Uniform $(0,\phi)$	$\left[\frac{x}{\phi}\right]1_{x>0}$	$\frac{1}{1-e^{-\lambda\phi}} < \left(\frac{1}{3}\frac{\lambda\phi}{2+\lambda\phi} + \frac{1}{\lambda\phi} + \frac{1}{2}\right)$, approx. when $\lambda\phi < 2.356$
Weibull (μ, k)	$\left[1 - e^{-(x/\mu)^k}\right]1_{x>0}$	$\frac{\rho^2 w_k^2}{2(1+\rho w_k^1)} + (1 + \rho w_k^1) > \frac{1}{w_\rho^k}$
		$w_k^i = \Gamma(1 + \frac{i}{k}) \quad \rho = \lambda\mu$ $w_\rho^k = E_k[e^{-\rho T}]$
Exponential (μ)	$\left[1 - e^{-\mu x}\right]1_{x>0}$	all μ
Hyperexpo($\{\mu_i, p_i\}$)	$\left[1 - \sum_{i=1}^{n} p_i e^{-\frac{x}{\mu_i}}\right]1_{x>0}$	all $\{\mu_i, p_i\}_i$

Note that $E[T^2] = \text{Var}(T) + (E[T])^2$ and we have the following important conclusions:

- DNP is the best for large update rates: as the update rate $\lambda \to \infty$, with distribution of T fixed and with $E[1/T] < \infty$, the above condition is satisfied (RHS converges to 0). Note $\lambda\gamma = E[\lambda e^{-\lambda T}] \to 0$ using L'Hopital's rule (applied point-wise) and dominated convergence theorem.
- DOP is the best for small update rates: as the update rate $\lambda \to 0$, with distribution of T fixed, the above condition is negated (RHS is approximately $1 + \rho$).
- The range of λ for which DNP is optimal is influenced by the variance. *'DOP scheme becomes optimal as the variance of the transfer times increases, for bigger range of λ'.*

For uniform transfer times we derived the conditions under which DOP performs better than DNP, using (24), and the condition is tabulated in the first row of Table 1. Approximately, DOP is optimal if $\lambda\phi < 2.35$. Weibull is also tabulated in the second row.

Based on this theoretical and numerical case studies we have the following:

- AAoI is (almost) optimized either by DOP scheme or by DNP scheme. No other threshold policy performs significantly better than the best among these two static policies.
- If expected renewal cycle with DOP is smaller than that with DNP, DOP scheme optimizes AAoI over all SMR policies.
- When DNP has smaller renewal cycle, DOP may still be the optimal (in some test cases).

The Fig. 3 also plots the Monte-Carlo estimates of AAoI along with formula (17). For Monte-Carlo estimates we generate several random sample paths and compute the time average of AAoI. As anticipated, the formula well matches the estimates (see Fig. 3).

Future Directions: We consider multiple sources in [12] and have some initial results; with multiple sources the drop decision should also include the differential priority that needs to be given to different sources. We also consider the case of multiple resources transferring information to single destination using ALOHA type protocol ([12]).

As of now our analysis considers memory less packet arrivals, it would be interesting to consider more general arrival processes (e.g., renewal processes). It would be interesting to investigate if the static policies (dropping always the new/old packets) are again optimal.

One can also think of more general decision epochs, one can think of dropping at maximum K packets and K can also be controlled etc. One can consider one storage option along with DNP protocol.

4   Conclusions

We considered problems related to freshness of information, for scenarios in which the destination is regularly updated with a certain information. In such cases, old information can become completely obsolete once a new update is available. The systems naturally become lossy, in the sense that, some packets would be discarded. We developed a methodology to study the freshness of information, using average age of information (AAoI) as performance metric, for lossy systems. A packet at destination can automatically be discarded once a new update is available. However a new packet at source, while the source is transferring an older packet, demands an important decision: which packet to be discarded. Older packets can be transferred faster to the destination, while the new packet may have fresh information but may require more time to reach the destination. It may be better to base these decisions on the state of the system, the age of the previous update of the same information at destination. However two static policies, drop always the new packets (DNP) or drop always the old packets (DOP), are optimal among a class of stationary Markov policies, for many scenarios.

Appendix A: Proofs

Proof of Lemma 1: By conditioning on $\xi_{k,1}$, $T_{k,0}$:

$$
\begin{aligned}
E[\Gamma_k] &= E\left[\Gamma_k \ ; \ \xi_{k,1} > T_{k,0}\right] + E\left[\Gamma_k \ ; \ \xi_{k,1} \le T_{k,0}\right] \\
&= E\left[T_{k,0} \ ; \ \xi_{k,1} > T_{k,0}\right] + E\left[\xi_{k,1} + \tilde{\Gamma} \ ; \ \xi_{k,1} \le T_{k,0}\right] \\
&= E\left[T_{k,0}; \xi_{k,1} > T_{k,0} + \xi_{k,1} \ ; \ \xi_{k,1} \le T_{k,0}\right] + E\left[\tilde{\Gamma} \ ; \ \xi_{k,1} \le T_{k,0}\right],
\end{aligned}
$$

where $\tilde{\Gamma}$ is an IID copy of Γ_k, which is independent of $T_{k,0}$ and $\xi_{k,1}$. By independence, $E[\Gamma_k] - E\left[\tilde{\Gamma} \ ; \ \xi_{k,1} \le T_{k,0}\right] = E[\Gamma](1 - P(\xi \le T))$, and thus by further conditioning on T we have the following:

$$
E(\Gamma) = \frac{E\left[T; \xi > T + \xi \ ; \ \xi \le T\right]}{P(T \le \xi)} = \frac{E[Te^{-\lambda T}] + (1 - E[e^{-\lambda T}])/\lambda - E[Te^{-\lambda T}]}{P(T \le \xi)} = \frac{1 - \gamma}{\lambda \gamma}. \quad (25)
$$

Using exactly similar logic:

$$
E[\Gamma^2] = E[\min\{T_0, \xi_1\}^2] + E[\tilde{\Gamma}^2](1 - \gamma) + 2E[\tilde{\Gamma}]E[\xi_{k,1} \ ; \ T_{k,0} > \xi_{k,1}].
$$

Using (25),

$$
\begin{aligned}
E[\Gamma^2] &= \frac{E[\min\{T_0, \xi_1\}^2] + 2E[\Gamma]E[\xi_{k,1} \ ; \ T_{k,0} > \xi_{k,1}]}{\gamma} \\
&= \frac{2(1-\gamma)}{\lambda^2 \gamma} - \frac{2E[Te^{-\lambda T}]}{\lambda \gamma} + \frac{2E[\Gamma]}{\gamma}\left(\frac{1-\gamma}{\lambda} - E[Te^{-\lambda T}]\right) \\
&= \frac{2(1-\gamma)}{\lambda^2 \gamma} - \frac{2E[Te^{-\lambda T}]}{\lambda \gamma^2} + \frac{2(1-\gamma)^2}{\lambda^2 \gamma^2} = \frac{2(1-\gamma)}{\lambda^2 \gamma^2} - \frac{2E[Te^{-\lambda T}]}{\lambda \gamma^2}. \quad (26)
\end{aligned}
$$

Using (25) and (26), the first two moments of the renewal cycle are:

$$
E[R_c] = \frac{1}{\lambda} + \frac{1-\gamma}{\lambda \gamma} = \frac{1}{\lambda \gamma} \ \text{and}
$$

$$
E[R_c^2] = E\left[\xi_{k,0}^2 + 2\xi_{k,0}\Gamma_k + \Gamma_k^2\right] = \frac{2}{\lambda^2 \gamma^2} - \frac{2E[Te^{-\lambda T}]}{\lambda \gamma^2}.
$$

■

Proof of Theorem 1: As a first step, one can easily observe that the coefficients of the lower bound function f_o depend upon θ only via the stationary distribution π_θ, in particular only via $\pi_\theta(0)$, i.e., $f_o(\theta) = f_o(\pi_\theta(0))$. Further the function $\theta \mapsto \pi_\theta(0)$ is ONTO (see (15)) and hence one can equivalently optimize f_o using $\pi := \pi_\theta(0)$:

$$f_o(\theta) = f_o(\pi) = \frac{d_o\left((b_n - b_o)\pi + b_o\right) + 0.5(c_n - c_o)\pi + 0.5c_o}{(d_n - d_o)\pi + d_o}.$$

The first derivative for the lower bound function is:

$$f_o'(\pi) = \frac{0.5(c_n d_o - c_o d_n) + d_o(b_n d_o - b_o d_n)}{(\pi(d_n - d_o) + d_o)^2}. \tag{27}$$

From (28) of Appendix B:

$$c_n d_o - c_o d_n = \frac{E[T^2]}{\lambda\gamma} + \left(\frac{1}{\lambda} + E[T]\right)\left(E[Te^{-\lambda T}] - \frac{(1-\gamma)}{\lambda}\right)\frac{2}{\lambda\gamma^2}.$$

Thus the numerator of the derivative (27) is proportional to,

$$c_n d_o - c_o d_n + 2d_o(b_n d_o - b_o d_n) = \frac{E[T^2]}{\lambda\gamma} + \left(\frac{1}{\lambda} + E[T]\right)E[Te^{-\lambda T}]\left(\frac{2}{\lambda\gamma^2} - 2\frac{1}{\lambda\gamma^2}\right)$$
$$- \frac{1}{\lambda\gamma}\left(\left(\frac{1}{\lambda} + E[T]\right)\frac{2(1-\gamma)}{\lambda\gamma} - E[T]\frac{2}{\lambda\gamma}\right)$$
$$= \frac{E[T^2]}{\lambda\gamma} + \frac{2}{\lambda^2\gamma}(E[T] - E[\Gamma]) > 0, \text{ when } d_n \geq d_o.$$

Thus the derivative $f_o'(\theta) > 0$ for all θ, hence the lower bound f_o is increasing with π, and thus the unique minimizer of f_o is at $\pi^* = 0$. This implies the DOP scheme (see (15)) is optimal for AAoI $\bar{a}(.)$. ∎

Proof of Theorem 2: As before it suffices to show that the numerator of derivative of f_n (with respect to π) is negative. Recall the following:

$$c_n d_o - c_o d_n = \frac{E[T^2]}{\lambda\gamma} + \left(\frac{1}{\lambda} + E[T]\right)\left(E[Te^{-\lambda T}] - \frac{(1-\gamma)}{\lambda}\right)\frac{2}{\lambda\gamma^2},$$
$$d_o = \frac{1}{\lambda\gamma}, \quad b_o = \frac{E[Te^{-\lambda T}]}{\gamma}, \quad d_n = \frac{1}{\lambda} + E[T]$$

The numerator of derivative of f_n is proportional to,

$$c_n d_o - c_o d_n + 2d_n(b_n d_o - b_o d_n)$$
$$= \frac{E[T^2]}{\lambda\gamma} + \left(\frac{1}{\lambda} + E[T]\right)E[Te^{-\lambda T}]\left(\frac{2}{\lambda\gamma^2} - \frac{2}{\gamma}\left(\frac{1}{\lambda} + E[T]\right)\right)$$
$$- \frac{2}{\lambda\gamma}\left(\frac{1}{\lambda} + E[T]\right)\left(\frac{1-\gamma}{\lambda\gamma} - E[T]\right)$$
$$= \frac{E[T^2]}{\lambda\gamma} - \frac{2}{\lambda\gamma}\left(\frac{1}{\lambda} + E[T]\right)\left(1 - \lambda E[Te^{-\lambda T}]\right)(d_o - d_n).$$

Thus the theorem follows from hypothesis. ∎

Appendix B: Some Useful Terms Used in the Proofs

The estimate of the term $c_n d_o - d_n c_o$:

$$c_n d_o - d_n c_o = \left(\frac{2}{\lambda^2} + E[T^2] + \frac{2E[T]}{\lambda} \right) \left(\frac{1}{\lambda} + \frac{1-\gamma}{\lambda\gamma} \right)$$

$$- \left(\frac{1}{\lambda} + E[T] \right) \left(\frac{2}{\lambda^2} + \frac{2(1-\gamma)}{\lambda^2\gamma} - \frac{2E[Te^{-\lambda T}]}{\lambda\gamma^2} + 2\frac{(1-\gamma)^2}{\lambda^2\gamma^2} + \frac{2(1-\gamma)}{\lambda^2\gamma} \right)$$

$$= \left(\frac{2}{\lambda^2} + E[T^2] + \frac{2E[T]}{\lambda} \right) \left(\frac{1}{\lambda\gamma} \right)$$

$$- \left(\frac{1}{\lambda} + E[T] \right) \left(\frac{2}{\lambda^2\gamma} - \frac{2E[Te^{-\lambda T}]}{\lambda\gamma^2} + 2\frac{(1-\gamma)^2}{\lambda^2\gamma^2} + \frac{2(1-\gamma)}{\lambda^2\gamma} \right)$$

$$= E[T^2] \left(\frac{1}{\lambda\gamma} \right) - \left(\frac{1}{\lambda} + E[T] \right) \left(-\frac{2E[Te^{-\lambda T}]}{\lambda\gamma^2} + 2\frac{(1-\gamma)^2}{\lambda^2\gamma^2} + \frac{2(1-\gamma)}{\lambda^2\gamma} \right)$$

$$= \frac{1}{\gamma} \left(\frac{E[T^2]}{\lambda} + \left(\frac{1}{\lambda} + E[T] \right) \left(\frac{2E[Te^{-\lambda T}]}{\lambda\gamma} \right) \right)$$

$$- \frac{1}{\gamma} \left(\left(\frac{1}{\lambda} + E[T] \right) \left(\frac{2(1-\gamma)}{\lambda^2\gamma} ((1-\gamma) + \gamma) \right) \right)$$

$$= \frac{1}{\gamma} \left(\frac{E[T^2]}{\lambda} + \left(\frac{1}{\lambda} + E[T] \right) \left(\frac{2E[Te^{-\lambda T}]}{\lambda\gamma} \right) - \left(\frac{1}{\lambda} + E[T] \right) \frac{2(1-\gamma)}{\lambda^2\gamma} \right)$$

$$= \frac{E[T^2]}{\lambda\gamma} + \left(\frac{1}{\lambda} + E[T] \right) \left(E[Te^{-\lambda T}] - \frac{(1-\gamma)}{\lambda} \right) \frac{2}{\lambda\gamma^2}. \tag{28}$$

References

1. Joo, C., Eryilmaz, A.: Wireless scheduling for information freshness and synchrony: drift-based design and heavy-traffic analysis. In: Modeling and Optimization in Mobile, Ad Hoc, and Wireless Networks (WiOpt) (2017)
2. Huang, L., Modiano, E.: Optimizing age-of-information in a multi-class queueing system. In: Information Theory (ISIT), 2015 IEEE International Symposium on IEEE (2015)
3. Kaul, S., Yates, R., Gruteser, M.: Real-time status: how often should one update? In: INFOCOM, 2012 Proceedings IEEE. IEEE (2012)
4. Talak, R., Karaman, S., Modiano, E.: Distributed scheduling algorithms for optimizing information freshness in wireless networks. arXiv preprint arXiv:1803.06469 (2018)
5. Altman, E., El-Azouzi, R., Ros, D., Tuffin, B.: Loss strategies for competing TCP/IP connections. In: Mitrou, N., Kontovasilis, K., Rouskas, G.N., Iliadis, I., Merakos, L. (eds.) NETWORKING 2004. LNCS, vol. 3042, pp. 926–937. Springer, Heidelberg (2004). https://doi.org/10.1007/978-3-540-24693-0_76
6. Altman, E.: Parallel TCP sockets: simple model, throughput and validation. In: IEEE Infocom (2006)

7. He, Q., Yuan, D., Ephremides, A.: Optimizing freshness of information: on minimum age link scheduling in wireless systems. In: Proceedings of the 14th International Symposium on Modeling and Optimization in Mobile, Ad Hoc, and Wireless Networks (WiOpt) (2016)
8. Yates, R.D., Kaul, S.K.: Status updates over unreliable multiaccess channels, pp. 331–335. In: Information Theory (ISIT), 2017 IEEE International Symposium on IEEE (2017)
9. Meyn, S.P., Tweedie, R.L.: Markov Chains and Stochastic Stability. Springer Science & Business Media, Heidelberg (2012)
10. Tse, D., Viswanath, P.: Fundamentals of Wireless Communication. Cambridge University Press, Cambridge (2005)
11. Menache, I., Shimkin, N.: Fixed-rate equilibrium in wireless collision channels. In: Chahed, T., Tuffin, B. (eds.) NET-COOP 2007. LNCS, vol. 4465, pp. 23–32. Springer, Heidelberg (2007). https://doi.org/10.1007/978-3-540-72709-5_3
12. Kavitha, V., Altman, E., Saha, I.: Controlling packet drops to improve freshness of information. arXiv preprint arXiv:1807.09325 (2018)
13. Wang, B., Feng, S., Yang, J.: When to preempt? Age of information minimization under link capacity constraint. J. Commun. Netw. **21**(3), 220–232 (2019)
14. Yates, R.D., Kaul, S.K.: The age of information: real-time status updating by multiple sources. IEEE Trans. Inf. Theory **65**(3), 1807–1827 (2018)

Maximizing Amount of Transferred Traffic for Battery Powered Mobiles

Eitan Altman[1,2]($\boxtimes$), Ghilas Ferrat[1,2], and Mandar Datar[1,2]

[1] INRIA Sophia Antipolis - Méditerranée, Valbonne, France
eitan.altman@sophia.inria.fr
[2] CERI/LIA, University of Avignon, Avignon, France

Abstract. There is a fast growing demand for mobile telephones. These rely on batteries to provide the power needed for transmission and for reception (up and downlink communications). Considering uplink, we analyse how the characteristics of the battery affect the amount of information that one can draw out from the terminal. We focus in particular on the impact of the charge in the battery on the internal resistance which grows as the battery depletes.

1 Introduction

In the design of power control algorithms, often one takes into account the level of depletion of the battery. When the battery is almost empty, a power saving mode is often applied. While this allows to extend the battery life, it is of interest to compare the benefits of this algorithm with one obtained from mathematical formulation of the problem as an optimisation one. In this paper we propose such a model that takes into account dynamic behavior of the battery and the fact that the internal battery resistance changes as a function of the battery's charge. Indeed, it is stated in that

"Li-ion has higher resistance at full charge and at end of discharge with a big flat low resistance area in the middle. Alkaline, carbon-zinc and most primary batteries have a relatively high internal resistance, and this limits their use to low-current applications such as flashlights, remote controls, portable entertainment devices and kitchen clocks. As these batteries deplete, the resistance increases further". We shall use the latter behavior in our modeling.

Our goal is to combine the dynamic behavior of the battery as function of its charge with capacity limits on transmission throughput from information theory to obtain limits on the amount of data that can be transmitted by mobiles that are powered with a battery.

2 Model

We assume that a fully charged battery has F Coulombs The figures in [1] reference suggests that the battery internal resistance R is linear decreasing in the battery's charge c and is thus given by

© Springer Nature Switzerland AG 2021
S. Lasaulce et al. (Eds.): NetGCooP 2021, CCIS 1354, pp. 78–84, 2021.
https://doi.org/10.1007/978-3-030-87473-5_8

$$R(t) = R_0 - \rho c(t)$$

for some constants ρ and R_0. The current i satisfies

$$\frac{dR(t)}{dt} = \rho i(t)$$

since $dc/dt = -i$.

We model the battery by a source of V volts with the internal resistance of R. Another resitor r is then connected in series to the two other elements and it represents the terminal equipment (TE).

$$i = \frac{V}{R+r}$$

Thus

$$\frac{dR(t)}{dt} = \frac{\rho V}{R(t) + r}$$

3   Analysis and Results

The solution of this differential equation is

$$R(t) = \sqrt{2\rho tV + (r + R_0)^2} - r$$

for some constants ρ and R_0. We could easily verify that $R(0) = R_0$ and that $\frac{dR(t)}{t} = \frac{\rho V}{R(t)+r}$. And so, $i(t) = \frac{V}{\sqrt{2\rho tV+(r+R_0)^2}}$.

We can then compute the time $T(x)$ the battery takes to discharge from F to x as the solution of $F - x = \int_0^{T(x)} i(t)dt$. We finally get the closed form expression of $T(x) = \frac{1}{2V}(F - x)[2(r + R_0) + \rho(F - x)]$. The power spent at the TE is given by

$$P(t) = ri(t)^2 = \frac{rV^2}{(r + R_0)^2 + 2\rho Vt}$$

Assume that all this power is transmitted and that the channel gain to the base station is h. Then assuming a single user and that the throughput Θ is given by Shannon capacity, we have for some constants W and N,

$$\Theta(t) = W \ln(1 + \frac{P(t)h}{N}) = W \ln(1 + \frac{rhV^2}{N((r + R_0)^2 + 2\rho Vt)})$$

and the total amount of data that can be transmitted is $\int_0^{T(0)} \Theta dt$, denoted TA.

In practice a terminal has to use some constant power Δ for its electronic circuit which reduces the amount of data transferred.

$$TA(0) = \int_0^{T(0)} \Theta(t)dt - \Delta T(0) = \int_0^{T(0)} W \ln(1 + \frac{rhV^2}{N((r + R_0)^2 + 2\rho Vt)})dt - \Delta T(0)$$

$$= W[\int_0^{T(0)} \ln[N(r + R_0)^2 + 2N\rho Vt + rhV^2]dt - \int_0^{T(0)} \ln[N(r + R_0)^2 + 2N\rho Vt]dt] - \Delta T(0)$$

$$(1)$$

Using $\frac{d}{dx}\frac{1}{a}((ax+b)\ln(ax+b)-(ax+b)) = \ln(ax+b)$ and computing the TA, we get that

$$
\begin{aligned}
TA(0) \;=\; W\Bigg(& \frac{1}{2N\rho V}N(r+R_0)^2\ln\left(1-\frac{2N\rho VT(0)rhV^2}{(rhV^2+N(r+R_0)^2)(N(r+R_0)^2+2N\rho VT(0))}\right) \\
& + \frac{1}{2N\rho V}rhV^2\ln\left(1+\frac{2N\rho VT(0)}{rhV^2+N(r+R_0)^2}\right) \\
& + T(0)\ln\left(1+\frac{rhV^2}{2N\rho T(0)+N(r+R_0)^2}\right)\Bigg) - \Delta T(0)
\end{aligned}
$$

4   Maximization of the Average Throughput

Previously we had a closed form expression of the time $T(x)$ that the battery takes to discharge from an initial charge F to a x level.

$$
T(x) = \frac{(F-x)[2(r+R_0)+\rho(F-x)]}{2V}
$$

$$
TA(x) = \int_0^{T(x)} (W\ln(1+\frac{rhV^2}{N(r+R_0)^2+2\rho NVt}) - \Delta)dt
$$

We also had a closed form expression of the total amount of data transmitted until we reach a x level of battery, $TA(x)$. Here the objective is to maximize the average throughput before charging the device, denoted $AP(x)$. Mathematically, the program can be written as

$$
\max_{x\in[0;F]} AP(x) = \frac{TA(x)-TC(x)}{T(x)} \tag{2}
$$

Let $\mathcal{L}(x,\lambda) = \frac{TA(x)-TC(x)}{T(x)} + \lambda(F-x)$ be the Lagrangian of the program. Solving it leads to

$$
\begin{cases}
\frac{\partial}{\partial x}\mathcal{L}(x,\lambda) = 0 \\
\frac{\partial}{\partial \lambda}\mathcal{L}(x,\lambda) = 0
\end{cases}
$$

We consider a case in which the total cost of charging the device is the same whether the battery is empty or almost full, denoted $TC(x) = \gamma$.

$$
\iff
\begin{cases}
\frac{T'(x)\left(\Theta(T(x)-\Delta-TC'(x))T(x)-(TA(x)-TC(x))\right)}{T(x)^2} - \lambda = 0 \\
F - x = 0
\end{cases}
$$

Considering the case of a border solution where $\lambda > 0$, we have $x^* = F$. Meaning that an agent charges the device, every time his battery full. In the case of an internal solution, where $\lambda = 0$, the first equation leads to either $T'(x) = 0$ or $\Theta(T(x))T(x) - \Delta T(x) - TA(x) + \gamma = 0$. The corresponding solution to $T'(x) = 0$ is $x = F + \frac{r+R_0}{\rho} > F$ therefore impossible considering the program. Using the software $Maple(Maplesoft)$ for the second equation, the time necessary to reach an optimal level of battery before charging the device $T(x^*)$ given by,

$$T(x^*) = \frac{e^{Root}}{2\rho N V}$$

Where *Root* is the solution of the following equation in terms of z,

$$2W(N(r+R_0)^2 + 2\rho N V)\ln(rhV^2 + N(r+R_0)^2)$$
$$+ W(e^z - 4\rho N V)\ln(rhV^2 + N(r+R_0)^2 + e^z)$$
$$- 2W(N(r+R_0)^2 + e^z)\ln(rhV^2 + e^z) + 2WN(r+R_0)^2 z$$
$$- We^z \ln(N(r+R_0)^2 + e^z) + 2(Wz - \Delta)e^z$$
$$- 2WN(r+R_0)^2 \ln(N(r+R_0)^2) + 2\gamma\rho N V = 0$$

In order to compute the internal solution x^*, the inverse function of $T(x)$ must be computed, and finally, the optimal level of battery is given by,

$$x^* = \frac{\rho F + r + R_0}{\rho} - \frac{\sqrt{e^{Root} + N(r+R_0)^2}}{\rho\sqrt{N}}$$

This solution has sense since it belongs to $[0; F]$ and that the agent charges his device only once his battery is at a lower level than the full level.

5   Numerical Applications

We have looked up real values for our various parameters, which make sense. We consider a battery life of 24 h, to simplify future computations. In general, a mobile phone's battery is a source of ca. 3, 7 V, has an initial internal resistance of ca. 105 mΩ, and when fully charged, has 7200 Coulombs. The second resistor has often a resistance of ca. 20.9 Ω (Fig. 1).

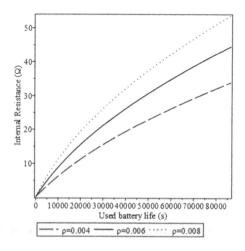

Fig. 1. Internal Resistance as a function of the remaining battery life

As stated in our model, a high level of charge means the internal resistance is low. We consider ρ as the restriction coefficient of the battery. As ρ is small as the power delivered by the battery is less constrained, the battery encounters less restrictions in delivering power, as shown in Fig. 2. In general, the restriction coefficient is ca. 0.006 (between 0.004 and 0.008) for most mobile phone battery's.

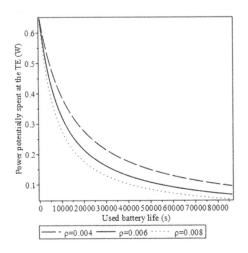

Fig. 2. Power potentially spent as a function of the remaining battery life

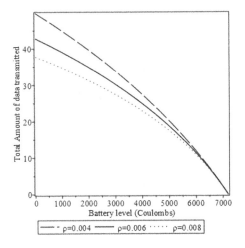

Fig. 3. Total amount of data transmitted as a function of the level of battery

Similarly, the total amount of data that can be transmitted in a full cycle of depletion is higher for smaller restriction coefficients, as shown in Fig. 3. And so, in a full cycle of depletion, the total amount of data that can be transmitted (i.e. TA(0)) is 49.78 for $\rho = 0.004$, 42.82 for $\rho = 0.006$, and 37.80 for $\rho = 0.008$.

Furthermore, assuming that recharging the battery is costly. We have optimized the average throughput over the level of remaining charge in the battery at which recharging would begin. We have assumed that the cost of charging the device was the same whether the battery was empty, almost empty, almost full or full. Here, with no loss of generality, we suppose that the cost is around 10 to 20% of the total amount of data that can be transmitted.

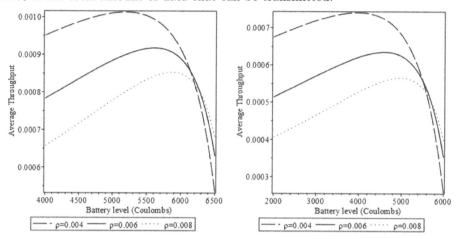

The figure on the left (respectively right) represents the average throughput, with a cost of 10% (resp. 20%) of the total amount of data that can be transmitted. We observe that, all other things being equal, a higher cost of charging drives the optimal battery level at which charging should begin to a lower level and so, leads to less cycles of charge.

6   Rising Internal Resistance

It is of interest to compare the amount of data that can be transmitted considering a constant resistance and a resistance as a function of the remaining battery life, like in our model. In [1] and [2], it is suggested that the battery's internal resistance R is an increasing function of used battery life and that a high level of resistance causes heating of the battery and less power can be spent, which ultimately leads to reducing the battery's life. As the battery's life is reduced, the total amount of data that can be transmitted decreases, the phenomenon through which rising internal resistance reduces performance of the device. As to show the phenomenon of rising internal resistance, we have plotted the difference between the throughput we would have if there was no rising effect and the average flow of data transmitted. Mathematically,

$$\int_0^{7200} \Theta(0) - \frac{TA(x)}{T(x)} dx.$$

In our model with our data, as shown in Fig. 4, we notice that the throughput with a constant resistance (i.e. $\int_0^{7200} \Theta(0) dx$) is of approximately 14 arbitrary

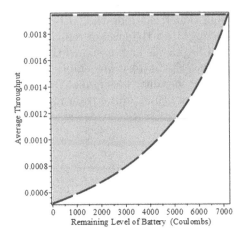

Fig. 4. Effect of Rising Internal Resistance

units and the average flow of data transmitted (i.e. $\int_0^{7200} \frac{TA(x)}{T(x)} dx$) is of approximately 7.01 in the same units. This difference is the physical phenomenon of rising internal resistance.

7   Conclusions

We have derived closed form solutions to the amount of information that a battery powered terminal can deliver. We made simplifying modelling assumptions, namely that the antenna can be represented as a resistor. In practice, it may be represented as an impedance that may vary in the frequency used. Also note that we did not model Alternative Current powers and the losses due to Direct Current/Alternative Current transformation. The next thing to do is to add a cost function that would penalize using the battery as a function of its charge so as to penalize operating it when the internal capacity of the battery is large. Indeed, as R increases, the efficiency of the battery decreases: most of its remaining power is spent on heating the battery as it is spent on R. Assuming that recharging the battery is costly, we have optimized over the level x of remaining charge in the battery at which the charging would begin so as to minimize the cost per cycle of depletion and recharging. We also plan to investigate interference and competition that may arise when more that one terminal is connected with a common base station. This is the objective of our future work.

References

1. https://batteryuniversity.com/learn/article/rising_internal_resistance
2. https://batteryuniversity.com/learn/article/tweaking_the_mobile_phone_battery
3. https://en.wikipedia.org/wiki/Electric_current
4. https://en.wikipedia.org/wiki/Shannon-Hartley_theorem

Game Theory in Mobile and Wireless Networks

An Ascending Implementation of the Vickrey-Clarke-Groves Mechanism for the Licensed Shared Access

Ayman Chouayakh[1,2]([⊠]), Aurélien Bechler[1]([⊠]), Isabel Amigo[2]([⊠]),
Loutfi Nuaymi[2]([⊠]), and Patrick Maillé[2]([⊠])

[1] Orange, Paris, France
aurelien.bechler@orange.com
[2] IMT Atlantique, Nantes, France
{isabel.amigo,loutfi.nuaymi}@imt-atlantique.fr,
patrick.maille@imt.fr

Abstract. *Licensed shared access* is a new sharing concept that allows Mobile Network Operators (MNOs) to share the 2.3–2.4 GHz bandwidth with its owner. This sharing can be done after obtaining a license from the regulator. The allocation is made among groups such that two base stations in the same group can use the same spectrum simultaneously. In this context, different auction schemes were proposed, however they are all one-shot auctions. In this paper, we propose an ascending implementation of the well-known Vickrey-Clarke-Groves mechanism (VCG) when the regulator has K identical blocks of spectrum to allocate. The implementation is based on the clinching auction. Ascending auctions are more transparent than one-shot auctions because bidders see the evolution of the auction. In addition, ascending auctions preserve privacy because bidders do not reveal necessarily their valuations.

1 Introduction

In order to accommodate data traffic for 5G networks, Mobile Network Operators (MNOs) need more radio spectrum. At the same time many holders of exclusive licenses –which we call incumbents– might not utilize all of their spectrum resources: usage varies with respect to time and location. Therefore the idea of Licensed Shared Access (LSA) has emerged. LSA is a new concept of spectrum sharing in which the holder of the 2.3–2.4 GHz bandwidth can share his spectrum with MNOs. This concept was proposed by the radio spectrum policy group (RSPG) in November 2011 [1]. Sharing is done after obtaining a license from the regulator. This license guarantees a certain quality of service to both the incumbent and the LSA licensees (MNOs). This differs from the traditional concept of sharing in which MNOs have no guarantees on finding the spectrum free for their own usage and have to use some techniques (such as cognitive radio) before accessing the spectrum.

In this context, since the regulator ignores the value of the LSA spectrum, then by auctioning that spectrum he can have an idea about the valuations of

S. Lasaulce et al. (Eds.): NetGCooP 2021, CCIS 1354, pp. 87–100, 2021.
https://doi.org/10.1007/978-3-030-87473-5_9

MNOs for that spectrum. A well designed auction mechanism should be truthful i.e., each player should not be able to play the system by bidding strategically. Also, in the case of LSA, it has to take into account the spacial reusability of spectrum i.e., two base stations can use the same spectrum if they do not cause interference to each other.

In [2–4], authors designed mechanisms which could be applied in the case where there is one and only one block to allocate. On the other hand, in [5] we have designed and analyzed a truthful scheme when spectrum is infinitely divisible. A more realistic assumption is to suppose that spectrum can be split in several sub-bands or blocks, that have a predetermined size. Therefore, in this paper we suppose that the regulator has K identical blocks to allocate. Identical means that besides being of the same width, there is no preference over blocks from the point of view of base stations [6,7].

All the previous mentioned mechanisms are one-shot auctions in which bidders reveal all their valuations. Contrary to one-shot auctions, ascending auctions preserve the privacy of the winning bidder(s) because the winner(s) do(es) not need to reveal all his valuations. Hence in this paper we focus on ascending auctions. Since the objective of the regulator (the auctioneer) is to optimize the use of spectrum, we propose to implement an ascending version of Vickrey-Clarke-Groves mechanism (VCG). To the best of our knowledge, this is the first ascending mechanism for LSA context. In fact, we propose to implement VCG using two approaches: the first approach is by adding representatives so that the auction will be between the auctioneer and those representatives and the second approach is by removing those representatives i.e., base stations communicate directly with the auctioneer. In this paper we use player, bidder and base station interchangeably.

The rest of this paper is organized as follows: Sect. 2 presents the system model. In Sect. 3, we present the clinching mechanism and show how it can be adapted for the LSA context using two approaches. In Sect. 4, we evaluate the performances of our proposition. Section 5 concludes the paper.

2   System Model

We consider N base stations of different MNOs in competition to obtain K blocks of spectrum at a certain time and geographical area. We suppose that blocks are identical. As it was presented in [4,8], the problem can be modeled using an interference graph. In order to optimize the use of spectrum, the regulator constructs from the interference graph M groups such that two base stations of the same group do not interfere with each other (and so they can use the same spectrum block simultaneously). So finally, the competition between the N base stations is transformed into a competition between M groups. In this paper we suppose that groups are constructed before the auction takes place.

2.1   Preferences of Base Stations

We assume that each base station i has a private valuation v_i vector of size K, each element $v_{i,k}$ representing the willingness-to-pay for the k-th extra resource block. The valuation of a block can be interpreted as the revenue from that block. As in [9], we suppose that the value of an extra block, for a base station, decreases with the number of blocks already obtained. This corresponds to a discretization of concave valuation function for spectrum [9], as illustrated in Fig. 1. Finally, we adopt a quasi-linear utility model: if a base station i obtains n_i blocks and pays p_i, its utility then is

$$u_i = \sum_{n=1}^{n_i} v_{i,n} - p_i.$$

In particular, a base station obtaining no block gets a utility equal to zero.

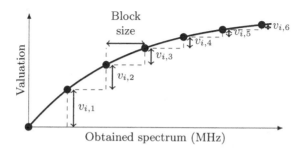

Fig. 1. An example of a concave valuation function of obtained spectrum, and the corresponding block valuations $v_{i,n}$ for a player i.

2.2   Why Implementing VCG?

In this paper, we suppose that the regulator wants to implement the ascending version of VCG motivated by the following features:

1. Efficiency: Efficiency is defined as the sum of the valuations served $\sum_{i=1}^{N} \sum_{n=1}^{n_i} v_{i,n}$ [10]. This means that the social value of the good being sold equals the maximum of the potential buyers' individual valuations.
2. Truthfulness: This property means that bidders' best strategy is to behave sincerely, i.e., lying about one's preferences is not beneficial. The strongest version is when truth-bidding is a dominant strategy, but it can also be a (weaker) ex-post Nash equilibrium strategy: when truthful bidding is an ex-post equilibrium, each player knows that bidding truthfully is a best strategy if all other players also bid truthfully and without knowing the other players' valuations [11].
3. Individual rationality: A mechanism is individually rational if each player has an incentive to participate in the auction, i.e., it has a strategy guaranteeing it a non-negative utility [12].

3   VCG Ascending Implementation for the LSA Context

In this section, we start by presenting the general framework of the clinching approach, then we show how to adapt it to the LSA context using two approaches.

3.1   Background

The clinching auction [13] is an ascendant auction for K homogeneous goods, where bidders have valuations as described in Sect. 2. At each round t, the auctioneer declares a price P^t and each bidder i responds by demanding a quantity, if demand is higher than supply, then the auctioneer increases the price at the next round $P^{t+1} = P^t + 1$. The auction ends when demand is not higher than supply. Bidder's payments are computed during the auction. We detail in the following how the clinching auction works. The cumulative clinch Cl_i^t of player i at round t is defined as:

$$\mathrm{Cl}_i^t = \max\{0, K - \sum_{j \neq i} d_j^t\}, \tag{1}$$

with d_j^t the demand of player j at round t. The current clinch at round t of player i (the number of blocks obtained at round t) is denoted by cl_i^t:

$$\mathrm{cl}_i^t = \mathrm{Cl}_i^t - \mathrm{Cl}_i^{t-1}. \tag{2}$$

When the auction ends, each bidder i obtains a quantity equal to its cumulative clinch Cl_i, and its payment p_i is:

$$p_i = \sum_{t=0}^{T} P^t \mathrm{cl}_i^t. \tag{3}$$

It was proven in [13] that the clinching auction achieves the outcome of VCG i.e., it ensures an efficient allocation, charges each player with its Vickrey payment and bidding truthfully is an ex post Nash equilibrium. Here bidding truthfully means that each player reports its demand with respect to its valuations: $d_i(P) = \max\{n \text{ such that } v_{i,n} > P\}$, for a given declared price P.

Remarks

1. We illustrate in the following example why truthful bidding is not a dominant strategy. We suppose we have two blocks and two players, where valuations of the first player are $\{3, 2\}$ and the second $\{2, 1\}$. Suppose that the second player uses the following strategy: if the first player demands two blocks at the first round then it will continue to demand 2 blocks until the end of the auction (even though it will obtain a negative utility), otherwise it demands one block. Clearly, given that strategy, player one has to demand only one block at the first round so at the second round ($P = 1$) player two demands one block. Thus the auction ends at the second round (since the total demand is two). Each player gets one block, the utility of player one is $3 - 1 = 2$.

2. We denote by $\{f_{i,1}, .., f_{i,K}\}$ the highest K valuations, extracted in a non decreasing order, of other players facing player i. There is a relation between clinching prices and those valuations: if player i clinches its nth block at price P then that price is the minimum price such that the sum of demands of all other players is $K - n$. This corresponds to the situation in which $P = f_{i,n}$.

Table 1 summarizes the notations used throughout the paper. In the following, in order to implement the ascending version of VCG, we propose at a fist time to add a representative part per group, that part will represent the members of that group. After that, we show how to implement VCG without those representatives.

Table 1. Notations

K	Number of blocks on sale
$v_{i,n}$	valuation of base station i for an nth block.
$d_i(P)$	demand of player i at price P
g_h	group h
n_h	number of players of group h
$D_h(P)$	demand of group h (or representative h) at price P
$(D_h(P))^{-i}$	demand of group h at price P when player i is absent
$B_{h,n}$	group-bid of representative h for its nth block
$B_{h,n}^{-i}$	group-bid of representative h for its nth block when player i is absent
$F_{h,n}$	The nth highest group-Bids of other representatives facing representative h

3.2   VCG Implementation with a Representative per Group

In this section, we show how to implement the clinching approach for the LSA concept when a representative part per group is introduced. It can be an interface between base stations and the auctioneer. There is no communication between base stations and the auctioneer. The auction will be between the M representatives and the auctioneer. Before the auction takes place, each base station i transmits to the representative of its group its bid vector b_i (which can be different from v_i), then each representative h constructs the group-bid vector B_h based on the received bids ($B_{h,n} = \sum_{i=1}^{N} b_{i,n} 1_{i \in g_h}$). At each round and for each price P each representative h demands a quantity $D_h(P)$ with respect to the group-bid vector (the demand corresponds to the number of components that are higher than P) as showing in the following equation .

$$D_h(P) = \max\{n, B_{h,n} > P\}$$

If a representative obtains a block then it charges each base station of that group an amount. The auction clears when the sum of demands of all representatives is equal or lower than K. The steps of the auction can be summarized as follows:

1. Each base station reports to the corresponding representative its bids vector.
2. The representative constructs the group-bid vector.
3. At each round, each representative reports its demand $D_h(P)$ to the auction-eer.
4. The auctioneer computes the cumulative clinch Cl_h^t of each representative h at round t which is defined as:

$$\text{Cl}_h^t = \max\{0, K - \sum_{j \neq i} D_j^t\}, \tag{4}$$

with D_j^t the demand of representative j at round t. The current clinch (the number of blocks obtained at round t) of representative h is denoted by cl_h^t:

$$\text{cl}_h^t = \text{Cl}_h^t - \text{Cl}_h^{t-1}. \tag{5}$$

5. If a representative clinches (obtains) a block at a price P then it charges each base station i of its group a price and sends that amount to the auctioneer.
6. If the demands of all representatives is higher than K, then the auctioneer increases P at the next round, otherwise the auction ends.

Clearly, if each base station reports to the corresponding representative its true valuations then the allocation is efficient (the procedure is similar to the original one). However, reporting true valuations depends on the payment that will be made, the question is how to charge each base station in a manner that guarantees truthful bidding? i.e., reporting its true valuations to the representative is a dominant strategy. We denote by $p_{i,n}$ the payment of base station i for its nth block and by F_h the vector of the K highest valuations of the other representatives (extracted in a non decreasing order) facing representative h. The following proposition proposes a payment rule ensuring truthful bidding.

Proposition 1. *The following payment rule ensures a truthful bidding (as dominant strategy):*

$$p_{i,n} = [P - B_{h,n}^{-i}]^+$$

Where $B_{h,n}^{-i} = B_{h,n} - b_{i,n}$.

Proof. Suppose that by reporting its true valuations, player i obtains n_i blocks and pays p_i. Any other reported bids may:

1. Increase the number of obtained blocks n_i', in this situation it will pay the same amount for the first n_i blocks (because its payment is independent of his bids), however it will pay an amount higher than its valuation for the other blocks: for each j in $\{n_{i+1}, .., n_i'\}$ we have $F_{h,j} > B_{h,j}^{-i} + v_{i,j}$ (otherwise group h would have obtained that block) thus $F_{h,j} - B_{h,j} > v_{i,j}$ i.e., $p_{i,j} > v_{i,j}$ which leads to reduce its utility.
2. Obtain the same number of blocks, player i obtains the same utility.
3. Decrease the number of obtained blocks. In this situation it will decrease its utility since those blocks are charged below its valuation for them.

We denote by B^S the vector composed of all the group-bid vectors sorted in a non increasing order. The size of B^S is $M \times K$. Please note that blocks are allocated to the groups with the first K components of B^S (efficiency is $\sum_{i=1}^{K} B_i^S$).

We now illustrate how the mechanism works for a given configuration.

Example 1. Suppose we have three group g_1, g_2 and g_3, and three blocks.

- g_1 is composed of three players with the following bids respectively $\{5,3,2\}$, $\{10,6,4\}$ and $\{10,6,3\}$.$B_1 = \{25,15,9\}$
- g_2 is composed of two players with the following bids respectively $\{10,6,3\}$ and $\{8,4,2\}$. $B_2 = \{18,10,5\}$
- g_3 is composed of one player with bids $\{17,11,4\}$. $B_3 = \{17,11,4\}$.

Applying the clinching approach with representatives leads to:

1. At $P = 11$, $D_1(P) = 2$, $D_2(P) = 1$ and $D_3(P) = 1$, thus group one clinches his first block. Player one pays $[11 - 20]^+ = 0$, similarly for player two and three, each one pays zero.
2. At $P = 15$ we have $D_1(P) = 1$, $D_2(P) = 1$ and $D_3(P) = 1$. The auction ends. The second and the third group obtain their first block each. The first player of the second group pays $15 - 8 = 7$, the second player pays $15 - 10 = 5$. The player of the third group pays 15.

Also, here $B^S = \{25,18,17,11,15,10,9,5,4\}$ and efficiency is $25 + 18 + 17 = 60$.

In the following we study the convergence rate (the number of rounds that the auction takes to end).

Efficiency and Convergence Rate Trade-Off. The convergence speed of the auction depends on the increment. A possible way to accelerate the convergence of the auction is by increasing the increment from round to another i.e., P will be increased by an amount $q > 1$. Before studying the impact of changing the increment from 1 to $q > 1$, let us first introduce the following proposition:

Proposition 2. *When the increment is equal to 1, the auction concludes after B_{K+1}^S rounds.*

Proof. At each price P the demand of each representative corresponds to the number of components that are higher than P. In particular if $P = B_{K+1}^S$ then the sum of demands of all representatives is exactly K, also $P = B_{K+1}^S$ is the first price at which the sum of demands is exactly K, for $P = B_{K+1}^S - 1$ the sum of demands is $K + 1$. Therefore, the auction ends after B_{K+1}^S rounds.

In the following proposition, we show that by increasing the increment we may accelerate the convergence rate but this may lead to loss in terms of efficiency. Please note that if the final price is higher than a component from the set $\{B_1^S, .., B_K^S\}$ then that component will not be demanded: suppose that we have two blocks and two bidders with valuations $\{7,5\}$ and $\{4,1\}$. Suppose that $q = 3$.

The auction ends at the second round when $P = 6$ and the total demand is 1 (bidder one will not demand two blocks because his valuation for the second block is lower than the final price) efficiency is therefore 7.

Proposition 3. *After introducing an increment $m > 1$, the auction ends after T_c rounds such that $T_c = \lceil * \rceil \frac{B_{k+1}^S}{q}$ (where $\lceil * \rceil x$ is the least integer greater than or equal to x) and the efficiency is $\sum\limits_{i=1}^{K} B_i^S \mathbb{1}_{B_i^S > T_c \times q}$*

Proof. The auction ends when P reaches a value higher than B_{K+1}^S. That value is reached after T_c rounds such that $T_c \times q > B_{K+1}^S$. Therefore $T_c = \lceil * \rceil \frac{B_{K+1}^S}{q}$. In terms of efficiency, since if the final price is higher than a component from the set $\{B_1^S, .., B_K^S\}$ then that component will not be demanded then $\sum\limits_{i=1}^{K} B_i^S \mathbb{1}_{B_i^S > T_c \times q}$. Note that A necessarily condition to obtain the optimal efficiency is to have $B_{K+1}^S < T_c \times q < B_K^S$

One may wonder if we can reduce the increment if the demands fall down rapidly from round t to another $t + 1$ i.e., we reduce P at $t + 2$ and ask bidders for their demands. By doing so the auction will not be truthful anymore because bidders may reduce their demands in order to reduce their payments. So the increment must be the same during the auction. Next, we set q to 1 in order to obtain an efficient implementation.

In the original version of clinching [13] (without groups), clinching prices represent also the payment of players. However, in our context, if a group clinches a block at a price P, then P is the maximum amount that it may pay. We prove that in the following proposition.

Proposition 4. *If a group (representative) clinches its nth block at price P then the sum of payments of players of that group can not be higher than P.*

Proof. We can distinguish two cases:

1. There exists a player such that $b_{i,n} > P$, then in this situation, each player j except i pays zero because $B_{h,n}^{-j} > P$, for player i it will pay $[P - B_{h,n}^{-i}]^+ < P$ thus the revenue in this situation is lower than P.
2. $\forall\, i,\ b_{i,n} < P$, we take any set S_h of group h such that the sum of bids of its members is higher than P and lower than P when removing any player of the set i.e., $\sum\limits_{i \in S_h} b_{i,n} \geq P$ and $\forall\, j \in S_h \sum\limits_{i \in S_h, i \neq j} b_{i,n} \leq P$, we can obtain that set as follows: we sort bids of group h in a non increasing order. In the beginning S_h is composed of the player with the highest bid. We keep extending S_h by adding players until both conditions hold. In this situation, each player of group h which does not belong to S_h pays zero, payment of group h is given by:

$$P_{h,n} = \sum_{i=1}^{|S_h|}(P - \sum_{j\neq i}^{n_h} b_{j,n}) \tag{6}$$

$$= |S_h|P - \sum_{i=1}^{|S_h|}\sum_{j\neq i}^{n_h} b_{j,n}) \tag{7}$$

Since

$$\sum_{i=1}^{|S_h|}\sum_{j\neq i}^{n_h} b_{j,n} = \sum_{i=1}^{|S_h|}(\sum_{j\neq i}^{n_h} b_{j,n} + b_{i,n} - b_{i,n}) \tag{8}$$

$$= |S_h|V_{h,n} - \sum_{i=1}^{|S_h|} b_{i,n} \tag{9}$$

We obtain

$$P_{h,n} = |S_h|P - |S_h|V_{h,n} + \sum_{i=1}^{|S_h|} b_{i,n} \tag{10}$$

$$\leq |S_h|P - |S_h|\sum_{i=1}^{|S_h|} b_{i,n} + \sum_{i=1}^{|S_h|} b_{i,n} \tag{11}$$

$$= (|S_h| - 1)(P - \sum_{i=1}^{|S_h|} b_{i,n}) + P \tag{12}$$

$$\leq P \tag{13}$$

This first implementation has the following advantages: first, truthful bidding is a dominant strategy. Second, the auctioneer could not have a precised idea about valuations of base stations, it may have only an idea about the total valuation of group h for an nth block but he can not see the valuation of each base station. In practice, it may be difficult to introduce those representatives because we may have "the black box effect": from the point of view of players, they can not see the evolution of the auction (they are just asked to pay an amount for an obtained block). For the auctioneer, he can not see how each base station is charged. Thus in the following we show how to implement the ascending version when removing those representatives so that the auction will be held between the auctioneer and base stations directly.

3.3   VCG Implementation Without Representatives

In this section, we propose to implement the ascending VCG auction when representatives are removed. In this scenario, the auction will be between the regulator and base stations. Similarly to what was presented before, the auctioneer fixes a price P and keep increasing P until demands of groups is no higher than supply.

The question here is how to compute the demand of groups? We propose to introduce a price p_h per group, for each price P, the auctioneer keep increasing p_h and ask each player of group h its demand $d_i(p_h)$, until he can compute the demand of group h $D_h(P)$ which is defined as:

$$D_h(P) = \max\{n : \exists\, \omega \subset g_h \text{ and } (r_1, ..., r_{|\omega|}) \in \mathbb{R}^{|\omega|}$$
$$\text{s.t. } d_i(r_i) = n \text{ and } \sum_{i=1}^{|\omega|} r_i > P \; \forall i \in \omega\}$$

Intuitively this demand means that there is a set of players of group h which will pay P in order to obtain n blocks and there is not a set of players which are willing to pay P for $n + 1$ blocks. Here demands of group is computed directly by the auctioneer instead of the representatives. Also this demand is the same as it was presented before. Therefore, if we assume truthful bidding we obtain an efficient allocation. The question now, is how to compute the payment of each player? In fact, the introduction of p_h is not only to compute the demand of a group as a function of demands of its members but also to compute payments of players. We denote by $\left(D_h(P)\right)^{-i}$ the demand of group h without player i.

We propose to operate as follows: if group h clinches its nth block at price P then, for each player i, we keep increasing p_h until:

1. Either $\left(D_h(P)\right)^{-i} > 0$, in this situation player i pays zero.
2. Or achieving the maximum amount (could be computed from demands) that all players (without counting i) of group h can pay in order to obtain n blocks i.e., $\left(D_h(m)\right)^{-i} = n-1$ and $\left(D_h(m-1)\right)^{-i} = n$, i.e., $m = V_{h,n}^{-i}$. In this situation player i pays $P - m$.

Thus we can see that payment is the same as the one of the previous implementation (with representatives).

In the following proposition we show that truthful bidding (demanding a quantity with respect to the valuations) is an ex post Nash Equilibrium.

Proposition 5. *In the proposed auction mechanism, truthful bidding is an ex post-Nash equilibrium.*

Proof. Let us fix a base station i, suppose that all other base stations report their demand truthfully during the auction, by reporting its true demand player i will obtain the same utility as in the auction with the representatives. we denote by u_1 that utility. Now we have to show that any other strategy of demanding for player i will reduce his utility i.e., it obtains a utility $u_2 \leq u_1$. Suppose that strategy generates a higher utility, this means that player i could obtain the same utility in the first implementation (with representatives) by proposing a bids vector with respect to its reported demands. This is a contradiction because in the first implementation, proposing the valuation is dominant strategy. Thus $u_1 \geq u_2$.

We illustrate in the following example how the auction works.

Example 2. We take the same configuration of example 1, applying the clinching approach without representatives leads to:

1. At $P = 11$, $D_1(P) = 2$ (could be obtained when $p_1 = 5$, $r_1 = 2.5, r_2 = r_3 = 4.5$), $D_2(P) = 1$ ($p_2 = 6$) and $D_3(P) = 1$ ($p_3 = 11$), thus group one clinches its first block. The first player of that group pays zero because without it and for $r_2 = r_3 = 5.5$ (p_1 was increased until 6) the second and the third players can always obtain that block. Similarly for player two, it pays zero because without it player one and three can obtain that block for $r_1 = 3$ and $r_3 = 8$ (p_1 was increased till 8). Similarly for player three it pays zero.
2. At $P = 15$ we have $D_1(P) = 1$, $D_2(P) = 1$ and $D_3(P) = 1$. The second and the third group obtain their first block each group. For the second group we keep increasing p_2 till 10 in order to compute the payments of players of that group. The first player of the second group pays $15 - 8 = 7$, the second player pays $15 - 10 = 5$. The player of the third group pays 15.

In the following section, we evaluate the performance of our proposed ascending version VCG. Note that both approaches have the same performances but we prefer the second approach (without representatives) since it is more transparent. Also, we set the increment to 1.

4   Performance Evaluations

We compare VCG with TLSAA and TLSAA2 [14], two truthful variants of LSAA [4] that we have proposed in previous work. LSAA is the first auction mechanism which was proposed as candidate for the LSA context. LSAA outperforms other potential candidate mechanisms such as TAMES [2] and TRUST [3] in terms of efficiency. However LSAA is not truthful.

Note that for LSAA, TLSAA and TLSAA2 all the available spectrum is allocated as a one block for only one group via a single round auction. Each bidder submits a bid which represents the maximum amount that it is willing to pay in order to obtain all the available spectrum. Then the auctioneer computes the group-bid of each group which is a positive real obtained via a function. Spectrum is allocated to the group with the highest group-bid.

We compare our implementation with TLSAA and TLSAA2 in terms of revenue of the auctioneer, efficiency and fairness of the allocation. In order to quantify the fairness of the allocation, we use Jain's index [15] which is a continuous function on the closed interval $[\frac{1}{N}, 1]$ and measures the fairness of the allocation between N players. In particular, that index achieves its maximum (1) when all players obtain the same amount and achieves its minimum ($\frac{1}{N}$) when all the available spectrum is allocated to only one player.

4.1   Simulation Settings

We have fixed $M = 10$ groups, the number of players in each group is chosen randomly from the discrete uniform distribution of integer values in the interval

[1 ; 30]. We suppose that there is a quantity of LSA spectrum that could be divided into 100 blocks. For TLSAA and TLSAA2 that quantity is allocated as a single block. We create the bid vector which is composed of 100 elements: the first element is drawn from the uniform distribution over the interval $[0, 100]$ and the n-th element $(n > 1)$ is drawn from the uniform distribution $[0, b_{i,n-1}]$ (For TLSAA and TLSAA2, the bid is the sum of those components). The average fairness, revenue and efficiency are computed over 1000 draws.

4.2   Simulation Results

In terms of efficiency, by construction VCG is more efficient than the other mechanisms. This result is confirmed by Table 3. Also, the more blocks we divide the spectrum into, the more efficient the allocation is. In terms of fairness of the allocation, VCG is fairer than the other two mechanisms as its shown in Table 4, this is natural since blocks will be distributed among groups and will not be allocated to one and only one group. If the auctioneer wants only to increase his revenue then our proposition is not the best as it is shown in Table 2. However, our proposition offers a good trade-off between fairness, efficiency and revenue: by splitting spectrum into five blocks. Compared to LSAA2 we lose 62% of the revenue but we win more than 300% in terms of efficiency and more than 500% in terms of fairness. In addition, we add transparency, price discovery and privacy to the auction.

Table 2. Average revenue as a function of the number of blocks for $M = 10$

K	100	10	5	4	2	1
VCG	3.82	1.5	422	458	488	368
TLSAA	–	–	–	–	–	380
TLSAA2	–	–	–	–	–	1304

Table 3. Average efficiency as a function of the number of blocks for $M = 10$

K	100	10	5	4	2	1
VCG	15574	15487	11337	9962	5469	2945
TLSAA	–	–	–	–	–	2887
TLSAA2	–	–	–	–	–	2558

Table 4. Average fairness as a function of the number of blocks for $M = 10$

K	100	10	5	4	2	1
VCG	0.98	0.99	0.71	0.6	0.33	0.18
TLSAA	–	–	–	–	–	0.17
TLSAA2	–	–	–	–	–	0.11

5   Conclusion

In this paper we have proposed two ways in order to implement the ascending version of VCG for the LSA context. For the first approach, we have introduced a representative per group. At each round, each representative transmits to the regulator the demand of its group based on bids of its members. Each base station is charged a price computed by the representative of its group. There are two advantages of this implementation. First, truthful bidding is a dominant strategy and second we preserve privacy of valuations of base stations. However, it can be difficult to introduce those representatives in practice. Thus, at a second time, we have proposed another ascending implementation of VCG without those representatives and in which communication is directly between the auctioneer and base stations. We have introduced a price per group and show how to compute the payment of each player. In the second approach truthful bidding is an ex post Nash equilibrium. Transparency is the main advantage of the second approach because each base station sees the evolution of the auction. In addition, we have shown by simulations that our proposition offers a good trade-off between fairness, efficiency and revenue compared to other mechanism proposed for the LSA context.

References

1. Matinmikko, M., Okkonen, H., Malola, M., Yrjola, S., Ahokangas, P., Mustonen, M.: Spectrum sharing using licensed shared access: the concept and its workflow for LTE-advanced networks. IEEE Wirel. Commun. **21**, 72–79 (2014)
2. Chen, Y., Zhang, J., Wu, K., Zhang, Q.: Tames: a truthful auction mechanism for heterogeneous spectrum allocation. In: IEEE INFOCOM, pp. 180–184 (May 2013)
3. Zhou, X., Zheng, H.: Trust: a general framework for truthful double spectrum access. In: Proceedings of the IEEE INFOCOM (2009)
4. Wang, H., Dutkiewicz, E., Fang, G., Mueck, M.D.: Spectrum sharing based on truthful auction in licensed shared access systems. In: Vehicular Technology Conference (July 2015)
5. Chouayakh, A., Bechler, A., Amigo, I., Nuaymi, L., Maillé, P.: PAM: a fair and truthful mechanism for 5G dynamic spectrum allocation. In: Proceedings of the IEEE PIMRC (2018)
6. Zhou, X., Gandhi, S., Suri, S., Zheng, H.: eBay in the sky: strategy-proof wireless spectrum auctions. In: Proceedings of the ACM MobiCom, pp. 2–13 (2008)
7. Wang, W., Liang, B., Li, B.: Designing truthful spectrum double auctions with local markets. IEEE Trans. Mob. Comput. **13**(1), 75–88 (2014)
8. Nemhauser, G.L., Wolsey, L.A., Fisher, M.L.: An analysis of approximations for maximizing submodular set functions i. Math. Program. **14**(1), 265–294 (1978)
9. Enderle, N., Lagrange, X.: User satisfaction models and scheduling algorithms for packet-switched services in UMTS. In: Vehicular Technology Conference, 2003. VTC 2003-Spring. The 57th IEEE Semiannual, vol. 3, pp. 1704–1709. IEEE (2003)
10. Roughgarden, T., Sundararajan, M.: Is efficiency expensive. In: Third Workshop on Sponsored Search Auctions (2007)

11. Roughgarden, T.: Ascending and ex post incentive compatible mechanisms (2014). https://theory.stanford.edu/tim/w14/l/l21.pdf. Stanford Lecture notes CS364B: Frontiers in Mechanism Design
12. Krishna, V.: Auction Theory. Academic Press, Cambridge (2009)
13. Ausubel, L.M.: An efficient ascending-bid auction for multiple objects. Am. Econ. Rev. **94**(5), 1452–1475 (2004)
14. Chouayakh, A., Bechler, A., Amigo, I., Maillé, P., Nuaymi, L.: Designing lsa spectrum auctions: mechanism properties and challenges. Submitted. https://hal.archives-ouvertes.fr/hal-02099959/document
15. Jain, R.K., Chiu, D.-M.W., Hawe, W.R.: A quantitative measure of fairness and discrimination. Eastern Research Laboratory, Digital Equipment Corporation, Hudson, MA (1984)

Resource Orchestration in Interference-Limited Small Cell Networks: A Contract-Theoretic Approach

Maria Diamanti[1], Georgios Fragkos[2], Eirini Eleni Tsiropoulou[2],
and Symeon Papavassiliou[1]($\boxtimes$)

[1] School of Electrical and Computer Engineering, National Technical University
of Athens, 15780 Athens, Greece
`mdiamanti@netmode.ntua.gr, papavass@mail.ntua.gr`
[2] Department of Electrical and Computer Engineering, University of New Mexico,
NM 87131 Albuquerque, NM, USA
`{gfragkos,eirini}@unm.edu`

Abstract. Recently, non-orthogonal multiple access (NOMA) small cell networks (SCNs) have been studied to meet the stringent requirements for spectral efficiency and massive connectivity in the emerging 5G networks. This paper aims at addressing the overall resource orchestration issue in 5G SCNs, by considering the problem of joint user to small cell association and uplink power allocation, employing NOMA technology. In particular, the power allocation is performed under an incomplete information scenario, where the users' channel conditions are unknown to the small cell. To treat this issue in an effective manner, contract theory is adopted in order to incentivize each user to select the power level that optimizes its own utility, while each small-cell base station (SBS) rewards them inversely proportionally to their respective sensed interference. The proposed framework is complemented by a distributed user-cell association mechanism based on reinforcement learning (RL). Indicative numerical results are provided to validate the operation and effectiveness of the proposed contract-theoretic approach.

Keywords: Small cell networks (SCNs) · Non-orthogonal multiple access (NOMA) · Reinforcement learning (RL) · Contract theory · Incomplete information · User association · Power allocation

1 Introduction

The deployment of small cells co-existing with the legacy macrocell network, constitutes a straightforward and effective approach to support the deluge of data traffic induced in the uplink of 5G wireless networks. Though small cells are typically low power, low cost, short range wireless transmission systems (base stations), they have all the basic characteristics of conventional base stations and

S. Lasaulce et al. (Eds.): NetGCooP 2021, CCIS 1354, pp. 101–109, 2021.
https://doi.org/10.1007/978-3-030-87473-5_10

are capable of handling high data rate for individual users [1,2]. Accordingly, they can efficiently reuse the spectrum across different geographical areas, thus, improving spectrum efficiency and enhancing network coverage and capacity. Towards the same direction of contributing to improved spectral efficiency and massive connectivity, non-orthogonal multiple access (NOMA) technology has been adopted by 5G networks, allowing multiple users to multiplex in the power domain over the same time/frequency/code resources. The successive interference cancellation (SIC) technique is, then, applied at the receiver to mitigate part of the interference, by sequentially decoding the received signals based on their signal strength [3].

One of the key problems and challenges in this emerging environment, is the inter-user interference caused by different users transmitting over the same resources. The latter heavily depends on the joint user to small cell association and corresponding power allocation problem. Therefore, the development of efficient resource allocation schemes that deal with this problem in NOMA small cell networks (SCNs) is of significant research and practical importance.

1.1 State of the Art and Motivation

Several research works (e.g., [4,5]) have addressed the emerging resource orchestration problem in NOMA SCNs, and in particular the joint problem of user association and power allocation. For instance, in [4], the problem is formulated as a coalition formation game that associates users with the small cell, for which the total power is minimized in the optimal partition of users. In [5], a similar problem focusing on the joint user clustering and power allocation, is treated in two iterative stages to avoid the exhaustive search of user clusters. Nevertheless, the majority of existing works on the topic of user association and/or interference mitigation, presume perfect knowledge of the global channel and/or network information (e.g., channel state information (CSI)), which is either impossible to have or impractical to obtain. Considering the situation when only statistical CSI is available at the small base stations (SBSs), the private information of the user devices regarding their experienced channel conditions could be utilized to ease the resource allocation procedure. In this context, contract theory, enables the modeling of an incentive mechanism in 5G SCNs under a practical scenario of incomplete information, where the users' private information (i.e., transmission power, wireless channel characteristics) are not known by the SBSs. Based on contract theory, the negotiations among the SBSs and the users are modeled under a network economics framework aiming to identify the SBSs' optimal rewards provided to the users in order the latter to determine their optimal transmission power. Indicatively, we note that the work in [6] has attempted to deal with the specific CSI incompleteness problem based on contract theory principles, for heterogeneous long-term evolution advanced (LTE-A) networks. Nonetheless, the proposed approach assumes a central entity determining the optimal user association and contracts, while it is targeting orthogonal frequency division multiple access (OFDMA) environments in order to mitigate or eliminate interference, which in turn reduces the spectral efficiency due to the requirement of channel access orthogonality.

1.2   Contribution and Outline

Targeting both spectral efficiency and application feasibility, it is practically beneficial to assume that the total system available bandwidth is subdivided into orthogonal frequency chunks (i.e., channels). In this regard, adjacent small cells are allocated to different frequency chunks that do not interfere with each other. Within such communication environment, our contribution aims at introducing a resource orchestration framework through a distributed user association and power allocation scheme which reduces computational complexity, while accounting for the CSI incompleteness. Specifically, we propose a contract-based approach, in which the small cell, acting as a monopolist, determines the users' transmit power level so as to be able of decoding their signals, and rewards them inversely proportionally to their interference in order to incentivize them to accept the proposed contract. The key idea is to offer the right contract item to each user, so that all users have the incentive to truthfully reveal their CSI. Furthermore, the determination of the small cell to which each user is associated with, is performed by the users individually based on the evaluation of the contract items, through a reinforcement learning (RL) algorithm. In a nutshell, our work offers a resource orchestration framework applicable in SCNs that differs from previous existing efforts in the literature, in the sense that it: a) considers an interference-limited wireless environment, adopting the use of NOMA technology that has arisen as a promising access technology in 5G networks, b) formulates the overall resource orchestration problem under an incomplete CSI scenario, and c) promotes the use of distributed approaches in the decision making, eliminating the need for centralized decision making entities.

The rest of the paper is organized as follows. Section 2 contains the considered system model and the introduced SBSs' and users' utility functions. In Sect. 3, we present the optimal contract design under incomplete information, while we determine the SBSs' optimal rewards to incentivize the users to transmit with their optimal uplink transmission powers. The users' association to the small cells based on reinforcement learning is highlighted in Sect. 4. In Sect. 5, indicative numerical results validating the operation of the contract theoretic framework are presented, while Sect. 6 concludes the paper.

2   System Model

We consider a heterogeneous dense wireless network consisting of a set of users $U = \{1, \ldots, |U|\}$ and a set of SBSs $C = \{1, \ldots, |C|\}$. A set $\mathbb{U}_c$ of cardinality $|\mathbb{U}_c|$ users represents the users associated with cell c. The channel gain of user u communicating with the SBS c is denoted by $G_u^c = k_u/(d_u^c)^a$, where k_u is a lognormal distributed random variable with mean 0 and variance σ^2, d_u^c [m] is the distance between user u and SBS c and a is the corresponding path loss exponent. To consider realistic scenarios, the SBS lacks specific information of users' transmission characteristics, i.e., CSI, type of user. Adopting the principles of contract theory [7], each SBS aims at incentivizing the users to exhibit their improved transmission characteristics by providing them some

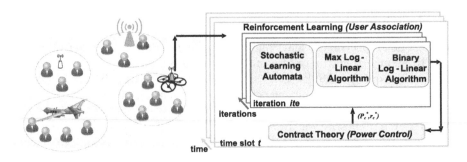

Fig. 1. System model & operation framework.

transmission-related rewards. The goal is to determine an equilibrium point, where both the SBS and the users maximize their utilities, in terms of collecting and transmitting information, respectively. The type of the user is defined as $t_u^c = (G_u^c \cdot \sum\{1/G_u^c\})^{-1/2}, t_u^c \in (0,1]$, and without loss of generality, we consider $|U_c|$ types of users with $t_1^c < \cdots < t_u^c < \cdots < t_{|U_c|}^c$. Adopting the NOMA technique and by applying SIC at the receiver in the uplink, the users with worse channel gain conditions are alleviated by the interference caused by the users with better channel conditions, as the signals of the latter are decoded first at the receiver and excluded from the interference sensed by the users with worse channel conditions. Thus, the SBS c rewards the user u with $r_u^c = \rho/I_u^c$, where $\rho \in \mathbb{R}^+$ is the reward factor and I_u^c is the interference that user u senses, while being associated with SBS c. The physical meaning and interpretation of the reward is that the SBS provides a greater reward to users that experience less interference. Thus, a user of higher type t_u^c (i.e., worse channel conditions), senses less interference and it is rewarded more by the SBS (for fairness purposes), as it is expected to transmit with higher power (i.e., invest greater effort).

Figure 1 summarizes the overall operation of the proposed framework. At the beginning of each time slot, the users select an SBS to be associated with in a distributed manner following a reinforcement learning approach (Sect. 4). Then, the optimal contracts are determined by each SBS for the corresponding users residing within each small cell (Sect. 3). The overall nested procedure is executed over the time guaranteeing the smooth operation of the 5G SCNs.

3   Optimal Contract Design

The considered heterogeneous dense network is characterized by asymmetry of information, as the exact users' types and transmission powers are unknown to each SBS. Instead each SBS c only knows the probability Pr_u^c that the user u is of type t_u^c and $\sum_{u=1}^{|U_c|} Pr_u^c = 1$. The utility that SBS c experiences from user u is defined as $U_c^u = P_u^c - \mathcal{C} \cdot r_u^c$, expressing the SBS's satisfaction in terms of its operation by receiving a signal with high power strength P_u^c, while being charged with the cost $\mathcal{C} \cdot r_u^c$ of providing incentives through the rewards to the user. Thus, the overall SBS's utility is $U_c = \sum_{u=1}^{|\mathbb{U}_c|} [Pr_u^c \cdot (P_u^c - \mathcal{C} \cdot r_u^c)]$. On the other hand, the user's utility is defined as $U_u^c(P_u^c) = t_u^c \cdot e(r_u^c) - \mathcal{C}' \cdot P_u^c$, which expresses the user's satisfaction $t_u^c \cdot e(r_u^c)$ from receiving the reward r_u^c from the SBS c, while considering its personal transmission cost $\mathcal{C}' \cdot P_u^c$. It is noted that the user's satisfaction depends on its type t_u^c and on the evaluation function $e(r_u^c)$ (with $e(0) = 0, e'(r_u^c) > 0, e''(r_u^c) < 0$), which captures the user's perception and personal satisfaction from the reward.

Following the principles of contract theory, the SBS establishes a personalized contract (r_u^c, P_u^c) with each user in the small cell, where the user invests its personal effort P_u^c (i.e., uplink transmission power) and the SBS rewards the user with r_u^c. A contract is feasible if the following conditions hold true: (i) the *Individual Rationality (IR)*, i.e., the contract should guarantee that the user's utility is non-negative $U_u^c(P_u^c) \geq 0, \forall u \in \mathbb{U}_c$; and (ii) the *Incentive Compatibility (IC)*, i.e., each user must select that contract which is designed for its type $t_u^c \cdot e(r_u^c) - \mathcal{C}' \cdot P_u^c \geq t_u^c \cdot e(r_{u'}^c) - \mathcal{C}' \cdot P_{u'}^c, \forall u, u' \in \mathbb{U}_c, u \neq u'$. Additionally, the following three conditions must hold true in order the contract to be feasible.

Proposition 1. *For any feasible contract (r_u^c, P_u^c), the following must hold true:* $r_u^c > r_{u'}^c \Longleftrightarrow t_u^c > t_{u'}^c$ *and* $r_u^c = r_{u'}^c \Longleftrightarrow t_u^c = t_{u'}^c$.

Proposition 1 can be proven by arguing as follows. We can prove separately the sufficiency and the necessity of the described condition. Regarding the sufficiency, we can write the incentive compatibility condition for two types of users, e.g., $t_u^c > t_{u'}^c$, add the inequalities, and exploit the strictly increasing property of the evaluation function $e(r_u^c)$ to conclude that $r_u^c > r_{u'}^c$. Regarding the necessity, we can work backwards by considering the strictly increasing property of $e(r_u^c)$, build the summation of the incentive compatibility constraints for two users, and conclude that $t_u^c > t_{u'}^c$. Similarly, we can show $r_u^c = r_{u'}^c \Longleftrightarrow t_u^c = t_{u'}^c$.

Proposition 2. *A user of higher type, i.e., $t_1^c < \cdots < t_u^c < \cdots < t_{|\mathbb{U}_c|}^c$ (i.e., worse channel conditions), will receive a greater reward from the SBS c to be incentivized to be served by this SBS, i.e., $r_1^c < \cdots < r_u^c < \cdots < r_{|\mathbb{U}_c|}^c$, and it will transmit with greater power (i.e., effort), i.e., $0 < P_1^c < \cdots < P_u^c < \cdots < P_{|\mathbb{U}_c|}^c$.*

The proof of this proposition intuitively stems from Proposition 1, given that $t_1^c < \cdots < t_u^c < \cdots < t_{|\mathbb{U}_c|}^c$.

Proposition 3. *A user of higher type, i.e., $t_1^c < \cdots < t_u^c < \cdots < t_{|\mathbb{U}_c|}^c$, receives higher utility to be incentivized by the SBS c, i.e., $U_1^c < \cdots < U_u^c < \cdots < U_{|\mathbb{U}_c|}^c$.*

Proposition 3 proof can be concluded based on the following steps. We consider the incentive compatibility constraint for one user, and we analyze the inequality by considering a user of lower type, i.e., $t_u^c > t_{u'}^c$, and sequentially we conclude that $U_u^c > U_{u'}^c$. Due to space limitations only the key arguments required for the proofs of Propositions 1–3 are provided here, while the detailed intermediate steps of the proofs are presented in [8,9].

Each SBS aims at maximizing its overall utility in order to be able to collect and properly decode the users' transmitted signals (as dictated by the received SINR). In parallel, each user should satisfy all of its personal constraints, as they have been described above, in order to be willing to be associated with the specific cell. Thus, the following distributed optimization problem is solved at each SBS, by jointly considering the SBS and corresponding users' sides.

$$\textbf{P1:} \quad \max_{(r_u^c, P_u^c)_{\forall u \in \mathbb{U}_c}} U_c = \sum_{u=1}^{|\mathbb{U}_c|} [Pr_u^c \cdot (P_u^c - \mathcal{C} \cdot r_u^c)] \tag{1a}$$

$$\textbf{s.t.}\ t_u^c \cdot e(r_u^c) - \mathcal{C}' \cdot P_u^c \geq 0, \forall u \in \mathbb{U}_c \tag{1b}$$

$$t_u^c \cdot e(r_u^c) - \mathcal{C}' \cdot P_u^c \geq t_u^c \cdot e(r_{u'}^c) - \mathcal{C}' \cdot P_{u'}^c, \forall u, u' \in \mathbb{U}_c, u \neq u' \tag{1c}$$

$$0 \leq r_1^c < \cdots < r_u^c < \cdots < r_{|\mathbb{U}_c|}^c \tag{1d}$$

The optimization problem **P1** is non-convex, thus, in order to solve it, we reduce its constraints. By performing appropriate derivations we can easily show that the constraints (1b) and (1c) can be reduced to (2b) and (2c) [8,9]. In particular, to show that the constraint (1b) can be reduced to (2b), we consider the incentive compatibility constraint for a user and by considering the strictly increasing property of the evaluation function, as well as the monotonicity of the users' types (proposition 2), we can sequentially rewrite the incentive compatibility constraint to be reduced to the inequality $t_u^c \cdot e(r_u^c) - \mathcal{C}' \cdot P_u^c \geq t_1^c \cdot e(r_1^c) - \mathcal{C}' \cdot P_1^c$. Then, by exploiting the individual rationality condition, we can conclude to the constraint (2b). Moreover, in order to show that the constraint (1c) is reduced to (2c), we consider the downward (i.e., $u, u', u' \in \{1, \ldots, u-1\}$), the upward (i.e., $u, u', u' \in \{u+1, \ldots, |\mathbb{U}_c|\}$), and the local downward (i.e., $u, u-1, \forall u, u-1 \in \mathbb{U}_c$) incentive compatibility constraints. Then, we can show that all the downward incentive compatibility constraints can be represented by the local downward incentive compatibility constraint, while all the upward incentive compatibility constraints can be equivalently captured by the local downward incentive compatibility constraints. Thus, the optimization problem **P1** can be rewritten as:

$$\textbf{P2:} \quad \max_{(r_u^c, P_u^c)_{\forall u \in \mathbb{U}_c}} U_c = \sum_{u=1}^{|\mathbb{U}_c|} [Pr_u^c \cdot (P_u^c - \mathcal{C} \cdot r_u^c)] \tag{2a}$$

$$\textbf{s.t.}\ t_1^c \cdot e(r_1^c) - \mathcal{C}' \cdot P_1^c = 0 \tag{2b}$$

$$t_u^c \cdot e(r_u^c) - \mathcal{C}' \cdot P_u^c = t_u^c \cdot e(r_{u'}^c) - \mathcal{C}' \cdot P_{u'}^c, \forall u, u' \in \mathbb{U}_c, u \neq u' \tag{2c}$$

$$0 \leq r_1^c < \cdots < r_u^c < \cdots < r_{|\mathbb{U}_c|}^c \tag{2d}$$

We can easily prove that **P2** is a convex programming problem by checking the Hessian matrix. Thus, **P2** can be solved by applying the KKT conditions, and accordingly the optimal users' transmission power vector $\mathbf{P_c^*} = [P_1^{c*}, \ldots, P_{|\mathbb{U}_c|}^{c*}]$ and the optimal SBS's rewards vector $\mathbf{r_c^*} = [r_1^{c*}, \ldots, r_{|\mathbb{U}_c|}^{c*}]$, can be determined.

4 Users' Association Based on Reinforcement Learning

In order to support the operation of the aforementioned framework, users are enabled to autonomously select the specific cell to be associated with, based on a reinforcement learning (RL)-based approach. In this work, we adopted the reinforcement learning mechanism of the stochastic learning automata (SLA) to realize this process in a distributed manner, though other alternatives may be considered as well (e.g., Max-logit, Binary-logit) [10]. In particular, each user acts as a learning agent, where at each iteration of the SLA algorithm makes a probabilistic-based selection of an SBS to be associated with, following a simple probabilistic rule. To realize this, each cell is characterized by a reputation function REP_c depending on socio-physical characteristics, e.g., average users' distance from an SBS, average users' transmission power invested to communicate with an SBS, average users' received rewards, etc., similarly to [11]. Accordingly, the cell's reputation functions are incorporated within the RL algorithms to enable the users to probabilistically learn their best cell association.

5 Numerical Results

In this section, we present some numerical results that validate the operation and performance of the contract-theoretic component of our proposed framework, obtained through modeling and simulation. Given that adjacent cells do not interfere with each other, we focus on the operation of one indicative cell, assuming that users have already performed their association with each SBS. In the following, we consider a small cell of 500-meter radius with an SBS placed in the center of the cell and $|U_c| = 10$ users placed in increased distance from the SBS (with a step of 50-meters). Specifically, we define one type for each user as analyzed in Sect. 2, presuming that all users have the same probability Pr_u^c of belonging to each type. As mentioned earlier, the channel gain of user u communicating with the SBS c is modeled as $G_u^c = k_u/(d_u^c)^a$, where k_u is assumed to be a lognormal distributed random variable with mean 0 and variance $\sigma^2 = 8dB$, and the corresponding path loss exponent is $a = 4$. Furthermore, the maximum uplink transmission power of the users is $P_{u,max}^c = 0.7$ [W], while the reward factor is $\rho = 10^{-15}$, the SBS's unit cost is $C = 0.7$, the user's unit power cost is $C' = 0.4$, and the user's evaluation function is assumed $e(r_u) = \sqrt{r_u}$.

Figure 2a presents the users' effort, i.e. optimal uplink transmission power, and reward vs. their index, while similarly Fig. 2b shows the achievable utilities for both the users and the cell. Logarithmic scale is used to better visualize the curves' trend and differences in the values. The results in both Fig. 2a and 2b, validate the monotonicity behavior in the offered contracts. That is the higher the user type, the more effort is required and thus, the more the reward it receives, leading to larger utility for the user itself and the cell. This is well aligned with the fact that the measured interference at the receiver after performing the SIC technique for the user with the lowest channel gain (i.e., the highest type user) is impacted only by the background noise, which is very low. Moreover, from Fig. 2b it can be seen that all types of users receive a non-negative utility, being consistent with the individual rationality constraint imposed. To complement the evaluation of the contract feasibility, the utilities of two selected users (i.e. users with type 5 and 8) are examined in Fig. 2c, for different possible contracts offered by the SBS (represented in the horizontal axis by user types). Observing these results we note that following the proposed approach, each user achieves equal or higher utility from the other types, if and only if selects the contract item intended for its own type (dictated by the red dashed vertical line in the graphs), demonstrating the satisfaction of the incentive compatibility constraint.

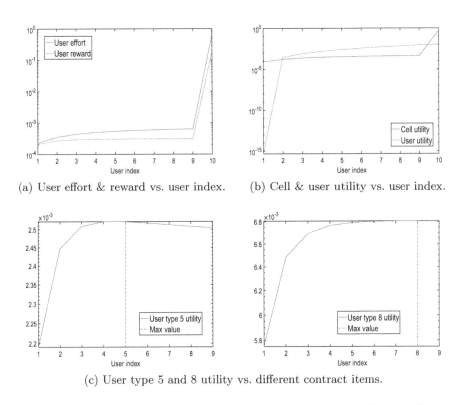

(a) User effort & reward vs. user index.   (b) Cell & user utility vs. user index.

(c) User type 5 and 8 utility vs. different contract items.

Fig. 2. Operation validation of the proposed contract-theoretic framework.

6   Conclusions and Future Work

In this paper, the problem of resource orchestration, in terms of user to small cell association and power allocation, in the uplink of 5G NOMA-based small cell networks is studied. A reinforcement learning technique is adopted to enable the users to select the optimal small cell to be connected with, in an autonomous and distributed manner. Thereafter, a contract-theoretic approach is introduced to design user specific contracts in terms of determining the users' optimal uplink transmission power and the small cell's optimal rewards provided to the users, to incentivize them to perform in an interference limited manner in the network.

Our current and future work contains the extension of this framework by considering multiple reinforcement learning approaches for the user association and comparing them in terms of efficiency and computation complexity. Finally, a natural extension of this work focuses on the use of the proposed contract-theoretic framework for the end-to-end study of the network operation, by including the backhauling communication of the SBSs to the macro base station.

Acknowledgments. The research work was supported by the Hellenic Foundation for Research and Innovation (H.F.R.I.) under the "First Call for H.F.R.I. Research Projects to support Faculty members and Researchers and the procurement of high-cost research equipment grant" (Project Number: HFRI-FM17-2436). The research of Dr. Tsiropoulou was conducted as part of the NSF CRII-1849739.

References

1. Muirhead, D., Imran, M.A., Arshad, K.: A survey of the challenges, opportunities and use of multiple antennas in current and future 5g small cell base stations. IEEE Access **4**, 2952–2964 (2016)
2. Tsiropoulou, E.E., Vamvakas, P., Papavassiliou, S.: Supermodular game-based distributed joint uplink power and rate allocation in two-tier femtocell networks. IEEE Trans. Mob. Comput. **16**(9), 2656–2667 (2016)
3. Dai, L., Wang, B., Yuan, Y., Han, S., Chih-Lin, I., Wang, Z.: Non-orthogonal multiple access for 5g: solutions, challenges, opportunities, and future research trends. IEEE Commun. Mag. **53**(9), 74–81 (2015)
4. Qian, L.P., Wu, Y., Zhou, H., Shen, X.: Joint uplink base station association and power control for small-cell networks with non-orthogonal multiple access. IEEE Trans. Wirel. Commun. **16**(9), 5567–5582 (2017)
5. Ali, M.S., Tabassum, H., Hossain, E.: Dynamic user clustering and power allocation for uplink and downlink non-orthogonal multiple access (noma) systems. IEEE Access **4**, 6325–6343 (2016)
6. Asheralieva, A., Miyanaga, Y.: Optimal contract design for joint user association and intercell interference mitigation in heterogeneous lte-a networks with asymmetric information. IEEE Trans. Veh. Technol. **66**(6), 5284–5300 (2017)
7. Bolton, P., Dewatripont, M.: Contract Theory, MIT, Cambridge (2005)
8. Fragkos, G., Patrizi, N., Tsiropoulou, E.E., Papavassiliou, S.: Socio-aware public safety framework design: a contract theory based approach, pp. 69–79. IEEE ICC, 2020. (preprint available at shorturl.at/CPRU9) (2020)
9. Zhang, Y., Song, L., Saad, W., Dawy, Z., Han, Z.: Contract-based incentive mechanisms for device-to-device communications in cellular networks. IEEE J. Sel. Areas Commun. **33**(10), 2144–2155 (2015)
10. Xu, Y., Wang, J., Wu, Q.: Distributed learning of equilibria with incomplete, dynamic, and uncertain information in wireless communication networks. In: IGI Global, Game Theory Framework Applied to Wireless Communication Networks, pp. 63–86 (2016)
11. Sikeridis, D., Tsiropoulou, E.E., Devetsikiotis, M., Papavassiliou, S.: Wireless powered public safety iot: a uav-assisted adaptive-learning approach towards energy efficiency. J. Netw. Comput. Appl. **123**, 69–79 (2018)

Unlicensed Spectrum for Ultra-Reliable Low-Latency Communication in Multi-tenant Environment

Ayat Zaki-Hindi[1,3(✉)], Salah-Eddine Elayoubi[2], and Tijani Chahed[3]

[1] Orange Labs, 44 Avenue de la Republique, 92320 Chatillon, France
ayat.zakihindi@orange.com
[2] CentraleSupelec, 3 Rue Joliot Curie, 91190 Gif-sur-Yvette, France
salaheddine.elayoubi@centralesupelec.fr
[3] Institut Polytechnique de Paris, Telecom SudParis, 19 Place Marguerite Perey, Palaiseau, France
tijani.chahed@telecom-sudparis.eu

Abstract. We study in this paper the transport of Ultra-Reliable Low-Latency Communications (URLLC) in the presence of multiple tenants in a system composed of unlicensed and licensed spectrum, the former being shared by the tenants. Due to the stringent reliability and latency constraints of URLLC, the more expensive 5G licensed spectrum is used to serve the traffic that is not served by the unlicensed spectrum within a certain time budget. In a competitive scenario, where each tenant tries egoistically to minimize its share of the licensed spectrum, this may result in the tragedy of the commons like situation. We formulate the problem using a game-theoretic approach to model the non-cooperative multi-tenant scenario. We model the medium access of the combined unlicensed and licensed system to quantify the performance of the system, in terms of the overall probability of failure, and validate the model against simulations. We then derive the strategies that minimize individual cost functions. Our work gives insights about the existence of Nash equilibria and identify them numerically. Finally, we quantify the so-called price of anarchy, i.e., ratio of the utility yielded by the competitive setting to the outcome of a cooperative scenario.

Keywords: URLLC · Unlicensed spectrum · Multi-tenant

1 Introduction

The use of unlicensed spectrum for mobile communications is possible since 3GPP release 13, when Licensed-Assisted Access (LAA) LTE was first proposed in the downlink only, to be followed by the enhanced LAA (eLAA) for uplink and downlink in release 14 and the de-facto Multefire standard [1]. Afterwards, 3GPP worked on the definition of 5G New Radio (NR) which includes several unlicensed bands, illustrating the importance of unlicensed spectrum for 5G [2]. However,

© Springer Nature Switzerland AG 2021
S. Lasaulce et al. (Eds.): NetGCooP 2021, CCIS 1354, pp. 110–124, 2021.
https://doi.org/10.1007/978-3-030-87473-5_11

a main drawback for unlicensed spectrum is that operators cannot guarantee the Quality of Service (QoS) for their users when multiple tenants are operating in the same area due to interference among them. A tenant may be a mobile network operator or a vertical operating a network for its users. This limitation is especially true for Ultra-Reliable Low Latency Communications (URLLC) which transport critical information with stringent latency and reliability requirements, on the order of 1 to 5 ms end-to-end and 99.999% respectively [3]. Nonetheless, unlicensed spectrum is being discussed for some URLLC services [4], notably for some smart factory use cases that are to be deployed in industrial areas where the environment can be controlled, e.g., by reducing the sources of outside interference. This may be true in environments managed by a single operator, but not in multi-tenant environments where several verticals manage plants in direct proximity. These co-existing networks operating in unlicensed spectrum create interference which degrades the QoS and compromises the value-add of unlicensed spectrum for verticals.

In this paper, we study the usage of unlicensed spectrum for URLLC services and advocate the combined usage of it along with licensed spectrum to ensure the stringent latency and reliability targets. In particular, a generated packet attempts transmission in unlicensed spectrum during a time budget shorter than the delay constraint; if it does not succeed then it is redirected to the licensed 5G spectrum. This scheme drastically decreases the need for licensed spectrum resources compared to a classical licensed-only system, as illustrated in [5]. However, in a multi-tenant environment, each of the verticals wants to maximize its economic gain from using unlicensed spectrum, which increases the overall system interference and decreases the value-add of unlicensed spectrum for all tenants, leading to a tragedy of the commons like situation. We model this situation as a game between tenants, where each tenant strategy consists in using the unlicensed resources more or less aggressively and the objective is to minimize its demand for the expensive licensed ones.

The literature on URLLC is becoming rich. There has been early works which study the transport of URLLC over LTE, for instance [6], but the majority of papers deal with URLLC on 5G's licensed NR, considering grant-free fast uplink access, where neither issuing a scheduling request nor waiting for a scheduling grant are required [7]. This approach is often associated with the blind replication of packets, where the packet is sent several times within the delay budget without waiting for negative acknowledgement (NACK) to increase reliability [8,9]. However, only few papers considered the use of unlicensed spectrum, such as [4], mainly because of the existence of other technologies such as WiFi on the same unlicensed bands, which decreases the reliability of the system. This paper is based on our work in [5], where we proposed a new model for unlicensed medium access and quantified the cost of deploying the joint unlicensed/licensed transmission scheme in terms of licensed bandwidth. In this paper, we focus on another aspect of the problem which is the existence of multiple tenants and the impact of each one's strategy on their and others' performance.

In the remainder of this paper, Sect. 2 describes the system model and formulates the corresponding non-cooperative game. We analyze in Sect. 3 the performance of the system in terms of loss rate and requested licensed spectrum resources for a given strategy from the viewpoint of one tenant. Based on this analysis, Sect. 4 derives the Nash equilibrium of the system and Sect. 5 is dedicated for numerical results and illustrates the existence of pure-strategy Nash equilibria and the associated price of anarchy of the system. Section 6 eventually concludes the paper.

2   System Description and Problem Formulation

We consider an industrial area containing several smart factories operated by different tenants. Each tenant deploys an unlicensed Access Point in its premises connected to a central controller, and uses a Base Station of a mobile network operator covering the factory as a relay for the packets back to the controller. In the sequel, we denote by 5G-U and 5G-L the parts of the system which use unlicensed and licensed resources, respectively. We denote the transmitting machine by station and assume that all stations are equipped with both 5G transmission systems. We focus on uplink URLLC traffic generated by machines where URLLC packets share the same latency and reliability requirements, denoted by T and R, respectively. We also assume the absence of foreign communications in 5G-U band in this area, such as personal Wi-Fi networks (only networks deployed by tenants exist) and if other types of services such as eMBB or mMTC exist, then they would use different slices from the ones dedicated for URLLC in 5G-L.

The proposed transmission mechanism is that when a packet is generated, it is transmitted through 5G-U during a time budget $T_U < T$. If it is successfully transmitted then the process stops, if not, the packet is switched to 5G-L and is transmitted within the remaining time budget $T_L = T - T_U$. With this method, we decrease the load on 5G-L and hence the amount of licensed bandwidth (BW) that the tenant has to buy from the operator.

Assuming now the existence of M tenants operating in proximity, from the viewpoint of one tenant there is a non-negligible interference in unlicensed spectrum, and if every tenant tries to use it selfishly (without considering neighbouring interfering stations) then the overall interference could increase and the gain is reduced. We denote a given tenant by v_1 (for vertical) to which we evaluate the system performance under interference from other tenants, denoted by $v_i : i \in \{2, ..., M\}$. v_1 deploys a URLLC transmission strategy with time division T_U^1, likewise other tenants deploy other strategies with equal or different time divisions; $T_U^i : i \in \{2, ..., M\}$.

This situation can be represented by a non-cooperative game with triplet $G = (V, \{S_i\}_{i\in V}, \{u_i\}_{i\in V})$ where $V = \{v_1, v_2, ..., v_M\}$ is the finite set of players, S_i is the set of strategies of v_i represented by the set of possible values of $T_U^i \in]0, T[$, and u_i is the utility function of v_i which is the inverse of its cost function represented by the required bandwidth BW_i on the licensed access to satisfy the reliability and latency requirements, which we determine next.

3   Performance Analysis of Medium Access

3.1   Performance Analysis of Unlicensed Medium Access

In 5G-U, channel occupancy is managed by sensing the medium before transmission according to a random backoff procedure, called Listen-Before-Talk (LBT).

We consider $N = N_1 + N_2 + ... + N_M$ stations inside a coverage area transmitting in the uplink, where N_i denotes the number of v_i stations in this area, and $i \in \{1, ..., M\}$. Without loss of generality, we focus on a transmitted packet belonging to v_1. v_1's coverage includes all its N_1 stations and a partial number of other tenants' stations. In non-cooperative games, players do not communicate directly, so interference can be assessed from sensing the medium. When a packet from v_1 is generated, it is associated to a backoff counter chosen randomly from the integer set $\{0, ..., W_0 - 1\}$ where W_0 is the maximum contention window size and identical for all tenants. Then the station senses the medium during one time slot, which is the smallest period required to sense the medium. If the medium is sensed idle then the backoff counter is decremented by one, else it is halted. This process is repeated until the counter reaches zero and the packet is sent without sensing, a positive or negative acknowledgment ACK/NACK is expected within a given time. Its absence is considered as a NACK. If the transmission is successful then the process ends here, else it is repeated for a number of attempts m_1, called stages.

We denote the number of time slots needed for transmitting a URLLC packet by x and it is identical for all tenants, it comprises the time of packet transmission until the reception of ACK/NACK (or its absence). Since no collision avoidance mechanism is considered in our case to limit the delay, the duration of a collision is equal to a successful transmission. Assuming a perfect channel, stations sense x consecutive busy slots every time the medium is sensed busy. In our case, we let the packet attempts as long as its delay respects the time constraint T_U^1, hence we can deduce the maximum number of stages when $b(t) = 0$ in all stages:

$$m_1 = \left\lfloor \frac{T_U^1}{x + 1} \right\rfloor$$

where $\lfloor . \rfloor$ is the floor function. We consider $x + 1$ time slots for every collision because according to LBT, all stations in backoff need to sense at least one idle time slot before decrementing their $b(t)$ and start transmission (if $b(t) = 0$).

We note that we deploy LBT cat3 with fixed contention window size in every stage instead of LBT cat4 deployed in most Wi-Fi-like systems which adapts the contention window according to collisions. This makes LBT cat3 more suitable to delay-constrained applications.

Timer-Based Modified Bianchi Model. Many mathematical models were proposed to model LBT, the most famous one is based on discrete time Markov chains, proposed by Bianchi [10]. We modify the latter to suit our context where we add: i. a timer to track time evolution until T_U^1 and ii. two novel states:

Success and Failure; the latter refers to unsuccessful transmission within time budget T_U^1. This Markov chain is transient and describes the lifecycle of a packet from the moment it enters the system until it is either successfully transmitted or handed to the licensed band. This model is introduced in Fig. 1 for the case of $T_U^1 = 3x$ and $W_0 = 4$.

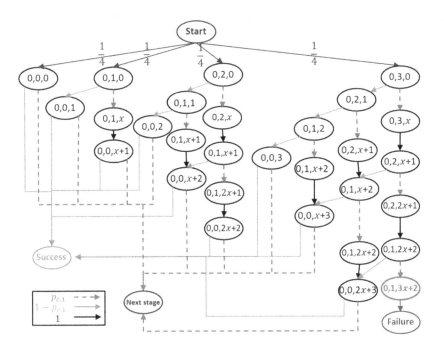

Fig. 1. Example of Timer-based modified Bianchi model

Each state of the 3-dimensional Markov chain is composed of three stochastic processes $\{s(t), b(t), d(t)\}$, representing the stage, the backoff counter and the delay at time t, respectively. We denote by $p_{c,1}$ the probability of collision seen by the transmitting station in one slot, in other words $p_{c,1}$ is the probability that at least one of the other $N-1$ stations is transmitting during the current time slot. $p_{c,1}$ is considered independent of the state in every time slot.

Figure 1 illustrates the fact that every busy period is followed by at least one idle slot. The chain is built dynamically depending on the values of T_U^1 and x, i.e., in every state we test if the constraint T_U^1 is still respected and generate next states by adding 1 or x to the current delay, hence the number of stages is determined by the last possible state with $d(t) = T_U^1$.

The chain begins from stage 0 and a random backoff counter, generating the first row of states $\{0, b(0), 0\} : b(0) \in \{0, ..., W_0 - 1\}$, then every state has two possibilities to proceed depending on the sensed medium: idle or busy, except for states with $b(t) = 0$ where the packet is immediately transmitted without medium sensing. If the medium is sensed idle, then $b(t)$ is decremented by one and $d(t)$ is incremented by one, otherwise, the medium is busy for a duration of x consecutive time slots and $b(t)$ remains unchanged. The chain is terminated when all its paths reach one of the absorbing states: Success or Failure. The state Success is reached when $b(t) = 0$ and the medium is idle having $d(t) < T_U^1$, while Failure is reached when $d(t) \geq T_U^1$ for any value of $s(t)$ and $b(t)$. We assume in this example that $x > W_0 - 1$ to get the illustrated chain, otherwise we would obtain another set of states. Due to difficulty of displaying a large number of states, we illustrate in Fig. 1 the states of first stage only and gather the rest of the chain in one state: Next stage.

Approximate Timer-Based Modified Bianchi Model. The existence of two possible increments of $d(t)$ complicates the problem at hand because this generates a huge number of states for practical values of W_0, T_U^1 and x, making the solution prohibitive. If we neglect the change in $d(t)$ after sensing an idle slot, then the states of different branches can be combined and the chain becomes more compact. This approximation affects the precision in calculating the probabilities of the chain, but the complexity reduction tips the balance in its favor. The approximate model for the previous example is shown in Fig. 2.

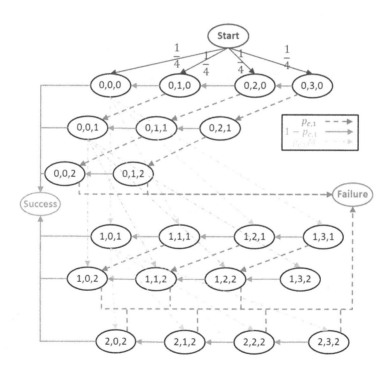

Fig. 2. Example of approximate timer-based modified Bianchi model

In Fig. 2, we simplify the notation of $d(t)$ to express now the multiples of $x+1$ slots when the medium is sensed busy followed by an idle slot. This chain is no longer built dynamically state by state, where it can be generated by knowing m_1 and W_0 only.

The sporadic nature of URLLC implies that the stations are not saturated, we denote by q the probability of having a URLLC packet to transmit (identical for all tenants). The interval between two consecutive packet arrivals for the same station is larger than T_U^1 then q is small enough to consider that the packets are not enqueued. In this case, we can assemble states Start, Success and Failure in one state: Inactive, represented in Fig. 3.

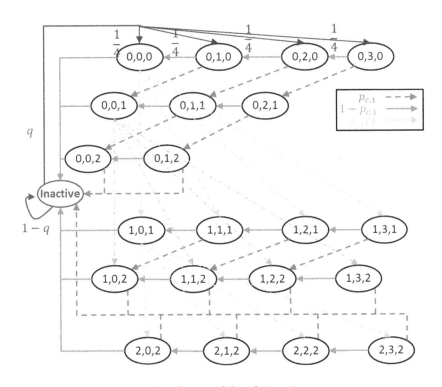

Fig. 3. Complete model with inactive state

We note that when $m_1 > W_0$, T_U^1 is not always attained in first stages because we have W_0 busy periods at most. With the help of Fig. 3, we can derive the balance equations of the Markov chain for general values of m_1 and W_0. We arrange the balance equations in a three-dimensional matrix Π with dimensions of $m_1 \times W_0 \times m_1$, where nonexistent states are replaced with zeros. We denote the probability of state Inactive by Π_{in}. We start by filling the elements of the matrix in a recursive manner row by row starting from states with smaller $d(t)$ and higher $b(t)$ in every stage, as follows:

$$\Pi_{0,W_0-1,0} = q\frac{\Pi_{in}}{W_0}$$

$$\Pi_{0,W_0-j,0} = q\frac{\Pi_{in}}{W_0} + (1-p_{c,1})\Pi_{0,W_0-j+1,0}, \quad 2 \le j \le W_0$$

$$\Pi_{0,W_0-j,k} = p_{c,1}\Pi_{0,W_0-j+1,k} + (1-p_{c,1})\Pi_{0,W_0-j+1,k-1}, \quad \begin{array}{l} 1 \le k \le m_1 - 1 \\ k+1 \le j \le W_0 \end{array}$$

We note that the delay in one stage cannot be less than its number of stage, the first rows in the next stages remain zeros.

$$\Pi_{l,W_0-1,k} = \frac{p_{c,1}}{W_0}\Pi_{l-1,0,k-1}, \quad \begin{array}{l} 1 \le l \le m_1 - 1 \\ l \le k \le m_1 - 1 \end{array}$$

$$\Pi_{l,W_0-j,k} = \frac{p_{c,1}}{W_0}\Pi_{l-1,0,k-1} + (1-p_{c,1})\Pi_{l,W_0-j+1,k}$$
$$+ p_{c,1}\Pi_{l,W_0-j+1,k-1}, \quad \begin{array}{l} 1 \le l \le m_1 - 1 \\ l \le k \le m_1 - 1 \\ 2 \le j \le W_0 \end{array}$$

Π_{in} is obtained by applying the normalization condition: the sum of all state probabilities equal to 1. Based on the definition of $p_{c,1}$ we can write the following equation:

$$p_{c,1} = 1 - (1-\tau_1)^{N_1-1}(1-\tau_2)^{N_2}...(1-\tau_M)^{N_M}$$

where τ_i is the probability of transmission in one time slot for a station of tenant v_i which is equal the sum of all states with $b(t) = 0$ of the corresponding chain of tenant v_i and depends on its policy T_U^i, given by:

$$\tau_i = \sum_{l=0}^{m_i-1}\sum_{k=0}^{m_i-1}\Pi_{l,0,k} \tag{1}$$

For tenant v_i, we get:

$$p_{c,i} = 1 - (1-\tau_i)^{N_i-1}\prod_{j\ne i}(1-\tau_j)^{N_j} \tag{2}$$

The solution requires solving numerically the set of fixed-point equations of every tenant: (1) and (2), until all $p_{c,i}$ values converge to their solution.

Finally, back to Fig. 2, we quantify the reliability of the system under delay constraint T_U^1 by evaluating the hitting probability of state Failure starting from state Start, which can be calculated directly from the balance equations by replacing $\Pi_{in} = 1$ after obtaining the value of $p_{c,1}$, where the reliability is $\Pi_{Success} = 1 - \Pi_{Failure}$:

$$\Pi_{Failure} = p_{c,1}\Pi_{m_1-1,0}$$

We denote $\Pi_{Failure}$ by $P_1^U(T_U^1, ..., T_U^M)$ the probability of failure of tenant v_1 in transmitting the packet through 5G-U within a delay budget T_U^1, as a function of transmission policies of all tenants T_U^i.

$$P_1^U(T_U^1, ..., T_U^M) = p_{c,1}(T_U^1, ..., T_U^M) \sum_{l=0}^{m_1(T_U^1)-1} \sum_{j=0}^{W_0-1} \Pi_{l,j,m_1(T_U^1)-1}$$

3.2   Performance Analysis of Licensed Medium Access

Existing methods to access the licensed medium in uplink are Grant-based (GB) scheduling and Grant-free (GF) on a common pool. GB scheduling is time consuming and does not meet URLLC constraint on latency. This leaves us with the less reliable solution, GF, where resources are accessible without prescheduling. The GF system we consider in our study deploys a simple replica-transmission mechanism, as for the standardized Transmission Time Interval (TTI) bundling for URLLC [11]. The stations are required to be synchronized at the beginning of TTIs. A reserved frequency bandwidth for tenant v_i: BW_i is divided into K_i blocks, each of width w, where we call the combination of TTI-w a Resource Block (RB) which is the required resource for one URLLC packet transmission. When the station has a packet to transmit on 5G-L, as advocated by [8], it chooses uniformly at random from the K_i available RBs in each TTI to transmit the packet. Depending on the available delay budget on 5G-L T_L^i and the length of TTI, the station can transmit δ_i replicas of the packet over δ_i consecutive TTIs, where δ_i corresponds to T_L^i/TTI, without waiting for any ACK/NACK, limiting by that the delay and increasing reliability. A collision occurs if two or more packets are transmitted on the same RB, and the packet is lost if and only if all its replicas are in collision.

Our aim here to evaluate the reliability of 5G-L for tenant v_1, as we did for 5G-U; the maximum number of stations here is N_1, contrary to 5G-U where it is N, since each tenant has its own licensed access. The random medium access mechanism is as follows: the packet under study chooses its RB in one TTI with probability $1/K_1$, if another packet in the system chooses to transmit in a different RB from the latter, it is chosen with probability $1 - 1/K_1$. Considering the number of other packets in the system is $n \in \{1, 2, ..., N_1 - 1\}$ (excluding the case of $n = 0$ because it is collision free), the probability of collision with at least one other packet is shown in Eq. (3), which represents the probability of the complementary event to choosing different RBs from the one under study, where A_n is the event of having n active packets simultaneously.

$$\mathbb{P}(C|A_n) = 1 - (1 - \frac{1}{K_1})^n \tag{3}$$

The probability of having n packets from the $N_1 - 1$ other stations is given by Equation (4), where P_a is the probability that a packet arrives to 5G-L, calculated in next sub-section depending on 5G-U's performance.

$$\mathbb{P}(A_n) = \binom{N_1 - 1}{n} P_a^n (1 - P_a)^{N_1 - 1 - n} \tag{4}$$

The packet is lost when all its δ_1 replicas collide with other transmissions: $\mathbb{P}(F|A_n) = \mathbb{P}(C|A_n)^{\delta_1}$. The probability of failure in 5G-L for v_1 is thus as follows:

$$P_1^L = \sum_{n=1}^{N_1-1} \mathbb{P}(F|A_n)\mathbb{P}(A_n)$$

$$P_1^L(\delta_1, ..., \delta_M) = \sum_{n=1}^{N_1-1} \binom{N_1-1}{n} P_a(\delta_1, ..., \delta_M)^n (1 - P_a(\delta_1, ..., \delta_M))^{N_1-n-l}(1$$

$$- (1 - \frac{1}{K_1})^n)^{\delta_1(T_U^1)}$$

3.3 Combined Unlicensed and Licensed Transmission

When combining 5G-U and 5G-L systems, we have to be careful about their time units. 5G-U operates in T_s unit which is considerably smaller than TTI used in 5G-L. To unify the units of the combined system, we assume that $TTI = zT_s$ where z is an integer. To adapt the aforementioned arrival probability q per T_s to TTI, and considering that a station generates one packet at most during δ_1 TTIs, the packet arrival probability to 5G-L is given by $P_a = 1 - (1 - q)^{\delta_1 z}$. For the case of combined 5G-U then 5G-L transmission, packets arrive at 5G-L after failing the transmission on 5G-U with probability $P_1^U(T_U^1, ..., T_U^M)$, which can be written equivalently as $P_1^U(\delta_1, ..., \delta_M)$ since $T_U^i = T - \delta_i z$, resulting:

$$P_a(\delta_1, ..., \delta_M) = 1 - (1 - qP_1^U(\delta_1, ..., \delta_M))^{\delta_1 z} \tag{5}$$

Since TTI bundling requires synchronization among stations, and since packets arrive randomly in time, if a packet reaches 5G-L amid the TTI, then it is postponed till the beginning of the next one, which may cause a maximum delay of the packet of $(z-1)T_s$ but its impact is small and is neglected in our analysis for simplification.

We derive now the formula of total probability of loss for the combined 5G-U and 5G-L system for tenant v_1, denoted by $P_1(\delta_1, ..., \delta_M)$, which quantifies the reliability R of the system under delay constraint T:

$$P_1(\delta_1, ..., \delta_M) = \sum_{n=1}^{N_1-1} \binom{N_1-1}{n} P_a(\delta_1, ..., \delta_M)^n (1 - P_a(\delta_1, ..., \delta_M))^{N_1-n-l}[1 \tag{6}$$

$$- (1 - \frac{1}{K_1})^n]^{\delta_1}$$

Using this formula, we determine numerically the minimum bandwidth cost K_1 which satisfies $P_1(\delta_1, ..., \delta_M) \leq R$ for a given δ.

4   On Nash Equilibrium

In non-cooperative games, each player aims to maximize its own utility over its strategy set, thus player i chooses the strategy s_i which maximizes its utility u_i for a given vector of strategies $\vec{s} = (s_1, ..., s_M)$. Thereafter, player v_i waits for others to change/keep their strategies, and then it changes/keeps its strategy accordingly. If there exists a vector of strategies $\vec{s}^* = (s_1^*, ..., s_M^*)$ which satisfies $\forall i \in V, \forall s_i' \in S_i, u_i(s_i^*, \vec{s}_{-i}^*) \geq u_i(s_i', \vec{s}_{-i}^*)$ where $\vec{s}_{-i}^*$ refers to the set of strategies for all players except for player i, then the game has Nash equilibria [12]. In the previous section, we determined $u_i(s_1, ..., s_M)$ in the form of $BW_i(T_U^1, ..., T_U^M)$ for our game.

The existence of Nash equilibria is a very important criterion for games, because it indicates that the system is able to operate in multi-player environment. Our game has a finite number of players $\{v_1, ..., v_M\}$ and corresponding set of transmission policies $\{T_U^1, ..., T_U^M\}$, where T_U^i can take values from an infinite range $]0, T[$. T_U^i is equivalent to δ_i which takes values from a finite set $\{1, 2, ..., \delta_{max} - 1\}$ where $\delta_{max} = T/TTI$. Therefore, our game is a finite game and it has been proven by Nash in [13] that it has at least one mixed strategy equilibrium.

To determine whether our game has pure Nash equilibria or not, we propose to investigate next this question numerically, because our model has a fixed point that cannot be solved analytically.

5   Numerical Results

5.1   Model Validation

In our scenario, 5G-U properties are considered similar to those defined in the latest IEEE 802.11 systems operating on the 5GHz unlicensed band, regarding the time slot duration T_s, $SIFS$, $DIFS$ and bit rate R_b. The transmission uses the whole available unlicensed band of spectrum, hence only one transmission can take place in a given time. The data packet size including all headers is denoted by L_{data} and the acknowledgment packet size by L_{ack}. The station receives the ACK/NACK after a duration of SIFS then all stations backoff during a period of DIFS before starting to contend again for medium access. The duration x is then calculated as: $x = \lceil \frac{(L_{data}+L_{ack})/R_a+SIFS+DIFS}{T_s} \rceil$, where $\lceil . \rceil$ is the ceiling function.

The system deploys LBT-cat3 with a fixed contention window size of W_0. We consider that the station generates a packet every $10ms$ following a Poisson distribution, then we can estimate the probability of packet arrival q per time slot. The latency and reliability requirements are set to $T = 1ms$ and $R = 1 - 10^{-5}$, respectively. Table 1 shows the considered numerical values.

Transmission in licensed spectrum has become more flexible in 5G than 4G. Depending on the application, it is now possible to choose TTI length from a range of values. For delay constrained applications, like URLLC, we prefer to

Table 1. Numerical values of the system

T_s	$9\mu s$	R_b	100 Mbps	x	7
$SIFS$	$16\mu s$	L_{data}	$32 Bytes$	W_0	16
$DIFS$	$34\mu s$	L_{ack}	$14 Bytes$	q	0.001

choose the smallest length of TTI defined in the standards: $TTI = 0.125\,ms$, albeit larger bandwidth needed for the same transmission.

We assume that a URLLC packet fits in one conventional LTE RB of 0.5 ms duration and 180 KHz bandwidth (12 subcarriers with carrier spacing of 15 KHz), having that $TTI = 0.125\,ms$, the bandwidth of our RB is then $w = 720\,KHz$.

We consider two tenants in the area and $N_1 = 0.75N$ because it is logical to have the majority of stations belonging to the tenant itself. We compare the results obtained from the analytical model with an event-driven simulation realised using MATLAB. The simulation output is calculated from the viewpoint of a designated station which is always active, the other $N-1$ stations generate a packet every T_s with probability q. Every packet in the system is tagged with a contention window and a timer T_U^i which depends on the tenant. Every station with a packet performs LBT-cat3, and in every time slot all timers are checked for time-out. In our analysis we considered that $p_{c,1}$ is independent in every time slot, which is not the case in the simulation since we are tracking all packets, which will lead eventually to a difference in the results. For the rest of the simulation, if the designated packet times-out, it is transferred to 5G-L in the remaining time budget. $p_{c,1}$ is calculated by enumerating all collisions of the designated packet divided by the total number of transmissions (success or collision). P_1 is also calculated by enumerating the number of lost packets (timed-out in 5G-U and all replicas collided in 5G-L) over the total number of generated packets.

We illustrate in Fig. 4 $p_{c,1}$ and P_1 for different values of N, where we fix the transmission policies for the tenants to $\delta_1 = 5$ and $\delta_2 = 3$. The number of available RBs is also fixed to $K_1 = 5$.

Fig. 4 shows a good match between analysis and simulation, the difference is due to the correlation of $p_{c,1}$ as stated above.

5.2 Nash Equilibria Illustration

For the case of two tenants, we illustrate Nash equilibria by evaluating the cost of all possible combinations of the pair (δ_1, δ_2). We first consider v_1 as the tenant of interest with $N = 180$: $N_1 = 135$ and $N_2 = 45$. We show in Table 2 the corresponding cost for every pair (δ_1, δ_2), then we consider v_2 as the tenant of interest with $N_1 = 45$ and $N_2 = 135$, which yields to a symmetrical scenario for both tenants. Note that in our example we limit the number of RBs to 99.

We observe from Table 2 that we have multiple equilibrium points, which correspond to the set of pure-strategy Nash equilibria: $(\delta_1, \delta_2) = (4,4), (4,5), (5,4)$. If the game begins at any of these strategies, it is not in the interest of either

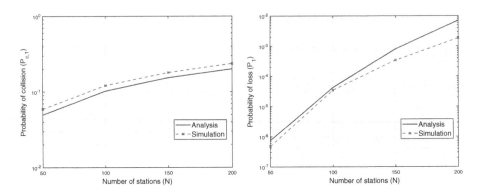

Fig. 4. Model validation in two-tenant case

Table 2. Nash equilibrium illustration

δ_1							
δ_2	1	2	3	4	5	6	7
1	99, 99	56, 99	24, 99	18, 99	19, 99	23, 99	34, 99
2	99, 56	56, 56	24, 56	18, 55	19, 54	23, 51	34, 46
3	99, 24	56, 24	24, 24	18, 24	18, 23	23, 23	34, 21
4	99, 18	55, 18	24, 18	**18, 18**	**18, 17**	23, 17	34, 16
5	99, 19	54, 19	23, 18	**17, 18**	18, 18	23, 18	34, 17
6	99, 23	51, 23	23, 23	18, 23	18, 23	22, 22	32, 21
7	99, 34	46, 34	21, 34	16, 34	17, 34	21, 33	32, 32

players to change their strategy because it does not improve its payoff (mini-mizes its cost). Neighbouring values of δ_i can lead to the same cost due to the quantization granularity where multiple close values of P_1 lead to the same value of K. Table 2 illustrates the fact that decreasing the time budget in 5G-U for one player (increasing its δ_i) improves the performance for the other player.

By this, we conclude that our game has pure-strategy Nash equilibria.

5.3   Price of Anarchy

It is interesting to discuss the notion of *price of anarchy* in non-cooperative games, which measures the efficiency deterioration of the system in the presence of multiple non-cooperative players, compared to a centralized cooperative sys-tem. We evaluate in Fig. 5 the cost in terms of number of RBs in the licensed band for the case of one player with N stations versus the case of two players with $N/2$ stations each; the cost in the second case is $K_1 + K_2$.

Fig. 5 illustrates the difference of cooperative games versus non-cooperative ones, which confirms that a centralized system achieves higher gain than a decen-tralized one.

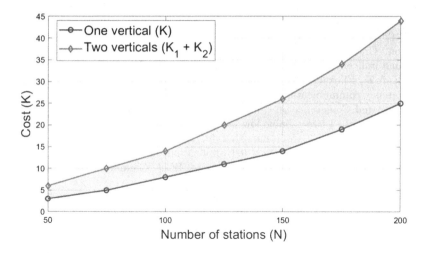

Fig. 5. Price of anarchy

6   Conclusion

We studied in this paper the transport of URLLC traffic in the uplink in a multi-tenant scenario, where the transmission is considered on both unlicensed and licensed spectrum. The unlicensed spectrum is shared among the tenants, which degrades its value-add in case of over solicitation; on the licensed spectrum however, each tenant reserves a certain bandwidth to reach a target reliability.

We modeled the MAC access of the combined unlicensed and licensed system incorporating a timer corresponding to the delay budget for URLLC traffic using Markov chains and obtained the overall performance of the system, in terms of overall probability of failure.

We modeled the multi-tenant system as a non-cooperative game where players are the tenants, the strategies are their usage policies of the unlicensed system and the objective of each tenant is to minimize its usage of the licensed spectrum. We showed that the game has pure-strategy Nash equilibria and illustrated these points numerically.

Our model gives insights on the usage of unlicensed spectrum for URLLC and shows that it is still valuable for verticals even in a multi-tenant scenario. However, as indicated by our analysis for the price of anarchy, a situation where a common operator manages the unlicensed spectrum access for all tenants would lead to better overall utilities.

References

1. Alliance, M.: Multefire release 1.0 technical paper: a new way to wireless, white paper (January 2017)
2. 3GPP: Physical layer procedures for data, 3GPP TR 38.214 v15.1.0, Technical report, March 2018

3. Maternia, M., et al.: 5G PPP use cases and performance evaluation models, 5G PPP (2016)
4. Sutton, G.J., et al.: Enabling ultra-reliable and low-latency communications through unlicensed spectrum. IEEE Net. **32**(2), 70–77 (2018)
5. Hindi, A.Z., Elayoubi, S.E., Chahed, T.: Performance evaluation of ultra-reliable low-latency communication over unlicensed spectrum. In: Proceedings of the IEEE International Conference on Communications (ICC), pp. 1–7. IEEE (2019)
6. Chen, H., et al.: Ultra-reliable low latency cellular networks: use cases, challenges and approaches. IEEE Commun. Mag. **56**(12), 119–125 (2018)
7. Schulz, P., et al.: Latency critical IoT applications in 5G: perspective on the design of radio interface and network architecture. IEEE Commun. Mag. **55**(2), 70–78 (2017)
8. Singh, B., Tirkkonen, O., Li, Z., Uusitalo, M.A.: Contention-based access for ultra-reliable low latency uplink transmissions. IEEE Wirel. Commun. Lett. **7**(2), 182–185 (2018)
9. Deghel, M., Brown, P., Elayoubi, S.E., Galindo-Serrano, A.: Uplink contention-based transmission schemes for urllc services. In: Proceedings of the 12th EAI International Conference on Performance Evaluation Methodologies and Tools, pp. 87–94 (March 2019)
10. Bianchi, G.: Performance analysis of the IEEE 802.11 distributed coordination function. IEEE J. Sel. Areas Commun. **18**(3), 535–547 (2000)
11. 3GPP: Study on latency reduction techniques for LTE, 3GPP TR 36.881 v14.0.0, Technical report, June 2016
12. Lasaulce, S., Debbah, M., Altman, E.: Methodologies for analyzing equilibria in wireless games. IEEE Signal Process. Mag. **26**(5), 41–52 (2009)
13. Nash, J.: Non-cooperative games. Ann. Math. 286–295 (1951)

Spectrum Sharing Secondary Users in Presence of Multiple Adversaries

Sayanta Seth[1(✉)], Debashri Roy[2(✉)], and Murat Yuksel[1(✉)]

[1] Electrical and Computer Engineering, University of Central Florida,
Orlando, FL 32826, USA
sayanta@knights.ucf.edu, murat.yuksel@ucf.edu
[2] Computer Science, University of Central Florida, Orlando, FL 32826, USA
debashri@cs.ucf.edu

Abstract. Cognitive radio technology brings a lot of interesting features which affect the transmission and reception properties of modern communication devices. Dynamic spectrum sensing, channel hopping, allocation and strategically nullifying adversarial attacks are among the few. In the presence of primary users (PUs), we inspect the secondary user (SU) pair behavior in decentralized, ad hoc cognitive radio networks, before and after adversarial attacks. We have taken into consideration the power radius of each SU pair to calculate the payoffs which decide if they should participate in a coalition or not. We propose algorithms for coalition formation for the SU pairs and the adversaries. We also propose two attack strategies for the adversaries: smart or naive. Overall, we propose a game-theoretic framework to study the multi-SU-pair multi-adversary scenario. We investigate the effect of adversarial attacks on the proposed framework. We show that the decrease in the average utilities of the SU pairs after attack varies from 66% to 75% with adversary count increasing by 14.2%. We study how the channels are allocated if there is an attack and how the payoffs of those SU pairs vary with increasing or decreasing number of channels. We also show that the payoffs decrease by 14.66%, if adversaries adopt a smart strategy instead of a naive one.

1 Introduction

A Cognitive Radio Network (CRN) typically consists of two types of users [9]: licensed primary users (PUs) and unlicensed secondary users (SUs). The PUs are the 'owners' of the spectrum, while SUs are smart, opportunistic users, exploiting unoccupied licensed spectrum. In the United States, according to Federal Communications Commission (FCC) [4,12], most of the radio spectrum is used inefficiently, resulting in over-utilization of some of its bands [10]. In a CRN, it is very important for an SU to predict with a high probability when the PU arrives so that it can vacate its band without much impact on the PU. In [1,3,7,13], the authors have focused on centralized and distributed spectrum sensing models, showing a novel power allocation scheme that uses dynamic sub-channel method based on a Nash Bargaining game. Saad *et al.* [11] studied cooperative spectrum

© Springer Nature Switzerland AG 2021
S. Lasaulce et al. (Eds.): NetGCooP 2021, CCIS 1354, pp. 125–135, 2021.
https://doi.org/10.1007/978-3-030-87473-5_12

sensing in CRNs with single PU. Here, the SUs increase their sensing accuracy by participating in a coalitional game. The authors explored the trade-off between the probability of actual detection of the PU and the probability of false alarm on the SU network topology and its dynamics. They reduced the interference on the PU through collaborative sensing. However, to our knowledge, the case of multiple users and multiple adversaries has not been considered.

The class of CRNs where an SU transmitter and an SU receiver work in tandem, is called Ad Hoc Cognitive Radio Networks (AHCRNs), which we mainly focus on. In AHCRNs, each user transmits based on its transmission power budget. For simplicity, we consider an SU transmitter and SU receiver to be a single entity, henceforth referred to as an 'SU pair'. We also consider PU activity in our system. There are a fixed number of PUs which are always present and are constantly accessing their own channels. The remaining channels are allocated to the *SU coalitions*, based on their payoffs. When the SU pairs work together to maximize their overall payoffs, we can say that they have formed a coalition. Also, the adversaries considered in the system, are capable of only attacking the SU pairs, and not the PUs, following the FCC mandated rules [5]. We consider the problem of tuning transmission power of such SU pairs under the presence of multiple adversaries. We use this transmission power criterion to decide the coalition formation of the SU pairs. We present a game-theoretic framework to study the multi-SU-pair multi-adversary scenario. The main contributions of this paper are:

1. We propose an intelligent coalition formation algorithm without overlapping transmission power radii, ensuring communication interference is avoided.
2. We devise an adversarial coalition formation algorithm, keeping in mind the smart and naive attack strategies of the proposed framework.
3. We present a stability criterion for the convergence of the coalition formation algorithm, so that the users maintain coalitions according to their payoffs.

Key insights from our study include:

1. We show that the decrease in the average utilities of the SU pairs after attack varies from around 66% to 75% with adversary count increasing by 14.2%. We also demonstrate that coalition utility increases by 28.5% with increasing channels, for constant number of SU pairs and adversaries.
2. For the same number of SU pairs and channels, we observe average coalitional utilities decrease with increasing adversary count.
3. Finally, we observe that smart attack strategy unleashes 14.66% more damage on the proposed framework than the naive one.

2   System Model and Assumptions

We consider a geographical area with radios of varying designated power budget. Each of the transmitter and the receiver of the SU pairs cover a portion of the area according to their power radii. We assume that each SU has a single transceiver which imposes the requirement of having pairs of SUs when forming a coalition. When two SUs are exchanging data they have to dedicate their transceivers to only one

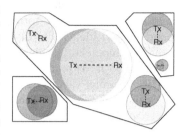

Fig. 1. SU pair coalition formation

wireless link. Overall, this requires the number of SUs that are actively transmitting/receiving data to be in multiples of 2. In Fig. 1, we show how SU pairs form coalitions. If an SU pair, e.g., with the biggest combined power radii in the figure, were to be a singleton, which is actually the smallest unit considered in this paper, then the cumulative power radius of that SU pair would be the outer boundary of the overlapped circles.

When an SU pair is singleton, then, the transmitter is communicating with the receiver on their own allocated channel. When multiple such SU pairs form a bigger coalition, it means that the member SU pairs will be allocated channels. When the SU pairs become a part of different coalitions, all of them still can access their Common Control Channel (CCC). This is necessary because, for decision making about coalition formation in the event of an attack, the SU pairs must have a communication medium. For example, a singleton SU pair might want to be part of a bigger coalition when it feels that it might be vulnerable to adversarial attacks. An adversarial attack could signify that the attacked SU pair might be left without any channel to communicate. If it joins the bigger coalition, then it is likely that the bigger coalition has more channels assigned to it. It is also likely that the newcomer finds a channel to transmit, which becomes an incentive to join a bigger coalition.

As described in [8], the SUs participate in spectrum sensing and access jointly, in a cooperative manner. A typical scenario of such cooperative behavior is presented in Fig. 2, where we show the SUs (devices) forming three different coalitions. Looking at the coalition marked in red, we see that there are three CR devices in it. According to the system model, this cannot happen, because even in a sin-

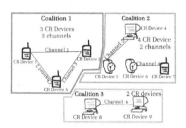

Fig. 2. Coalition formation among SU pairs (Color figure online)

gleton, there must be at least two SUs, as we have considered an SU pair to be a single entity. Hence, the red coalition is not feasible. The remaining blue coalitions are feasible because both of them have SUs in multiples of two.

3 Proposed Game-Theoretic Framework

Our system is deployed on a plane containing T PUs, N SU pairs (legitimate user pairs) and A adversaries. A total bandwidth of W is available, divided into C independent and identically distributed (i.i.d.) channels. The PUs stay on their designated channels. In this section, we discuss the proposed coalition formation games for both SU pairs and adversaries. The SU pairs work jointly to communicate their channel vacancy knowledge with their peers, such that they can improve the channel capacities accessed by them as a coalition. The goal of the SU pairs is to maximize their channel usage, while the goal of the adversaries is to block as many channels as possible, on which the SU pairs are working. The game that we design is a cooperative hedonic game, in which players may want to join or leave a coalition, or even stay alone. Let $\mathcal{N} = 1, 2, 3, ..., N$ be the set of players. If there is a subset S of $\mathcal{N}$, then the subset S is called 'a coalition'. A function $v(.)$ assigns a value to every subset of players. If all the members inside the subset S of $\mathcal{N}$ act in unison, then $v(S)$ is the payoff to all members of the subset. In other words, the value of the coalition is $v(S)$. At the starting of the game, there is no coalition, hence $v(\phi) = 0$. As the members start forming coalitions, based on their common interests, we have $v(S) > 0, \forall S \subseteq \mathcal{N}$. A detailed discussion in [2] presents the ideas of cooperative game theory. In characteristic form, an outcome of a game consists of: (i) A coalition structure, which is essentially a partition of $\mathcal{N}$ players into smaller coalitions, and (ii) a payoff vector to distribute the value of each coalition among its members. A non-empty collection of non-empty non-overlapping subsets can be referred to as a coalition structure (CS), $CS = S_1, S_2, S_3, ..., S_k$ where $S_i \subseteq \mathcal{N}$ represents coalition i and k is the total number of coalitions, which satisfies the followings:

$$\bigcup_{i=1}^{k} S_i = \mathcal{N}; \qquad S_i \cap S_j = \phi \text{ if } i \neq j \qquad (1)$$

3.1 Stability Criteria of Coalitions

For our proposed coalitional game, we consider two types of stability criteria: inner and outer. When no SU pair has incentive to leave its current coalition to become a singleton, we say that the players within that coalition have *inner stability*. Similarly, when no SU pair has any incentive to join another coalition, or in other words, no coalition in a CS has any incentive to merge with another coalition, we refer to it as *outer stability*. For example, consider a CS with two coalitions S_1 and S_2. The inner stability conditions will be:

$$v(S_1) > v(i), \forall i \in S_1 \qquad \text{and} \qquad v(S_2) > v(i), \forall i \in S_2 \qquad (2)$$

The outer stability conditions will be:

$$v(S_2) > v(S_1 \cup S_2) \qquad \text{and} \qquad v(S_1) > v(S_1 \cup S_2) \qquad (3)$$

For the rest of the paper, we will express the value function $v(.)$ of joining a coalition as the *payoff function*, quantifying the data rate achieved by joining that coalition.

3.2 Payoff Function Formulation

For a broader applicability, we consider Gaussian fading channel for our proposed framework. Hence, the transmission parameters are modelled based on Gaussian complex channel. The achievable data rate R in a Gaussian complex channel of bandwidth W is given by the Shannon's information capacity formula [6]:

$$R = W \log_2 \left(1 + \frac{P_t |h|^2}{N_0 W} \right) \tag{4}$$

where P_t is the transmission power, h is the channel gain, and $N_0 W$ is the cumulative noise of that channel. The payoff or the bit rate of a singleton is:

$$R_i^{singleton} = W \log_2 \left(1 + \frac{P_i |h|_{ii}^2}{N_0 W \sum_{j \in N, j \neq i} h_{ji} P_j} \right) \tag{5}$$

where h_{ii} is the channel gain between the transmitter and receiver of SU pair i. h_{ji} is the channel gain between SU pair j's transmitter and SU pair i's receiver. P_i is the transmission power of SU pair i. Extending this concept for a whole coalition S_i, we have:

$$R_i^{coalition} = \mu_i W \log_2 \left(1 + \frac{P_i |h|_{ii}^2}{N_0 W \sum_{j \in S^c, j \neq i} h_{ji} P_j} \right) \tag{6}$$

where μ_i is the number of channels of the total bandwidth W that is being used by coalition S_i. S^c is the complement of coalition S, which are the other coalitions in the CS. The term $\sum_{j \in S^c, j \neq i} h_{ji} P_j$ means the total interference power which is the summation of the power received from all members in other coalitions. After the coalition formation algorithm has converged, the set of coalitions will be $CoA = S_1, S_2, S_3, ..., S_\zeta$, where ζ is the total number of formed coalitions. When ζ is less than or equal to that of number of channels (C), then all coalitions will get at least one channel. Whereas, if $\zeta > C$ then, there will be at least one coalition which will not get any channel. Therefore, the second case may not converge. Further, ζ or the size of the set CoA should lie between 1 and N, i.e., $1 \leq \zeta \leq N$. Naturally, the second case of $\zeta > N$ is unrealistic.

3.3 Coalition Formation Algorithm

The power radius of each SU pair (a randomly generated positive real number in our case; a constant in reality, unique to a transmitter) means the geographical area that can be covered by each SU pair with its omni-directional antenna range. It is to be noted that the transmitter SU of the SU pair is at the center of the said area and the receiver SU can reside anywhere within that area. A pictorial representation of a certain coalition formation scenario is presented in Fig. 3.

We present the methodology for SU pair coalition formation in Algorithm 1. The coalition is formed based on the product of the probability of each SU pair

Algorithm 1 Algorithm for SU pair coalition formation

Inputs: total number of PUs (T), total number of SU pairs (N), total number of channels (N_C), probability of SU pairs picking up a channel (P).
Output: List of Coalitions of (CoA)

/*Phase 1: Finding the potential SU pairs for coalition formation or Initialization*/
/*Phase 2: Coalition Formation*/
while SU_i and SU_j do not change their coalitions **do**
  Check and compare their payoffs with each other;
  if (SU pair i's payoff is less than the combined payoff of SU pair i and SU pair j) AND (SU pair j's payoff is less than the combined payoff of of SU pair j and SU pair i) **then**
      max_pay = combined payoff of SU pair i and SU pair j **else**
        max_pay = payoff of individual SU pair i
      end
  end

end
Formation of intermediate coalition list CoA based on max_pay. /*Phase 3: Check stability criteria*/
$Maxpay$ = MAX(payoff of coalition k with all of its member SU pairs, payoff of coalition k with all of its member SU pairs along with one member SU pair from coalition j, payoff of singleton coalition k).
if $Maxpay$ = payoff of coalition k with all of its member SU pairs **then**
  Continue
else if $Maxpay$ = payoff of coalition k with all of its member SU pairs along with one member SU pair from coalition j **then**
  Remove the SU pair from coalition j and merge it with coalition k
else
  Remove the SU pair coalition j and maintain it as singleton.
end
Return the final stable coalition list (CoA)

Algorithm 2 Algorithm for modelling adversarial attack

Inputs: SU pair coalition list (CoA)
Output: SU pair coalition list after adversarial attack (CoA)

/*Phase 1: Initiating the adversaries in the same way as the SU pairs*/
/*Phase 2: Attack strategy*/
if Smart Strategy: **then**
  1. Communicate the potential target list to other adversaries through adversaries' dedicated CCC.
  2. Attack a SU pair from the potential target list.
  3. Broadcast the attacked target to other adversaries through CCC such that the other adversaries may remove the already attacked SU pairs from their attack list.
else
  4. Attack a SU pair from the potential target list without updating the other adversaries.
end
Return the updated coalition list (CoA)

Fig. 3. Randomly deployed SU pairs and adversaries

picking up a channel and the incentive or value function associated with each coalition. The initialization phase of the algorithm includes initializing the PUs and the SU pairs on the plane, randomly initializing the power radii of each SU pair, calculating the Euclidean distance between the given SU pair and other SU pairs, and based on these, forming the potential coalition list. This information is broadcasted over CCC and the number of available channels is updated as: $C = N - C_T$. Probability P is also updated after each of N SU pairs picks a channel from the available C channels. In most of the cases, the incentive for an SU pair to join the biggest coalition is the highest as there are already more SU pairs in a bigger coalition and more channels are assigned. Then, again, it is imperative that all the SU pairs will judge their own incentive and decide to join the biggest coalition, resulting in a grand coalition.

Ideally, the communication between an SU pair of a coalition should not prevent another SU pair of the coalition from successful communication. This is the reason we decided to introduce the factor of power radius, so that the SU pairs only form an alliance with other pairs who are far enough from each other. This factor results in non-overlapping power radii, which in turn, creates the basis of a healthy communication mechanism. The point to be noted here is that the total combined power radius of an SU pair cannot overlap that of another SU pair. If the combined power radii of two or more SU pairs overlap with each other, then there will be probable communication loss. One could argue that if multiple channels are allocated to a coalition, then in spite of overlapping power radii, different SU pairs could choose different channels and the communication could be carried out successfully. But, we do not always have abundant channels and the proposed algorithm also helps in overcoming the problem of shortage of available channels.

3.4 Checking for Stability Criterion for the Coalition

Once initial coalition is formed by checking the max_pay value, we then try to evaluate the stability of a coalition k as follows:

1. Find the maximum payoff max_pay of coalition k with all of its members.
2. If max_pay of coalition k is the greatest, then the coalition is already stable.
3. If max_pay of coalition k is the greatest if one SU pair member comes in from coalition j, then the new member is incorporated into k, which then becomes stable.
4. If the max_pay value of the member SU pair of coalition j, alone is greater than that of coalition k itself, then we remove that member from coalition j and put it as a singleton.

These calculations are repeated for all the combinations of coalitions and their members, and the final combination with the maximum max_pay value is chosen as the stable coalition(s).

3.5 Adversarial Attack Model

The adversaries in our case are SUs, capable of transmitting white noise and disrupting the transmission of one SU pair at a time. For the adversaries, being successful means being able to block or jam any SU pair's communication, or in other words, forcing the legitimate SU pairs to adopt a different strategy to keep going on with their transmission. In Fig. 3, the yellow dots are the adversaries introduced into the system. We have followed the concept of Euclidean distance between an adversary and any one of the SU pairs. If the power radius of the said adversary is able to overlay (partially or fully) that of the said SU pair, then we assume that the adversary will be able to unleash some amount of devastating effect upon the SU pair. In Algorithm 2, we propose how the adversaries can effectively hamper the stable transmission. We present two types of attack strategies, where adversaries can act either smartly or naively rendering to different destructive effect on the proposed framework.

3.6   Channel Allocation

Next, we allocate the available channels to the formed coalitions where the number of allocated channels is directly proportional to the number of SU pairs present in that particular coalition. Ideally, if there are 10 SU pairs, 10 channels and 5 coalitions, each coalition consisting of 2 SU pairs, then each coalition will also get 2 channels. But in reality, such is not always the case, as the coalition formation depends on the value of the incentive function, which in turn depends on the terms in the $R_i^{coalition}$. The transmission power factor is randomly generated by a pseudo-random number generator, so the value of the incentive function is also random. All of this leads to the fact that channel allocation might not be as expected, and in some cases, coalitions can end up receiving a very small number of channels compared to the number of SU pairs present in them. However, if the number of channels is greater than or equal to the number of SU pairs, then we can expect a good channel allocation number for most of the coalitions. If the number of channels is fewer than the number of SU pairs, they will have to rely on techniques like Time Division Multiple Access (TDMA) to utilize the fewer channels in turn, in consecutive timestamps.

4   Simulation Results

The simulation for the multi-channel, multi-SU-pair and multi-adversary game has been performed under geographical boundary conditions. There are 10 PUs in the system, each using their own designated channel. We study how the average utilities of the coalitions vary when we introduce more adversaries into the system keeping the number of SU pairs and channels constant. The results that we discuss here are obtained through simulations run 10 consecutive times and then averaging them out. In Fig. 4(a), we show the general decrease in average utility of SU pairs after adversarial attack. In Fig. 4(b), we have shown how the average utility for 50 SU pairs decreases with increasing number of adversaries in the system from 20 to 50 (also represented in tabular form in Table 1). This set of simulations converged to generate 25 coalitions. Now, it is evident that when the number of adversaries deployed becomes large enough, they can render the entire system of CRN useless. We also investigate how the total utility of the 25 coalitions vary with increasing number of channels, keeping the number of SU pairs and the number of adversaries constant, represented in Fig. 4(c) and in tabular form in Table 2. As we increase the number of channels from 45 to 60, we calculate the total utility of the 25 coalitions increase by 28.5%. This happens because, now, under attack, the SU pairs have more leeway to switch to other channels and continue with their communication.

Smart vs. Naive Attack Strategies by the Adversaries. Now, we have done a comparative study of the adversarial behaviour. The adversaries can choose to be smart. If they act smart, they should be communicating among themselves through their dedicated CCC and broadcast their SU communication blocking information to others. The potential targets for an adversary are decided by its power radius. Any of the SU pairs falling under that power radius could be chosen by the adversary to block. Now, based on the random geographical position in which all the SU pairs and adversaries are deployed, it could so happen that another adversary might have the same SU pair in its potential blocking list as the previous one. If both of them end up blocking the same one, then they will be wasting their resources and their payoffs as a whole adversarial group will drop. Hence, a smart adversary should always choose an SU pair and communicate its choice to other adversaries (over dedicated CCC), so that they can concentrate on blocking others. In this way, the adversaries as a coalition will be able to wreck a bigger havoc. On the other hand, adversaries working naively without communication will not be able to do as much damage as compared to them working smartly. In Fig. 4(d), we have compared the payoffs of all the 25 formed coalitions, based on the attack strategy. From experimental results, the coalition utilities decreased additionally after attacking with the smart strategy when compared to naive strategy by 14.66%.

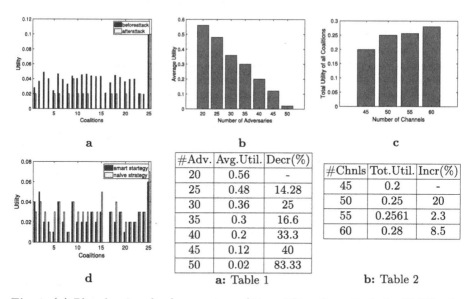

a

b

c

d

#Adv.	Avg.Util.	Decr(%)
20	0.56	–
25	0.48	14.28
30	0.36	25
35	0.3	16.6
40	0.2	33.3
45	0.12	40
50	0.02	83.33

a: Table 1

#Chnls	Tot.Util.	Incr(%)
45	0.2	–
50	0.25	20
55	0.2561	2.3
60	0.28	8.5

b: Table 2

Fig. 4. (a) Plot showing the decrease in coalition utility after attack, for 50 SU pairs and 60 channels (10 channels being used by the PUs), (b) Change in average Utilities for increasing adversaries, (c) Change in average utilities for increasing channels, (d) Utility comparison for smart attack strategy vs. naive strategy, **Table (a)** shows decrease in avg. utility with increasing adversaries, **Table (b)** shows increase in total utility with increasing channels.

5   Conclusions

In this paper, we showed how SUs in an ad-hoc CRN can create coalitions autonomously in the absence of base stations to increase their payoffs. We presented an intelligent coalition formation algorithm and formulated the payoff function for the calculation of the utilities of the SU pairs in terms of SNR. We have devised a coalition formation algorithm which can be used by SUs in the CRN to find potential partners for coalition formation. Using these potential coalitions, we have come up with the final coalition model. We also introduced adversaries in the proposed framework and modelled an algorithm for adversarial attack against the legitimate SUs. Next, we presented the data which shows percentage decrease in the average utilities with varying number of adversaries, keeping the number of SUs and channels constant. Then, we showed the increase in coalition utility with increasing number of channels, keeping the number of SUs and number of adversaries as constant. Finally, we presented the difference in coalition utilities for smart and naive adversarial attack strategies. We envision to extend this work by introducing significant PU activities and therefore to carry forward the idea of adversarial attack in the case of PUs as well.

Acknowledgement. This work was supported in part by NSF award 1647189. The authors offer their gratitude to Mostafizur Rahman for proofreading the paper.

References

1. Baharlouei, A., Jabbari, B.: Dynamic subchannel and power allocation using nash bargaining game for cognitive radio networks with imperfect PU activity sensing. In: 8th International Conference on Cognitive Radio Oriented Wireless Networks (2013)
2. Chalkiadakis, G., Elkind, E., Wooldridge, M.: Cooperative game theory: basic concepts and computational challenges. IEEE Intell. Syst. **27**, 86–90 (2012)
3. Chen, F., Qiu, R.: Centralized and distributed spectrum sensing system models performance analysis based on three users. In: International Conference on Wireless Communications Networking and Mobile Computing (WiCOM) (2010)
4. Das, D., Das, S.: A survey on spectrum occupancy measurement for cognitive radio. Wirel. Pers. Commun. **85**, 2581–2598 (2015)
5. FCC ET Docket No. 08–260: Second Report and Order and Memorandum Opinion and Order - Unlicensed Operation in the TV Broadcast Bands/Additional Spectrum for Unlicensed Devices Below 900 MHz and in the 3 GHz Band. US Federal Communications Commission, Washington DC, FCC 08–260 (2008)
6. Goldsmith, A.: Wireless Communications. Cambridge University Press, Cambridge (2005)
7. Le, L.B., Hossain, E.: Resource allocation for spectrum underlay in cognitive radio networks. IEEE Trans. Wirel. Commun. **7**, 5306–5315 (2008)
8. Mitola, J., Maguire, G.Q.: Cognitive radio: making software radios more personal. IEEE Pers. Commun. **6**(4), 13–18 (1999)
9. Niyato, D., Hossain, E.: Competitive spectrum sharing in cognitive radio networks: a dynamic game approach. IEEE Trans. Wirel. Commun. **7**, 2651–2660 (2008)

10. Roy, D., Mukherjee, T., Chatterjee, M., Pasiliao, E.: Defense against PUE attacks in DSA networks using GAN based learning. In: IEEE Global Communication Conference (2019)

11. Saad, W., Han, Z., Debbah, M., Hjorungnes, A., Basar, T.: Coalitional games for distributed collaborative spectrum sensing in cognitive radio networks. In: IEEE INFOCOM 2009 (2009)

12. Valenta, V., Maršálek, R., Baudoin, G., Villegas, M., Suarez, M., Robert, F.: Survey on spectrum utilization in Europe: measurements, analyses and observations. In: 2010 Proceedings of the Fifth International Conference on Cognitive Radio Oriented Wireless Networks and Communications (2010)

13. Xing, X., Jing, T., Cheng, W., Huo, Y., Cheng, X., Znati, T.: Cooperative spectrum prediction in multi-PU multi-SU cognitive radio networks. Mob. Netw. Appl. **19**(4), 502–511 (2014)

Scheduling and Resource Allocation Problems in Networks

.

Dynamic Bus Dispatch Policies

M. Venkateswararao Koppisetti$^{(\boxtimes)}$ and Veeraruna Kavitha

IEOR, IIT Bombay, Mumbai, India
{venky12,vkavitha}@iitb.ac.in

Abstract. The time gap between two successive buses is called headway in transport systems. In moderate/high frequency routes, with moderate/small headways, the random perturbations (traffic conditions, passenger arrivals, etc.), can alter the headway along the route significantly which possibly leads to bunching of buses. Two or more (successive) buses may start travelling together. Bus bunching results in inefficient and unreliable bus service and is one of the critical problems faced by bus agencies. Thus it is imperative to reduce the bunching possibilities (probability). Another important aspect is the expected time that a typical passenger has to wait before the arrival of its bus. If one increases the headway, the bunching chances might reduce, however, may significantly increase the passenger waiting times. We precisely study this inherent trade-off and derive a bus schedule optimal for a joint cost related to all the trips, which is a weighted combination of the two performance measures.

We consider a system with Markovian travel times, fluid passenger arrivals and derive dynamic headways which control the bus frequency based on the observed system state. The observation is a delayed information of the time gaps between successive bus arrivals at various stops, corresponding to two earlier (previous to previous) trips. We solve the relevant dynamic programming equations to obtain near-optimal policies, and the approximation improves as the load factor reduces. The near-optimal policy turns out to be linear in previous headway and the (earlier) bus-inter-arrival times. Using Monte Carlo based simulations, we demonstrate that the proposed dynamic policies significantly improve (both) the performance measures, in comparison with the previously proposed partial dynamic policies that only depend upon the headways of the previous trips.

1 Introduction

Public transport plays an important role in any system. We consider public transport systems like that of buses, trams, metros, local trains etc., for brevity we refer them as bus transport systems. In fact in the later cases, bunching (buses travelling together) is a major issue. Bus agencies desire to provide the best service to the passengers due to heavy competition from other transport services and would strive hard to reduce/eliminate the bunching possibilities.

© Springer Nature Switzerland AG 2021
S. Lasaulce et al. (Eds.): NetGCooP 2021, CCIS 1354, pp. 139–153, 2021.
https://doi.org/10.1007/978-3-030-87473-5_13

We consider a bus-transport system, where the buses travel repeatedly along a fixed route consisting of a fixed number of stops. Each bus starts at the depot, traverses all the stops and returns to the depot, while facilitating the transfer of the (encountered) passengers from their origin to their destination. Typically successive buses are designed to depart at depot, according to a pre-designed time-table. The time period between two successive bus-departs is referred to as headway (at depot).

The randomness of travel times, load conditions, etc., leads to random headways at bus stops. These random delays can lead to bunching of two or more buses; the leading bus can get delayed excessively due to a large number of passengers and or due to heavy random traffic en-route, and the trailing bus may have relatively lesser load which can eventually lead to both of them coming close to each other somewhere along the path.

The larger headways at depot reduce the bus bunching, but leads to an increase in the passenger waiting times. Thus one needs to design the headways optimally, to ensure proper trade-off between these two important performance aspects. Hence, one needs to design T-successive depot headways (with $(T + 1)$-trips), and hence a need for a finite horizon headway policy. We considered such optimal headway policies in [2, 10], and these policies depend at maximum on the headways of the previous trips. Static/stationary policies (constant headways) are considered in [10] while a variety of non-stationary and dynamic policies (including the policies of this paper) are available in [2].

One can do much better if one has access to a better system state that is influenced by the random fluctuations governing the system. For example, if one can observe the number of passengers waiting in various stops (at bus arrival instances) or an equivalent information of the previous trips, a better headway policy can be designed using this knowledge. In this paper we consider that the bus-inter-arrival times between various stops (of previous trips) are observable, based on which the headway times of the future trips are decided.

The natural tool to design such policies is the theory of Markov decision processes [11]. However in this system, one will <u>have access only to delayed information</u>: the headway decision for the current bus has to be made immediately after the previous bus departs the depot, the information related to the previous bus trajectory (no delay) is obviously not available for this decision epoch. Further one may not even have the information about some of the trips, previous to the trip that just started. For notational simplicity we assume that 1-delay information is available and derive the optimal policies. One can easily extend this analysis to any arbitrary delay, we provide some initial suggestions regarding the same (the exact details in [2]).

We obtained closed form expressions for an ε-optimal policy, that is ε-optimal under small load factors. Interestingly the policy is linear in the previous trip headways and the bus-inter-arrival times at various stops of the previous trip (information about which is available). We showed numerically that this dynamic policy has significant improvement in comparison with respect to the optimal policy of [2] that dynamically adapts the headways only based on previous trip headways. This improvement is significant even for considerable load factors (up to 0.5). We used Monte-Carlo based simulations to estimate the two performance measures. Thus one can do much better, if there is a possibility to observe more details related to the previous trips. The observation process might be complicated, but, the complexity of proposed policy is negligible.

Related Literature

Bus bunching is a critical issue faced by bus agencies and this problem has been thoroughly investigated over past few decades. However to the best of our knowledge none of the literature studies the important trade-off between the bunching chances and the passenger waiting times (see [10] for details on this observation). Other than [2,10], none of the papers study/consider the probability of bunching. As already mentioned, work in [2,10] does not consider the fully dynamic policies.

There is a vast literature that studies other topics related to bus bunching and we discuss a few of them here. Existing control strategies are based on ideas like skipping some bus stops (e.g., [4,7,13]), limited boarding (e.g., [5,6]) or forcibly holding the buses at some stops (e.g., [4,5,14]) etc. Skipping some stops/ holding control at intermediate bus-stops are not passenger friendly policies. In this paper, we applied holding control only for the depot.

In [5], authors consider minimizing the total sojourn time (travel time between the boarding stop and the destination stop) of all the passengers. In papers like [5,6,12,14], authors control the error variance between ideal and proposed schedules when the number of buses and stops increases to infinity. They assume no bunching. Stochastic models that optimize the holding times to minimize passenger waiting times (defined via sum of squared headways), using the real-time information, are discussed in [3]. They avoid bus bunching, while maintaining the frequency of the buses as high as possible for the next few trips. In many scenarios with high randomness, it is not possible to completely avoid bunching nor is it possible to adhere to the ideal schedules. In such scenarios, it is rather important to reduce the probability of bunching, and we precisely consider this probability. Further we consider a more realistic definition of the passenger waiting times: as the time difference between their arrival and the arrival of their bus.

2   System and Problem Description

We consider a bus-transport system in which the buses travel in a loop, traversing the given path of M stops and repeating this for $(T+1)$ number of trips. We begin with the description of the details of the system considered and the required assumptions.

2.1   Bus Travel and (Stop) Inter-arrival Times

Let S_k^i be the time taken by the k-th bus to travel between the stops $(i-1)$ and i. We consider Markovian (correlated) travel times, between any two stops. To be precise we assume that

$$S_k^i = S_{k-1}^i + W_k^i, \text{ for any, } k \geq 0, \text{ and } S_0^i = s^i,$$

where W_k^i is the random difference between two successive travel times and $\{s^i\}_i$ are the sojourn times of the first trip. We assume $\{W_k^i\}_k$ are IID (Independent and Identically distributed) *Gaussian random variables with mean* 0 *and variance* ϵ^2 and this is true for all stops i. Further these are independent across the stops.

The boarding time of (all) the passengers at any stop majorly constitutes the dwell time of the bus. We assume the following:

A.1: Gated service: *Only the passengers that arrived to a stop before the arrival of the bus can board.*

A.2: Parallel boarding and de-boarding: *We neglect the time taken to de-board while computing the dwell times.*

A.3: Fluid arrivals and boarding: *The number of passengers arrived to a stop, during a period t equals λt, where $\lambda > 0$ is the arrival rate.* More details about this modelling is provided in the next section. *The time taken to board X number of passengers equals bX, where $b > 0$ is the boarding rate.*

These assumptions are not very restrictive, and are satisfied by most of the commonly used practices in bus transport systems. The fluid arrivals can be justified owing to Elementary Renewal theorem, and because typical (bus) inter-arrival times at any stop would be significant; passenger arrival (e.g., Poisson) process can be modelled as renewal process with rate λ and then the number of passenger arrivals in a large time interval $[0, t]$ (during one bus inter-arrival time), approximately equal λt. Further, usually negligible number of passengers arrive during the boarding process, thus gated service assumption is not very restrictive.

We require the following additional assumptions:

A.4: Surplus number of buses: *For any trip, there exists a bus (at depot) to start after the prescribed headway (without having to wait for the return of the previous buses).*

A.5: Order of buses *is maintained throughout the journey, i.e., even if the buses are bunched the next bus will board and depart the stop after the previous bus.* Thus overtaking is not allowed.

A.6: *There is no constraint on capacity of the bus.*

The system may require one or two additional buses to satisfy **A.4**, which is a normal practice to cater for any eventuality. The systems usually operate with small bunching probabilities, as such the event in assumption **A.5** is a rare event. Further this is a common practice in Tram, metro, local train etc., systems; the vehicles respect the order without overtaking. Even if not, our analysis will go through by interchanging the bus numbers whenever there is a overtaking. Assumption **A.6** can be restrictive, but is a commonly made assumption in literature [7–9,13,14].

Because of the above assumptions, the passengers boarding a bus in any trip k and at any stop i equals the ones that arrived during the bus-inter arrival time

$I_k^i := A_k^i - A_{k-1}^i$, where A_k^i is the arrival instance of k-bus at stop i. Thus the total number of passengers X_k^i waiting at stop i, at bus arrival instance, equals λI_k^i. Thus the dwell time of k-th bus at stop i equals[1]:

$$V_k^i = X_k^i b = b\lambda I_k^i = \rho I_k^i \text{ with } \rho := \lambda b. \tag{1}$$

In the above ρ represents the load factor of the stop[2].

Let h_k be the headway between $(k-1)$-th and k-th bus at depot. Then the inter-arrival times are given by:

$$I_k^1 = \left(h_k + S_k^1\right) - S_{k-1}^1 = h_k + W_k^1 \text{ for first stop, similarly}$$

$$I_k^i = h_k + \sum_{1 \le j \le i} S_k^j + \sum_{1 \le j < i} V_k^j - \left(\sum_{j \le i} S_{k-1}^j + \sum_{j < i} V_{k-1}^j\right)$$

$$= h_k + \sum_{1 \le j \le i} W_k^j + \rho \sum_{1 \le j < i} \left(I_k^j - I_{k-1}^j\right) \text{ for any stop } i. \tag{2}$$

The last equality follows by fluid arrival and gated service assumptions as in (1).

The analysis of these inter-arrival times are instrumental in obtaining the results of this paper. In Lemmas 2–3 provided in Appendix, it is shown that the inter-arrival times are Gaussian and their expectations, variances are computed.

We now describe the Markov decision process based problem formulation that optimizes a given weighted combination of the two performance measures, the bunching probability and the passenger-average waiting times.

3   Markov Decision Process (MDP)

3.1   Decision Epochs, State and Action Spaces

When the $(k-1)$-th bus leaves the depot, the system needs to determine the headway for the k-th bus. A decision at this epoch, can depend upon the available system state. One will have access only to delayed information: the headway decision for the current bus has to be made immediately after the previous bus departs the depot, the information related to the previous bus trajectory (no

---

[1] Since bunching is a rare event we neglect the affects of **A**.5 in this part of the modelling. There would be differences in the travel time and dwell time estimates when the buses bunch at a stop (with or without overtaking). But since the bus systems operate typically with small bunching probabilities, not considering these effects into the models does not introduce drastic errors in the performance measures; besides making it mathematically tractable. In sections dealing with numerical examples in [2], we established for many examples using exhaustive (and realistic) numerical simulations, that the theoretical performances well match with the corresponding Monte-Carlo estimates.

[2] One can easily consider the case with different load factors at different stops, for notational simplicity we consider the same load factor at all stops. We consider stop dependent load factors in [2].

delay) is obviously not available. Further one may not even have the information about some of the trips, previous to the trip that just started also.

We assume the availability of 1-delayed information[3]; the bus-inter-arrival times $\{I_{k-2}^j\}_{1\leq j\leq M-2}$ at various stops related to $(k-2)$-th trip are known at k-th decision epoch. One can also have access to the information about the headways of all the previous trips $\{h_{k-l}\}_{l\geq 1}$; there is no delay in this component of the information. We will observe later that one requires only the headways of the unobserved trips, that of 1-previous trip. Thus, in all, at $(k-1)$-th bus departure (i.e., at k-th decision epoch), we have access to the following state (see Eq. (2)):

$$Y_k = (h_{k-1}, \{I_{k-2}^j\}_{1\leq j\leq M}), \tag{3}$$

By fluid arrivals and gated boarding, this state is equivalent to the number of passengers boarding the bus at various stops corresponding to the latest available trip (see Eq. (1)).

Remark: In (3), it is interesting to observe that we require only the information related to first $(M-2)$ stops of the $(k-2)$-th trip; it becomes evident from Theorem 1 (given below) that such a state is sufficient if one has access (at maximum) to 1-delay information; from Lemmas 2–3 of Appendix the random components (I_k^i and $I_k^i - \rho I_{k-1}^i$), that define the required performance measures related to the k-th trip, depend at maximum upon the information related to stops $1, 2 \cdots i - 2$, when 1-delay information is available.

It is easy to observe (see Eq. (2)) that the random vector sequence $\{Y_k\}_k$ forms a controlled Markov Chain, whose evolution depends upon the headway (of the current trip that needs to be decided) and the previous state.

Trips Prior to the Controlled Trips. Initial trips may have light load conditions (passenger arrival rates) and can be subjected to small variations in traffic, load conditions. Alternatively, in some cases the initial trips can be subjected to high load conditions. Further, one may consider controlling some of the trips and not all, probably ones that have maximum fluctuations. It is clear that some of the trips (just prior) to the controlled ones also influence the performance.

By abuse of notation, we call all the previous trips that influence the controlled trips (the one whose headways are to be controlled) as initial trips. One can show that only some t_0 (to be more precise (at maximum) previous $(M+1)$-trips can influence) previous trips would influence. We assume that the buses operate during these initial trips (say t_0 of them) at some fixed headway h_0. We consider controlling the depot-headway starting from trip t_0+2, and to keep the notations simple, we refer $(t_0 + 2 + k)$-th trip by index k. Alternatively one can consider controlling the buses starting from the first trip as in [2].

---

[3] General d-delay information case is discussed briefly at the end of the paper and more details are in [2].

Table 1. Notations and constants

$$\theta^j = \tfrac{\rho^2}{2}\sum_{i=j+2}^{M}(i-j-1)(1+\rho)^{i-j-2}, \ \theta := \tfrac{1}{2}\sum_{i=1}^{M}(1+\rho)^{i-1} = \tfrac{(1+\rho)^M - 1}{2\rho},$$

$$\bar\theta := \tfrac{\rho}{2}\sum_{i=1}^{M}(i-1)(1+\rho)^{i-2}, \psi = (1+\rho)^{M-1}, \bar\psi := \rho(M+\rho)(1+\rho)^{M-2}$$

$$\psi^j = \rho^2\left((1+\rho)^{M-2-j}(M-j+\rho)\mathbb{1}_{j<M-1} + \mathbb{1}_{j=M-1}\right),$$

$$\omega^2 = \epsilon^2\left(\sum_{j=1}^{M}(1+\rho)^{2(M-j)} + \sum_{j=1}^{M-1}\rho^2(1+\rho)^{2(M-1-j)}[M-j+1+\rho]^2 + \rho^2\right)$$

$$\eta_{T-k} = \tfrac{(\theta+\eta_{T-k+1})\bar\psi}{\psi} - \bar\theta - \sum_{j=1}^{M}(1+\rho)^{j-1}\gamma_{T-k+1}^j,$$

$$a_{T-k} = \omega\sqrt{-2\log\left(\tfrac{(\theta+\eta_{T-k+1})\sqrt{2\pi}\omega}{\psi\alpha}\right)},$$

$$\gamma_{T-k}^j = \tfrac{(\theta+\eta_{T-k+1})}{\psi}\psi^j - \left(\theta^j + \rho\sum_{i=j+1}^{M-1}(1+\rho)^{i-1-j}\gamma_{T-k+1}^i\right)\mathbb{1}_{j<M-1},$$

$$\delta_{T-k} = \tfrac{(\theta+\eta_{T-k+1})a_{T-k}}{\psi} + \alpha\left[1 - \Phi(a_{T-k})\right] + \delta_{T-k+1}.$$

3.2   Performance Measures

Passenger Waiting Times. *The waiting time of a typical passenger is the time gap between its arrival instance at the stop and the arrival instance of its bus (to the stop).* Let $W_{n,k}^i$ be the waiting time of the n-th passenger that boards the k-th bus at stop i. The customer average of the waiting times corresponding to trip k and stop i equal (e.g., [2,10]):

$$\bar w_k^i \triangleq \frac{\bar W_k^i}{X_k^i} \text{ with } \bar W_k^i := \sum_{n=1}^{X_k^i} W_{n,k}^i.$$

Recall X_k^i is the number of customers that board the k-th bus at i-th stop and hence the above is the average of the waiting times of all the customers that belong to stop i and that correspond to trip k.

Fluid Approximation/Arrivals: The passengers are assumed to arrive at regular intervals (of length $1/\lambda$), with λ large. The waiting time of the first passenger during bus inter-arrival period (I_k^i) is approximately[4] $W_{1,k}^i \approx I_k^i$, that of the second passenger is approximately $W_{2,k}^i \approx I_k^i - 1/\lambda$ and so on. Thus as $\lambda \to \infty$, the following (observe it is a Riemann sum) converges:

$$\frac{\bar W_k^i}{\lambda} = \frac{1}{\lambda}\sum_{n=0}^{\lambda I_k^i}\left(I_k^i - \frac{n}{\lambda}\right) \to \int_0^{I_k^i}(I_k^i - x)dx = \frac{(I_k^i)^2}{2}. \tag{4}$$

---

[4] The residual passenger inter-arrival times at bus-arrival epochs get negligible as $\lambda \to \infty$.

Thus for large λ,

$$\bar{w}_k^i \approx \frac{I_k^i}{2} \text{ because } \bar{W}_k^i \approx \lambda\frac{(I_k^i)^2}{2} \text{ and } X_k^i \approx \lambda I_k^i. \tag{5}$$

We *refer this approximation, as the fluid approximation.*

The trip average of the waiting times is given by:

$$\frac{1}{T}\sum_{k=1}^{T}\sum_{i=1}^{M}E[\bar{w}_k^i] = \frac{1}{T}\sum_{k=1}^{T}\sum_{i=1}^{M}\frac{E[I_k^i]}{2},$$

which is one of the components to be optimized. By Lemma 2 (Appendix), the conditional expectation given the state Y_k and the depot-headway decision h_k equals (see (3)):

$$E\left[I_k^i\middle|Y_k, h_k\right] = h_k(1+\rho)^{i-1} - h_{k-1}(i-1)\rho(1+\rho)^{i-2}$$

$$+ \sum_{j=1}^{i-2} I_{k-2}^j(i-j-1)\rho^2(1+\rho)^{i-j-2} \text{ for any } k \geq 1. \tag{6}$$

Bunching Probability. Starting from the depot, the buses travel on a single route with some headway (time gap between successive arrivals to the same location) between successive buses. If these headways were maintained constant thought their journey, the successive buses would not meet each other. However, because of variability in load/traffic conditions, the above is not always true. A bus can get delayed (to some stop) significantly because of the random fluctuations. The delayed bus has larger number of passengers to board and hence is further delayed for the next stop. The trailing bus has lesser number of passengers and hence departs early from the stop. This continues in the subsequent stops, and there is a possibility of the headway between the two buses becoming zero. This is called *bus bunching.*

Bus bunching increases the waiting times (and their variability) of passengers, further and more importantly wastes the capacity of the trailing buses. Thus the system becomes inefficient. The larger depot headway times decreases the chances of bunching but, however, increases the passenger waiting times. Thus one needs an optimal trade-off.

The bunching probability *is the probability that a bus arrives to a stop before the departure of the previous bus.* It is easy to verify that this is the probability that the dwell time of $(k-1)$-th bus, V_{k-1}^i given by (1) is greater than the inter arrival time between $(k-1)$ and k-th buses, I_k^i given by (2):

$$P_{B_k}^i = P(V_{k-1}^i > I_k^i) = P(I_k^i - \rho I_{k-1}^i < 0). \tag{7}$$

We consider optimizing the bunching probability of the last stop, as this stop experiences maximum variations[5]. Conditioned on Y_k, h_k the value $I_k^M - \rho I_{k-1}^M$

---

[5] One can alternatively consider bunching probability of an initial stop or that of an important stop, and, the optimal policy could be derived in exactly the same manner as in this paper.

is Gaussian distributed (see (2)) and from Lemma 3 of Appendix, we have for any $k \geq 1$ (constants are given in Table 1):

$$P^\phi\left(I_k^M - \rho I_{k-1}^M < 0 \middle| Y_k, h_k\right) = 1 - \Phi\left(\psi h_k - \bar{\psi} h_{k-1} + \sum_{j=1}^{M-1} \psi^j I_{k-2}^M\right),$$

$$\Phi(x) := \int_{-\infty}^{x} \frac{1}{\sqrt{2\pi\omega^2}} \exp\left(\frac{-t^2}{2\omega^2}\right) dt, \tag{8}$$

where Φ is the cdf of a normal random variable with mean 0 and variance ω^2. Observe that the variance ω^2 does not depend upon headway policy, rather only depends upon the load factors of various stops and traffic variability. However, also observe that the bunching probability depends upon the headway policy.

3.3   The MDP Problem

Let $\phi = (d_1 \cdots, d_T)$ be any Markov policy, in that $d_k(y)$ represents the depot headway for the k-th bus if the system observes the state y. *We choose a headway in the range* $[0, \bar{h}]$ *for some* $\bar{h} < \infty$. The expected values of the above two cost components depend upon the policy ϕ and the initial trajectories specified by (t_0, h_0) (see Subsect. 3.1). To be more specific, given (t_0, h_0) one has probabilistic description of the system state Y_1. Let E_{t_0,h_0}^ϕ represent the expectation given the policy and the initial conditions, at times we omit the subscript and superscript to keep notations simple. We have multi-objective (two) optimization and a natural way is to optimize the following weighted combination of the two costs (6), (8):

$$J(\phi; h_0, t_0) = \sum_{k=1}^{T} E^\phi\left[\bar{w}_k^i\right] + \alpha P^\phi\left(I_k^M - \rho I_{k-1}^M < 0 \middle| Y_k, h_k\right)$$

$$= \sum_{k=1}^{T} E_{t_0,h_0}^\phi\left[r(Y_k, h_k)\right] \text{ with}$$

$$r(Y_k, h_k) = \sum_{i=1}^{M} E[I_k^i | Y_k, h_k] + \alpha P^\phi\left(I_k^M - \rho I_{k-1}^M < 0 \middle| Y_k, h_k\right)$$

$$= h_k \theta - h_{k-1}\bar{\theta} + \sum_{j=1}^{M-2} \theta^j I_{k-2}^j$$

$$+ \alpha\left[1 - \Phi\left(\psi_k h_k - \bar{\psi}_k h_{k-1} + \sum_{j=1}^{M-1} \psi_k^j I_{k-2}^j\right)\right],$$

where $\alpha > 0$ is the trade-off factor and the constants are in Table 1. Our objective is to obtain a policy that optimizes the following for any given (t_0, h_0):

$$v(t_0, h_0) := \inf_\phi J(\phi; t_0, h_0).$$

It is easy to verify that the above value function equals:

$$v(t_0, h_0) = E_{t_0, h_0}\left[v(Y_1)\right],$$

and this can be solved by solving the MDP problem for any given initial condition $y_1 = (h_0, \{I^j_{-1}\}_j)$, i.e., by deriving the value function $v(y_1)$ for any y_1 (e.g., [11]).

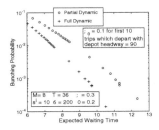

Fig. 1. Comparison between partial dynamic and dynamic policies.

Table 2. Performance for various configurations with initial trip details: $\rho_0 = 0.2$, $h_0 = 100$, $t_0 = 12$ and the controlled trip details: $M = 10$ $T = 36$ $s^j = 10$ $\lambda = 200$.

Configuration	Bunching probability		Waiting times	
	Dynamic	Partial	Dynamic	Partial
$\epsilon = 0.3$, $\rho = 0.3$, α big	1.95e−02	2.1e−02	18.42	25.29
$\epsilon = 0.3$, $\rho = 0.3$, α small	1.52e−01	1.51e−01	11.33	14.01
$\epsilon = 0.4$, $\rho = 0.3$, α big	1.78e−02	1.77e−02	24.41	34.15
$\epsilon = 0.4$, $\rho = 0.3$, α small	1.36e−01	1.37e−01	14.85	18.95
$\epsilon = 0.2$, $\rho = 0.5$, α big	2.64e−01	2.66e−01	144.23	185.81
$\epsilon = 0.2$, $\rho = 0.5$, α small	6.54e−02	6.51e−02	242.68	380.01

4   Optimal Policies

The optimal policy is obtained by solving dynamic programming (DP) equations using backward induction. The DP equations, for any $k < T$ are given by [11]:

$$v_k(Y_k) = \inf_{h_k \in [0,\bar{h}]} \left\{ r_k(Y_k, h_k) + E\left[v_{k+1}(Y_{k+1})|Y_k, h_k\right] \right\}, \text{ and,}$$

$$v_{T+1}(Y_{T+1}) = 0.$$

From the trip wise running costs (6)–(8), these equations are rewritten as (constants are given in Table 1):

$$v_k(Y_k) = \inf_{h_k \in [0,\bar{h}]} \left\{ h_k \theta - h_{k-1}\bar{\theta} + \sum_{j=1}^{M-2} \theta^j I^j_{k-2} \right.$$

$$\left. + \alpha \left[1 - \Phi\left(\psi h_k - \bar{\psi} h_{k-1} + \sum_{j=1}^{M-1} \psi^j I^j_{k-2} \right) \right] + E\left[v_{k+1}(Y_{k+1})|Y_k, h_k\right] \right\}. \quad (9)$$

One can derive optimal policies by solving these DP equations and there are many known numerical techniques to do the same (e.g., [11]). In the following we derive the structure of near optimal policies (closed form expressions) for the case with small load factors:

Theorem 1. *Assume $T > M + 1$. We define the coefficients $\{\eta_k\}_k$, $\{\gamma_k^j\}_{k,j}$ and $\{a_k\}_k$ backward recursively: first set $\eta_{T+1} = 0$, $\delta_{T+1} = 0$, $\gamma_{T+1}^j = 0$ for all $1 \leq j \leq M$ and then set the rest of them as in Table 1. There exists a $\bar{\rho} > 0$, such that for all $\rho \leq \bar{\rho}$: the following is an ε-optimal policy*[6] *with*[7] $\varepsilon = O(\rho)$:

$$h_{T-k}^*(Y_{T-k}) = \max \left\{ 0, \ \min \left\{ \bar{h}, \ h_{T-k}^{uc}(Y_{T-k}) \right\} \right\}, \tag{10}$$

$$h_{T-k}^{uc}(Y_{T-k}) := \frac{1}{\psi} \left[\bar{\psi} h_{T-k-1} - \sum_{j=1}^{M-1} I_{T-k-2}^j \psi^j + a_{T-k} \right].$$

The expected value function (for any k, Y_{T-k-1}, h_{T-k-1}) equals:

$$E[v_{T-k}(Y_{T-k})|Y_{T-k-1}, h_{T-k-1}] = E\left[\eta_{T-k} h_{T-k-1} \right.$$

$$\left. - \sum_{j=1}^{M-1} I_{T-k-2}^j \gamma_{T-k}^j + \delta_{T-k} \middle| Y_{T-k-1}, h_{T-k-1} \right] + O(\rho).$$

Proof: The complete proof is in [1] as well as in [2]. ∎

Remark: Thus the ε-optimal policy is affine linear in the previous trip headways and the bus-inter-arrival times. By the above theorem, the policy well-approximates the optimal one, as ρ the load factor reduces. We will notice that the policy works well even for nominal load factors (in some examples even up to $\rho = 0.5$) in the next section. One can have a much simplified approximate algorithm as in [2], which is constructed using some asymptotic arguments for partial dynamic policies (also described below).

5   Numerical Analysis

5.1   Partial Dynamic Policies [2]

As already mentioned, in [2] we derive policies that depend only on previous headways. We refer them as 'partial-dynamic' policies, as they do not consider the random component of the system state Y_k. It is obvious that one can improve with the 'fully' dynamic policies of Theorem 1. In this section, we study the extent of improvement provided by the extra information. Towards this we reproduce the optimal policies of [2], for the purpose of completion. For all load factor $\rho \leq \bar{\rho}$ (for some $\bar{\rho} > 0$, the optimal policy is given by ($\mathbf{h}_{T-k} := [h_{T-k-1}, \cdots h_{T-k-M}]$):

---

[6] The cost under this policy is within ε radius of the optimal cost.

[7] Big O notation: $f = O(\rho)$, as $\rho \to 0$, implies $f(\rho) \leq C\rho$ for some constant $C > 0$, for all ρ sufficiently small.

$$h^*_{T-k}(\mathbf{h}_{T-k}) = \left[-\sum_{l=1}^{M} h_{T-k-l} \psi^p_l + a^p_* \right], \quad \text{with} \qquad (11)$$

$$a^p_* = \frac{\sigma^M_M}{(1+\rho)^{M-1}} \sqrt{-2 \log \left(\frac{M\sqrt{2\pi}\sigma^M_M}{2(1-\rho)\alpha} \right)}. \qquad (12)$$

$$\psi^p_l = \frac{1}{(1+\rho)^{M-1}} \left((-1)^l \binom{M-1}{l} \rho^l (1+\rho)^{M-1-l} \right.$$
$$\left. -(-1)^{l-1} \binom{M-1}{l-1} \rho^l (1+\rho)^{M-l} \mathbb{1}_{l>0} \right).$$

In the above, the constant σ^M_M is given by [2]:

$$(\sigma^M_M)^2 = \epsilon^2 \left(\sum_{l=0}^{M-1} \sum_{j=1}^{M-l} \left(\tilde{\gamma}^{M-j+1}_l \right)^2 + \frac{\rho^2(1-\rho^{2M})}{1-\rho^2} \right) \quad \text{with}$$

$$\tilde{\gamma}^i_l \triangleq \gamma^i_l - \rho\gamma^i_{l-1} \mathbb{1}_{l>0}, \text{ and,}$$

$$\gamma^i_l \triangleq (-1)^l \binom{i-1}{l} \rho^l (1+\rho)^{i-1-l}, \text{ with } \binom{n}{r} := 0 \text{ when } n < r.$$

5.2 Experiments

We conduct many Monte Carlo based simulations to compare the proposed dynamic policies with the partial dynamic policies of [2]. We basically generate several sample paths of transport system trajectories, where each sample path is generated using a sample of the random walking times between the stops and the random passenger arrivals at various stops for all the T-trips. We dispatch the buses according to one of the two policies for different values of trade-off factors α and obtain the estimates of the bunching probability and the average passenger waiting times using the sample means.

In Fig. 1, we plot the estimates of average passenger waiting times versus the estimates of the bunching probability for different values of α and for both the policies. The details of the experiment are mentioned in the figure itself. We notice a significant improvement with fully dynamic policies. The curve of bunching probability versus expected waiting time obtained with fully dynamic policy is placed well below that with partial dynamic policies. This implies that one can simultaneously improve both the performance measures, when one has access to the more information about the system state. We conducted many more experiments and the observations are similar.

In Table 2 we consider various system configurations, which is described in the first column. We choose different values of α for the two policies such that the bunching probabilities are almost equal (under both the policies) and these values are reported in next two columns. We then tabulate the corresponding average passenger waiting times in the last two columns. These are the estimates averaged across all the T trips. The different configurations span across different levels of traffic variability (ϵ), different load factors during controllable trips (ρ) and or different level of α/trade-off factors. In all the configurations, we notice a good improvement with fully dynamic policies. Since α were chosen such that the bunching probabilities of both the policies are almost equal, one can study the improvement via the improvement in average passenger waiting times. We observe that improvements are in the range of 21% to 44%.

Extension to Arbitrary Delays

One can easily extend this analysis to arbitrarily delayed information, i.e., for the case when the observation is d-delayed we have, $Y_k = (h_{k-1}, h_{k-2} \cdots , h_{k-d}, \{I_{k-d}^j\}_j)$ with $1 \leq d < M$. Basically the Lemmas 2–3 can easily be extended and the rest of the proof can be completed after some changes. We conjecture the following would be an ε-optimal policy for some appropriate coefficients $\{\bar{\psi}_r\}$, $\{\psi^j\}$ and $\{a_{T-k}\}$:

$$h^*_{T-k}(y_{T-k}) = \min\left\{\bar{h}, \max\left\{0, \sum_{r=1}^{d} \bar{\psi}_r h_{T-k-r} - \sum_{j=1}^{M-d} I^j_{T-k-d-1}\psi^j + a_{T-k}\right\}\right\}.$$

Note that the above matches with the partial dynamic policy of [2] reproduced in (11) as well as the fully dynamic policy (10) proposed in this paper. The proof of this result is derived recently and is available in [2]. It would be further interesting to consider the case where one has partial information (only for some stops) related to some (delayed) trips.

6   Conclusions

Unlike the popular models considered in literature, we directly studied the inherent trade-off between the two most important aspects of any bus transport system, the bunching possibilities and the passenger waiting times. Further, we formulated a Markov decision processes based problem to derive optimal (depot) dispatch (i.e., headway) policies that depend upon the random state observed at various bus stops of the previous trips. The observation is that of the time gaps between arrivals of the successive buses at the same stop.

We consider systems with Markovian travel times, fluid passenger arrivals and with delayed (one delay) information. The objective function optimized is the sum of a weighted combination of the two performance measures, corresponding to all the trips of the given session. We obtained a near-optimal dynamic policy for small load factors by solving the corresponding finite horizon dynamic programming equations, using backward induction. This policy is linear in previous trip headway and the bus-inter-arrival times corresponding to the earlier trips. We conducted Monte-Carlo based simulations to plot the estimates of the average passenger waiting times and the bunching probability for various trade off factors. We also observed that the proposed dynamic policy performs significantly better than the previously proposed partial dynamic policies of [2]. These partial dynamic policies depend only upon the headways of the previous trips.

A Appendix

Only lemma 2 and lemma 3 statements are provided here. While the rest of the details including the proofs are in Technical report [1] or in [2].

Lemma 2. *The conditional expectation of inter arrival times given the state Y_k (from (3)) and h_k equals:*

$$E\left[I_k^i \middle| Y_k, h_k\right] = (1+\rho)^{i-1}h_k - (i-1)\rho(1+\rho)^{i-2}h_{k-1}$$

$$+ \sum_{j=1}^{i-2}(i-j-1)\rho^2(1+\rho)^{i-j-2}I_{k-2}^j. \tag{13}$$

∎

Lemma 3. *For any $2 \leq i \leq M$ we have:*

$$I_k^i - \rho I_{k-1}^i = (1+\rho)^{i-1}h_k - (i+\rho)\rho(1+\rho)^{i-2}h_{k-1} + \sum_{j=1}^{i}(1+\rho)^{i-j}W_k^j$$

$$- \sum_{j=1}^{i}\left((i-j+1+\rho)\rho(1+\rho)^{i-j-1}\mathbb{1}_{j<i} + \rho\mathbb{1}_{j=i}\right)W_{k-1}^j$$

$$+ \sum_{j=1}^{i-1}\left(\rho^2(1+\rho)^{i-2-j}(i-j+\rho)\mathbb{1}_{j<i-1} + \rho^2\mathbb{1}_{j=i-1}\right)I_{k-2}^j. \tag{14}$$

The bunching probability of $(k-1)$ and k-th bus at stop i given the state Y_k (from (3)) and h_k equals:

$$P\left(I_k^i - \rho I_{k-1}^i < 0 \middle| Y_k, h_k\right) = 1 - \Phi\left(\psi h_k - \bar{\psi}h_{k-1} + \sum_{j=1}^{i-1}\psi^j I_{k-1}^j\right),$$

where the constants are given in Table 1.

∎

References

1. Dynamic bus bunching policies. Technical report. http://www.ieor.iitb.ac.in/files/faculty/kavitha/bunchingTR.pdf
2. Stationary and finiate horizon bus schedules for optimal bunching and waiting. Manuscript under preparation. http://www.ieor.iitb.ac.in/files/faculty/kavitha/BunchJourn.pdf
3. Berrebi, S.J.Y.: A real-time bus dispatching policy to minimize headway variance. Ph.D. thesis (2014)
4. Cortés, C.E., Sáez, D., Milla, F., Núñez, A., Riquelme, M.: Hybrid predictive control for real-time optimization of public transport systems' operations based on evolutionary multi-objective optimization. Transp. Res. Part C Emerg. Technol. **18**(5), 757–769 (2010)

5. Delgado, F., Muñoz, J., Giesen, R., Cipriano, A.: Real-time control of buses in a transit corridor based on vehicle holding and boarding limits. Transp. Res. Rec. J. Transp. Res. Board **2090**, 59–67 (2009)
6. Delgado, F., Munoz, J.C., Giesen, R.: How much can holding and/or limiting boarding improve transit performance? Transp. Res. Part B Methodol. **46**(9), 1202–1217 (2012)
7. Liping, F., Liu, Q., Calamai, P.: Real-time optimization model for dynamic scheduling of transit operations. Transp. Res. Rec. J. Transp. Res. Board **48–55**, 2003 (1857)
8. He, S.-X.: An anti-bunching strategy to improve bus schedule and headway reliability by making use of the available accurate information. Comput. Ind. Eng. **85**, 17–32 (2015)
9. Hickman, M.D.: An analytic stochastic model for the transit vehicle holding problem. Transp. Sci. **35**(3), 215–237 (2001)
10. Venkateswararao Koppisetti, M., Kavitha, V., Ayesta, U.: Bus schedule for optimal bus bunching and waiting times. In: 2018 10th International Conference on Communication Systems and Networks (COMSNETS), pp. 607–612. IEEE (2018)
11. Puterman, M.L.: Markov Decision Processes: Discrete Stochastic Dynamic Programming. Wiley, New York (2014)
12. Sánchez-Martínez, G.E., Koutsopoulos, H.N., Wilson, N.H.M.: Real-time holding control for high-frequency transit with dynamics. Transp. Res. Part B Methodol. **83**, 1–19 (2016)
13. Sun, A., Hickman, M.: The real-time stop-skipping problem. J. Intell. Transp. Syst. **9**(2), 91–109 (2005)
14. Xuan, Y., Argote, J., Daganzo, C.F.: Dynamic bus holding strategies for schedule reliability: optimal linear control and performance analysis. Transp. Res. Part B Methodol. **45**(10), 1831–1845 (2011)

Multi Objective Decision Making for Virtual Machine Placement in Cloud Computing

Wissal Attaoui[1(✉)], Essaid Sabir[1,2(✉)], Halima Elbiaze[2(✉)],
and Mohamed Sadik[1(✉)]

[1] NEST Research Group, ENSEM, Hassan II University of Casablanca,
Casablanca, Morocco
{w.attaoui,e.sabir,m.sadik}@ensem.ac.ma
[2] Computer Science Department, University of Quebec at Montreal (UQAM),
Montreal, QC H2L 2C4, Canada
elbiaze.halima@uqam.ca

Abstract. Cloud computing is an innovative process that delivers on-demand services over the internet. Virtualization is considered as the key concept of cloud computing since it handles running multiple virtual resources in a single physical resource. Mapping the virtual machine (VM) to the appropriate physical machine (PM) is called virtual machine placement (VMP). In this context, the dilemma of placing VMs in the cloud environment presents a significant challenge that has been wholly addressed but not yet totally fixed. This paper provides a multi-objective decision-making approach for VMP in a cloud computing infrastructure. We propose a conic scalarization method to solve the optimization problem. Simulation results prove that the offline algorithm yields good results compared to online deterministic algorithms.

Keywords: Cloud computing · Virtual machine placement

1 Introduction

Nowadays, cloud computing is considered as an emerging technology that regularly evolves towards a significant field of computer science. Cloud services are continually being expanded to meet customer demands where the Infrastructure as a Service (IaaS) package is the most requested by cloud users. The cloud service provider delivers hardware services (i.e., CPU, memory, storage, network bandwidth, etc.) using virtualization technology. To execute or serve a task in the cloud, the first process to be performed is the Virtual Machine (VM) allocation, which is the process of allocating or mapping a VM with a specific configuration where the assigned VM must meet the Quality of Service (QoS). The next step is the VM placement process, which is the process of placing or mapping the VM to its best fit physical machine (PM), as presented in Fig. 1. By analogy,

ⓒ Springer Nature Switzerland AG 2021
S. Lasaulce et al. (Eds.): NetGCooP 2021, CCIS 1354, pp. 154–166, 2021.
https://doi.org/10.1007/978-3-030-87473-5_14

we can see this problem as a "Tetris" game whose goal is to place the arrived shapes in the right places properly. From the provider's point of view, energy consumption, cost, and resource wastage are the main objectives that need to be optimized. At the same time, the QoS and the quality of experience are the critical elements to be maximized from the customer perspective. In this context, many algorithms and policies were proposed to solve the VM allocation and placement problems.

The traditional VM mapping approaches were treated as single-objective, whereas the recent ones address the VMP as a multi-objective optimization problem. Moreover, this placement process can be accomplished either offline (static) or online (dynamic). On the one hand, for offline placement, the data center (DC) providers consolidate data and make placement decisions to meet the consumer requests considering different constraints. On the other hand, for online placement, the DC suppliers gather data periodically then decide whether a rescheduling of the placement procedure is required.

Virtualization gives the possibility to conveniently move a Virtual Machine (VM) from a specific host to another without turning it off; therefore, this can offer a dynamicity on VM placement optimization with a negligible impact on performance. Despite its numerous benefits, these dynamics may result in suboptimal or unstable configurations of the virtual networks. Moreover, VMs may experience some fluctuations within the resource utilization (e.g., a mobile application server and a web server may possess identical patterns of the incoming workload while using the same CPU). In this context, several challenges hinder the efficient placement of the virtual machine, which can be dealt with as multi-objective optimization approaches; Considering various trade-offs between energy consumption, reliability and performance degradation [7,10], power consumption and resource utilization [13], cost and QoS [1,21], network traffic and resource utilization [6,8,18], etc.

Accordingly, to handle these combinatorial optimization problems, we model the VM placement problem as a multi-objective decision making (MODM) approach aiming to simultaneously optimize five objective functions: energy consumption, cost, network traffic, resource utilization, and QoS. Therefore, to solve this optimization model, we transform the problem into a single objective function using a scalarization method.

This paper is structured as follows; the next section presents the related works. Section 3 presents the problem formulation analyzed as a multi-objective decision making (MODM) approach. In Sect. 4, we describe the scalarization method proposed to solve the MODM problem. Section 5 presents the test environment to evaluate our model and examine the most appropriate algorithm among proposals.

2   Related Works

Several approaches in the literature have studied the VMP problem in both static and dynamic environments. In the following, we cite some references that

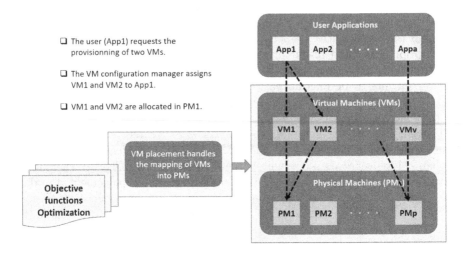

The user (App1) requests the provisionning of two VMs.

The VM configuration manager assigns VM1 and VM2 to App1.

VM1 and VM2 are allocated in PM1.

Fig. 1. Cloud computing architecture

have dealt with this placement problem in different ways, mono-objective or multi-objective optimization in online or offline settings. From 2007 to 2030, global energy consumption will increase by 76%, referring to *energy outlook 2030*. Therefore, we should think about the methods to be followed to reduce the energy consumption in the cloud; since cloud data centers included thousand of servers that consume an enormous amount of power. Therefore, to save energy consumption, it's preferable to place VMs on as few PMs as possible. However, if VMs are placed densely in a server, this can cause an occurrence of heat islands, which can affect the reliability of the device. In this context, [7] proposes a bi-objective optimization algorithm that considers energy-saving and server reliability. This algorithm aims to minimize the total power consumption of all servers as a function of resource utilization and simultaneously achieves reliability by adding backup servers when the number of working servers exceeds a redundancy ratio. Simulation results prove the performance of the recommended algorithm in terms of power consumption in both working and backup servers compared to online deterministic heuristics as First-Fit-Decreasing (FFD), Modified FFD (MFFD) and Thermal Aware Workload Scheduling Algorithm (TASA).

In the same regard, unbalanced resource usage may lead to resource wastage, SLA violation, and high power consumption. Authors in [13] propose a multi-objective virtual machine placement algorithm for IaaS cloud named (MOVMP). This model allows reducing the number of active PMs through migrating VMs and achieving a more balanced use of resources, which minimizes the energy consumption and resource wastage. The experiment setup of the proposed algorithm yields good results compared to First-Fit (FF) and Power-Aware Best Fit Decreasing (PABFD).

The quality of service (QoS) in cloud computing is another challenge for cloud providers to attract and satisfy users. Authors in [1] treated the VMP problem as a tradeoff between QoS and cost. They propose a two-layer model, where the first one considers the allocation problem in a cost-effective way using linear programming for load balancing. The second one regards the VM placement by proposing a Genetic Algorithm Based Virtual Machine Placement (GABVMP) algorithm as a solution to the optimization problem. Simulation results show that GABVMP is more performant than two greedy heuristics (Random placement and FF placement).

Several papers study the problem of minimizing network traffic [6,8,18] to enhance the performance of a DC by selecting the most suitable physical machines for virtual machines. Daniel et al. [9] present a VMP algorithm to reallocate virtual machines in DC Server contingent on the memory usage, the traffic matrix network, and the overall CPU. The first phase of this VMP algorithm considers collecting data from VMs and DC topology. The second one focuses on partitioning servers with a higher level of connectivity. The last one consists of clustering VMs by defining the amount of purchased traffic using graph theory to manage all the virtual servers. Simulation results prove that this solution improves network traffic quality and the availability of bandwidth at DC.

3   Multi-objective Decision Making: Problem Formulation

Based on our literature survey in [3], we analyze the different problems that may interrupt the VM placement. We classify the existing solutions into five primary objective functions based on multiple performance metrics such as energy consumption, quality of service, service level agreement, resource usage, and incurred cost.

In this paper, we form the problem of virtual machine placement as a multi-objective decision making (MODM) approach, by optimizing the five objective functions: (1) Energy consumption minimization, (2) cost optimization, (3) Network traffic minimization, (4) Resource utilization and (5) Quality of service maximization simultaneously. We consider the available PMs specifications, the requested VMs, the network traffic between VMs and their current placement as Inputs, and the new convenient placement as Output. This section presents a theoretical approach to solve the VMP problem (see Fig. 2).

Given a set of physical machines $S = S_1, S_2..., S_n$ and a set of virtual machines $V = V_1, V_2, .., V_m$. We are looking for new placement of VMs V on a set of PMs S while satisfying the constraints and simultaneously optimizing the five objective functions cited above in a pure multi-objective context.

Each physical machine S_i is characterized by a specific processing resources CPU, RAM memory, storage, and maximum power consumption represented as:

$$S_i = [S_i^{cpu}, S_i^{ram}, S_i^{stor}, S_i^{pmax}]; \quad \forall i \in \{1, 2, .., n\} \qquad (1)$$

Each virtual machine V_j needs processing resources of CPU V_j^{cpu}, RAM V_j^{ram}, and storage V_j^{stor}, providing economic revenue R_j and attributing SLA to each VM. Consequently, a virtual machine V_j will be represented as:

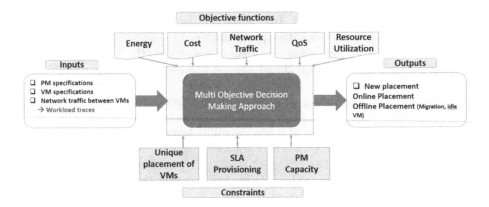

Fig. 2. Multi attribute decision making approach

$$V_j = [V_j^{cpu}, V_j^{ram}, V_j^{stor}, R_j, SLA_j]; \quad \forall j \in \{1, 2, .., m\} \qquad (2)$$

The network traffic between the requested VMs is represented as follows:

$$T_j = [T_{j_1}, T_{j_2}, .., T_{j_m}]; \quad \forall j \in \{1, .., m\} \qquad (3)$$

where T_{j_k} represents the average communication rate in [Kbps], between the virtual machine V_j and the virtual machine V_k. Note that we can consider $T_{jj} = 0$.

3.1   Objective Functions

The five objective functions are mathematically formulated as follows:

Energy Consumption Minimization

Based on [4], the energy consumption of a resource is defined as the sum of the power consumption of each PM considering a linear relationship with the CPU utilization:

$$\min \ g_1(x) = \sum_{i=1}^{n} \left[(P_i^{max} - P_i^{min}) \cdot U_i^{cpu} + P_i^{min} \right] \cdot A_i \qquad (4)$$

where

- $g_1(x)$: the total energy consumption of the PMs;
- P_i^{max}: is the power consumption at the peak load;
- P_i^{min}: is the minimum power consumption in the inactive mode;
- U_i^{cpu}: is the utilization ratio of CPU resources used by S_i;
- A_i: is a binary parameter equal to 1 if PM turned on, 0 otherwise.

Cost Optimization

Cloud service providers avoid economic loss generated by failures. Therefore, based on [22], we adopt the proposed approach of maximizing cloud provider's revenue regarding SLA violation cost. We have two cases:

- if V_j is finished without failure: provider will receive revenue:

$$R_j = c_j \cdot t_j - E_j \qquad (5)$$

   where c_j is the price of V_j; E_j is the execution cost of V_j that can be estimated through paying site infrastructure Capex and energy costs overall machines, and t_j is the lifetime of VM V_j.
- if V_j is failed: the customer is eligible to receive a refund from the provider, which is equal to $D_j = c_j \cdot t_j \cdot p_{SLA}$, where p_{SLA} is a constant parameter and indicates the part of the entire invoice that the provider has to deliver.

The expected revenue generated from placing V_j on S_i is defined as the following:

$$F_j = R_j(1 - f_j^i) - (E_j + D_j) \cdot f_j^i, \qquad (6)$$

where f_j^i is the probability of at least one failure occurring on S_i in t_j.

The problem of maximizing cloud provider's revenue (minimizing total cost) is expressed by:

$$\min\ g_2(x) = \sum_{j=1}^{n} \widehat{F_j} \cdot B_j; \qquad (7)$$

where:

- $g_2(x)$: is the total economical revenue for placing VMs; $\widehat{F_j} = -F_j$
- B_j: is a binary variable equals to 1 if V_j is located on PM, 0 otherwise.

Network Traffic Minimization

Based on [16], the total network traffic among VMs is defined as the sum of average network traffic T_{jk} generated by each VM V_j, which is located to run on any PM, with other VMs V_k that are located to run on different PMs.

$$\min\ g_3(x) = \sum_{j=1}^{n} \sum_{k=1}^{n} (T_{jk} \cdot X_{jk}) \qquad (8)$$

where:

- $g_3(x)$: is the total network traffic between VMs;
- X_{jk}: A binary variable equals to 1 if V_j and V_k are located in different PMs and 0 otherwise.

The traffic between two VMs V_j and V_k which are located on the same PM do not contribute to increase the total network traffic given by Eq. (8).

Resource Utilization

To make effective use of the resources in all dimensions and balance the resource utilization on each server along different dimensions, we adopt the resource wastage model proposed in [15]. The unused resources available on each server may change considerably with different VM placement solutions. The following equation is adopted to compute the implied cost of wasted resources:

$$W_i = \frac{|Y_i^p - Y_i^m| + \epsilon}{Z_j^p + Z_j^m} \tag{9}$$

where:

- W_i: denotes the resource wastage of the i-th server;
- Y_i^p and Y_i^m represent the normalized CPU and memory resource usage (i.e., the ratio of used resource to total resource);
- Z_j^p and Z_j^m represent the normalized remaining CPU and memory;
- ϵ is a very small positive real number and its value is set to be 0.0001;

The related objective function of resource wastage can be expressed by:

$$\min \ g_4(x) = \sum_{i=1}^{m} W_i \tag{10}$$

Quality of Service Maximization

Quality of Service maximization is achieved when locating maximum number of VMs with highest SLA level of priority. We use the same equation proposed in [16] aiming to minimize SLA violations by using the highest level of priority SLA_i:

$$\min \ g_5(x) = \sum_{j=1}^{n} C^{SLA_j} \cdot SLA_j \cdot B_j. \tag{11}$$

where:

- $g_5(x)$: is the total QoS function for a given placement;
- C: is a constant, sufficiently large, to give priority to services with a higher SLA over those with a lower SLA. Otherwise, if the constant C had small value, $g_5(x)$ might choose a large number of VMs with a lower priority, which is not correct, considering the intended purpose in Eq. (11).

3.2 Constraints

PMs Capacity. A PM must have sufficient available resources to meet the requirements of all VMs. The capacity constraints can be mathematically formulated as:

$$h_1(x) \quad : \quad \sum_{j=1}^{m} V_j^{cpu} \cdot P_{ji} - S_i^{cpu} \leqslant 0 \tag{12}$$

$$h_2(x) \quad : \quad \sum_{j=1}^{m} V_j^{ram} \cdot P_{ji} - S_i^{ram} \leqslant 0 \tag{13}$$

$$h_3(x) \quad : \quad \sum_{j=1}^{m} V_j^{stor} \cdot P_{ji} - S_i^{stor} \leqslant 0 \tag{14}$$

SLA Provisioning. A virtual machine V_j with decisive SLA (i.e., $SLA_j = 1$) must certainly be located to run on a physical machine S_i. Consequently, this restriction is expressed as:

$$h_4(x) \quad : \quad \sum_{i=1}^{n} S_{ji} = 1 \quad \forall j \quad such \quad that \quad SLA_j = 1 \tag{15}$$

Where:

- SLA_j: Service Level Agreement $SLA_j = 1$ if V_j is critical, or 0 otherwise.

Unique Placement of VMs. A VM V_j should be located to run on a single PM. Alternatively, it could not be located in any PM if the associated SLA_j is not the highest level of priority. Consequently, this constraint is expressed as:

$$h_5(x) \quad : \quad \sum_{i=1}^{n} S_{ji} - 1 \leqslant 0 \tag{16}$$

where S_{ji} is a binary variable that equals 1 if V_j is located to run on PM; otherwise, it is 0.

3.3   Output Parameters

Solution should indicate the placement of each virtual machine V_j on the necessary physical machine S_i, considering the multi-objective optimization criteria. A placement is represented as a matrix $S = S_{ji}$ of dimension $(m \times n)$, where $S_{ji} \in \{0,1\}$ indicates if the virtual machine V_j is located ($S_{ji} = 1$) or not ($S_{ji} = 0$) for execution on a physical machine S_i (i.e., $S_{ji}: V_j \rightarrow S_i$).

4   Solving MODM Problem

4.1   Scalarization Method

The general multi objective decision making problem can be presented as follows:

$$\begin{cases} \quad g(x) = [g_1(x), g_2(x), g_3(x), g_4(x), g_5(x)] \\ Subject to \quad h_i(x) \quad with \quad i \in \{1,2,3,4,5\} \end{cases} \tag{17}$$

The idea of finding a solution for (18) would be challenging since a single point that minimizes all objective functions simultaneously often doesn't exist. In this context, we normalize each objective function by computing the normalized objective function cost $\widehat{g}_i(x)$ as:

$$\widehat{g}_i(x) = \frac{g_i(x) - g_i(x)_{min}}{g_i(x)_{max} - g_i(x)_{min}} \qquad (18)$$

Based on previous works [12], different scalarization methods can be used to solve the multi-objective optimization approaches as Benson's method, Weighted Tchebyshev method, Pascoletti-Serafini method, Weighted sum method (WS), Euclidean Distance (ED), etc. In [5], the efficient method for VMP placement was ED, while in [12], authors recommend the Tchebychev method, and as per [19] and [16], the often-used method is Weighted sum. In this paper, we consider the Conic Scalarization (CS) method detailed in [14]. The concept of this method is straightforward: (i) choosing preference parameters which consist of weights of objective functions and a reference point for these objectives and (ii) solving the scalar optimization problem.

Consider the multi-objective virtual machine placement that aim to simultaneously minimize the five objective functions as follows:

Minimise $g(x) = [g_1(x), g_2(x), g_3(x), g_4(x), g_5(x)]$

First, choose preference parameters:

- Weight vector $\omega = \omega_i$: denotes the importance degree associated to each i-th objective function for decision maker, where $i \in \{1, .., 5\}$.
- Reference point $r = (r_1, r_2, .., r_5)$: identified by decision maker to compute the minimal elements and can be chosen arbitrarily.

Second, choose an augmentation parameter α such that $(\omega, \alpha) \in \mathbb{C}$, where:
$\mathbb{C} = ((\omega_1, .., \omega_5), \alpha) : 0 \leqslant \alpha \leqslant \omega_i, \omega_i > 0, i = \{1, .., 5\}$
The scalar problem for the given parameters (ω, α) and r is:

$$\min \sum_{i=1}^{5} \omega_i(\widehat{g}_i(x) - r_i) + \alpha \sum_{i=1}^{5} |\widehat{g}_i(x) - r_i| \qquad (19)$$

In this work, the weight ω_i is constant $(\frac{1}{5})$. In the case where $\alpha = 0$, the scalarization method (20) becomes that of the weighted sum method.

4.2   Algorithms

This subsection presents the proposed alternatives used in our experiments to solve the VMP problem comparing the four online algorithms (FF, FFD, BF, BFD) against the offline algorithm ACO based memetic algorithm with VM migration.

- **First-Fit (FF):** This algorithm places VMs according to the first in first out (FIFO) basis where requested VMs are allocated on the first host with available resources [11].

- **First-Fit-Decreasing (FFD):** The FFD algorithm works in a similar way to the FF algorithm presented above. It aligns VMs in the decreasing order, then finds and places servers with available resources to place VMs [2].
- **Best-Fit (BF):** This algorithm assigns the VMs required on the first PM with the available capacity from a sorted list of PMs in ascending order by a rating associated to each PM [2].
- **Best-Fit-Decreasing (BFD):** This algorithm is similar to BF. The difference is only on selecting VM lists in decreasing order by inquired CPU resources [2].
- **Memetic Algorithm (MA):** The term Memetic Algorithm describes population based hybrid evolutionary algorithms that are coupled with local refinement strategies, more details are presented in [20].
- **Ant Colony Optimization (ACO):** The ACO based algorithm is introduced as an instance of the multi-dimensional bin-packing problem [17].

5   Test Environment

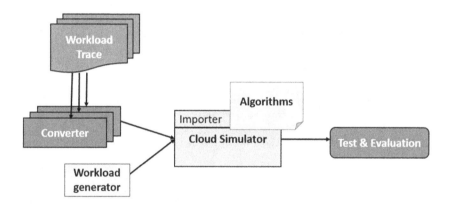

Fig. 3. Test environment

Experiments were conducted on a GNU Linux System with an Intel(R) Xeon(R) E3-1505M at 2.80 GHz CPU and 32 GB of RAM. Figure 3 shows an overview of the test environment. The proposed algorithms were developed by Java programming language. Unfortunately, it's not feasible (at least expensive) to test the VM placement based on real traces for a large-scale environment. In this paper, we use uncertain workload traces similar to real-world ones. We import data files of a particular format, to convert workload traces to the given format, and create test data in the given format. Physical resources comprise a diversified IaaS cloud, taking into account four categories of physical machines (i.e., small, medium, large, and xlarge), as presented in Table 1.

In this paper, we compare different algorithms in online and offline environments for various CPU load types. Simulation results presented in Table 2 prove the performance of BFD in both medium and high CPU load compared to other dynamic placement algorithms, while BF is the best one for low CPU load. On the other hand, one can see the ACO algorithm yields good results for all CPU load sizes. We can conclude that ACO based memetic algorithm is better then deterministic algorithms, otherwise the offline placement outperforms the online one. In Fig. 4, we compare the results of the conic scalarization method for MODM problem, considering different values of the augmentation parameter α (0.01, 0.05, 0.1, 0.15, 0.2) less than $\omega = \frac{1}{5}$. We see that the more α tends towards ω, the better results are obtained for all algorithms.

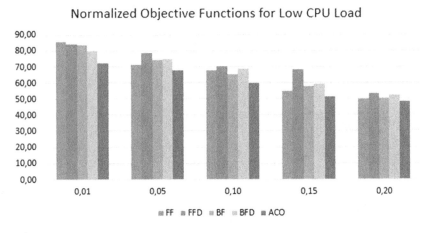

Fig. 4. Normalized objective functions for different values of augmentation parameter α (scale x100)

As a perspective, in the future work, we propose to combine the online and offline placement, and compare the CS method to other scalarization methods.

Table 1. Types of physical machines

PM characteristics	CPU	RAM	Network	$P_{max}(W)$
S.small	32	128	1000	500
S.medium	64	256	1000	1000
S.large	256	512	1000	2000
S.xlarge	512	1024	10000	4000

Table 2. Normalized objective functions of evaluated algorithms

Algorithm	Low CPU load	Medium CPU load	High CPU load
First-Fit	0.675	0.733	0.972
First-Fit decreasing	0.703	0.742	0.965
Best-Fit	**0.653**	0.723	0.846
Best-Fit decreasing	0.685	**0.721**	**0.837**
ACO	**0.595**	**0.654**	**0.683**

6   Conclusion

We consider the VMP problem as a multi-objective decision-making approach that aims to optimize simultaneously five objective functions: energy consumption, cost optimization, network traffic, resource utilization, and QoS. The optimization problem is solved based on a conic scalarization method. Simulation results show that ACO based-memetic gives good results compared to FF, BF, FFD, and BFD. However, offline algorithms can't be used in a pure dynamic environment. Therefore, our future will consider a combination of both offline and online placement algorithms.

References

1. Akintoye, S.B., Bagula, A.: Improving quality-of-service in cloud/fog computing through efficient resource allocation. Sensors **19**(6), 1267 (2019)
2. Atchukatla, M.S.: Algorithms for efficient VM placement in data centers: cloud based design and performance analysis. Master's thesis, Department of Computer Science and Engineering (2018)
3. Attaoui, W., Sabir, E.: Multi-criteria virtual machine placement in cloud computing environments: a literature review. arXiv:1802.05113 (2018)
4. Beloglazov, A., Abawajy, J., Buyya, R.: Energy-aware resource allocation heuristics for efficient management of data centers for cloud computing. Future Gener. Comput. Syst. **28**(5), 755–768 (2012). Special Section: Energy Efficiency in Large-Scale Distributed Systems
5. Chamas, N., López-Pires, F., Baran, B.: Two-phase virtual machine placement algorithms for cloud computing: an experimental evaluation under uncertainty. In: 2017 XLIII Latin American Computer Conference (CLEI), pp. 1–10, September 2017
6. Chen, T., Zhu, Y., Gao, X., Kong, L., Chen, G., Wang, Y.: Improving resource utilization via virtual machine placement in data center networks. Mob. Netw. Appl. **23**(2), 227–238 (2018)
7. Choi, J.Y.: Virtual machine placement algorithm for energy saving and reliability of servers in cloud data centers. J. Netw. Syst. Manage. **27**(1), 149–165 (2019)
8. Cohen, R., Lewin-Eytan, L., Seffi Naor, J., Raz, D.: Almost optimal virtual machine placement for traffic intense data centers. In: 2013 Proceedings IEEE INFOCOM, pp. 355–359, April 2013

9. Dias, D.S., Costa, L.H.M.K.: Online traffic-aware virtual machine placement in data center networks. In: 2012 Global Information Infrastructure and Networking Symposium (GIIS), pp. 1–8, December 2012

10. Dong, J., Wang, H., Jin, X., Li, Y., Zhang, P., Cheng, S.: Virtual machine placement for improving energy efficiency and network performance in IaaS cloud. In: 2013 IEEE 33rd International Conference on Distributed Computing Systems Workshops, pp. 238–243, July 2013

11. Fang, S., Kanagavelu, R., Lee, B., Foh, C.H., Aung, K.M.M.: Power-efficient virtual machine placement and migration in data centers. In: 2013 IEEE International Conference on Green Computing and Communications and IEEE Internet of Things and IEEE Cyber, Physical and Social Computing, pp. 1408–1413, August 2013

12. Gopu, A., Venkataraman, N.: Optimal VM placement in distributed cloud environment using MOEA/D. Soft. Comput. **23**(21), 11277–11296 (2019)

13. Gupta, M.K., Jain, A., Amgoth, T.: Power and resource-aware virtual machine placement for IaaS cloud. Sustain. Comput. Inform. Syst. **19**, 52–60 (2018)

14. Kasimbeyli, R., Ozturk, Z.K., Kasimbeyli, N., Yalcin, G.D., Icmen, B.: Conic scalarization method in multiobjective optimization and relations with other scalarization methods. In: Le Thi, H.A., Pham Dinh, T., Nguyen, N.T. (eds.) Modelling, Computation and Optimization in Information Systems and Management Sciences. AISC, vol. 359, pp. 319–329. Springer, Cham (2015). https://doi.org/10.1007/978-3-319-18161-5_27

15. Kumar, D., Mandal, S.K.: Multi-objective virtual machine placement using improved teaching learning based optimization in cloud data centers. Int. J. Appl. Eng. Res. **12**, 10809–10815 (2017)

16. López-Pires, F., Barán, B.: Many-objective virtual machine placement. J. Grid Comput. **15**(2), 161–176 (2017)

17. Malekloo, M., Kara, N.: Multi-objective aco virtual machine placement in cloud computing environments. In: 2014 IEEE GLOBECOM Workshops (GC Workshops), pp. 112–116, December 2014

18. Meng, X., Pappas, V., Zhang, L.: Improving the scalability of data center networks with traffic-aware virtual machine placement. In: 2010 Proceedings IEEE INFOCOM, pp. 1–9, March 2010

19. Mishra, S., Sangaiah, A.K., Sahoo, M.N., Bakshi, S.: Pareto-optimal cost optimization for large scale cloud systems using joint allocation of resources. J. Ambient Intell. Hum. Comput. (2019). https://doi.org/10.1007/s12652-019-01601-x

20. Urselmann, M.: Derivative-free chemical process synthesis by memetic algorithms coupled to aspen plus process models. In: Kravanja, Z., Bogataj, M. (eds.) 26th European Symposium on Computer Aided Process Engineering. Computer Aided Chemical Engineering, vol. 38, pp. 187–192. Elsevier (2016)

21. Zhang, J., He, Z., Huang, H., Wang, X., Gu, C., Zhang, L.: SLA aware cost efficient virtual machines placement in cloud computing. In: 2014 IEEE 33rd International Performance Computing and Communications Conference (IPCCC), pp. 1–8, December 2014

22. Zhao, L., Lu, L., Jin, Z., Yu, C.: Online virtual machine placement for increasing cloud provider's revenue. IEEE Trans. Serv. Comput. **10**(2), 273–285 (2017)

A Markovian Approach for Improving End-to-End Data Rates in the Internet

Marine Ségneré-Yter, Olivier Brun$^{(\boxtimes)}$, and Balakrishna Prabhu

LAAS-CNRS, Université de Toulouse, CNRS, Toulouse, France
brun@laas.fr

Abstract. We model the tradeoff between the monitoring costs and gain in throughput for overlay-based routing in the Internet. A Markovian model is shown to fit the real throughput traces quite well. The tradeoff problem is formulated as Markov decision process and it is observed that the myopic policy that maximizes the immediate utility is close to optimal on the real traces.

1   Introduction

More than two decades ago, it was observed that the performance of network flows could be improved by choosing other paths than those computed by IP routing protocols (see, e.g., [7]). Routing overlay networks were then proposed as a solution for achieving spectacular performance improvements, without the need to re-engineer the Internet (see [1] and references therein). An overlay network is composed of Internet end-hosts which can monitor the quality of Internet paths between themselves by sending probe packets. Since all pairs of nodes are connected, the default topology of a routing overlay is that of a complete graph. Although the monitoring cost is highly variable depending on the metric to be probed, it is usually not possible to discover an optimal path by probing all links in large overlay networks (see [2] for a graph-theoretic analysis of this issue). An alternative approach is to devise a parsimonious monitoring approach making the trade-off between the quality of routing decisions and the monitoring cost. Given a source and a destination node in the overlay, the idea is to probe only a small number of overlay paths between the two nodes at each measurement epoch, but to choose those paths so as to make the best routing decision.

Assuming known Markovian models for path delays, this trade-off problem was formulated as a Markov Decision Process (MDP) in [8]. Using delay data collected over the Internet, it was shown that the optimal monitoring policy enables to obtain a routing performance almost as good as the one obtained when all paths are always monitored, but with a modest monitoring effort.

In this paper, we adopt the theoretical framework introduced in [8], but focus on data throughput rather than RTT. We note that efficient parsimonious monitoring strategies are even more important for the throughput metric. Indeed, although lightweight methods for estimating the available bandwidth between

S. Lasaulce et al. (Eds.): NetGCooP 2021, CCIS 1354, pp. 167–171, 2021.
https://doi.org/10.1007/978-3-030-87473-5_15

two Internet end-hosts were proposed in [4,5], in practice the only accurate method is to transfer a large file between the two endpoints. It turns out that the MDP formulation for maximizing the data throughput is equivalent to the MDP formulation for minimizing the RTT. The contribution of the present paper is therefore not on the theoretical side, but rather to investigate the applicability of the approach proposed in [8] for optimizing throughput in overlay networks. To this end, we use we use throughput measurements that were made between 9 AWS (Amazon Web Services) data centres.

2   MDP Formulation

The problem formulation in this section is essentially the same as in [6,8] except that the quantity of interest is bandwidth instead of delay. Consider a single origin-destination pair and $\{1, 2, \ldots, P\}$ a set of P paths between the origin and the destination. The network topology is thus that of parallel links. At time step t, path i is assumed to be have a bandwidth $X_i(t)$, where $X_i(t)$ is a discrete-time Markov chain taking values in a finite set. The transition matrix for path i will be denoted by M_i.

At each time step, the routing agent has to decide on which path it should send data. For this, the agent has at its disposal the last observed bandwidth for each path. Further, it can choose to measure the bandwidth on one or more paths and update its state information before taking the routing decision. The agent incurs a cost of c_i for probing path i independently of time step. The decision-maker must find a compromise between paying to retrieve information from the system to get a higher bandwidth and not retrieving information leading to a lower bandwidth.

Let $u(t) \in \{0, 1\}^P$ whose ith component indicates whether path i is monitored in time step i or not. The total cost paid for action $u(t)$ is $\sum_{i|u_i(t)=1} -c_i = -\mathbf{c} \cdot \mathbf{u}(t)$ with $\mathbf{c} = (c_1, \cdots, c_P)$. Let $r(t)$ be the path chosen in time step t. A policy θ can be defined by the sequence $\{(\mathbf{u}(t), r(t))\}_{t \geq 0}$. Just as in [8], it can be seen that knowing only the last observed bandwidth for a link is not enough to determine the distribution the bandwidth that will be obtained in a given step. The state can be made Markovian by incorporating the age of the last observed bandwidth as well. That is, the pair $(y_i(t), \tau_i(t))$, where $y_i(t)$ is the last observed bandwidth of link i at time t and $\tau_i(t)$ is the age of the last observation is sufficient as the state variable for a Markovian representation of path i. All this information is summarized in a vector $\mathbf{s}(t) = (s_1(t), s_2(t), \ldots, s_P(t))$ where $s_i(t) = (y_i(t), \tau_i(t))$.

Since the state is now Markovian, the problem can be formulated as a Markov Decision Process (MDP). This MDP can be further simplified by noting that in the model, the routing decision does not have any impact on the evolution of the state. Thus, a locally greedy routing decision conditioned on $\mathbf{u}(t)$ and the current state will be optimal. In other words, for a given $\mathbf{u}(t)$, it will be optimal to choose the path that maximizes the expected bandwidth. With this in mind, the decision problem can be reduced to determining which paths to monitor in each time step. For a given state $s \equiv (y, \tau)$ of path i, define the belief on the bandwidth being z of this path as follows: $b_i(z|s) := \mathbb{P}(X_i(\tau) = z|X_i(0) = y)$, which is just the probability of path i transitioning from y to z in τ steps, and can be computed by choosing the corresponding element of M_i^τ.

If path i is measured, then its actual bandwidth, $X_i(t)$, will be known and can be used in the routing decision. Otherwise, it is its expected conditional bandwidth $\mathbb{E}[X_i|s_i] = \sum_{x \in \mathcal{X}_i} x \cdot b_i(x|s_i)$ that will be used. The locally greedy routing decision will be to choose $r(t)$ that maximizes $(u_i X_i + (1 - u_i)\mathbb{E}[X_i|s_i])$. Note that this decision is taken after performing the action of monitoring the subset of selected links. This leads to maximum bandwidth conditioned on $\mathbf{s}$ and $\mathbf{u}$ of $B(\mathbf{X}|\mathbf{s}, \mathbf{u}) = \max_i (u_i X_i + (1 - u_i)\mathbb{E}[X_i|s_i])$, and an expected maximum bandwidth of:

$$\bar{B}(\mathbf{s}; \mathbf{u}) = \sum_{\mathbf{x}} \left(\prod_{i=1}^{P} b_i(x_i|s_i) \right) B(\mathbf{x}|\mathbf{s}; \mathbf{u}). \tag{1}$$

Here the product measure is used because X_is evolve independently.

Now that the routing decision is known, the final MDP takes the form:

$$\max_{\theta} \mathbb{E}_{\mathbf{s}_0}^{\theta} \left\{ \sum_{t=0}^{\infty} \rho^t \left[\bar{B}(\mathbf{s}(t); \mathbf{u}(t)) - \mathbf{c} \cdot \mathbf{u}(t) \right] \right\}. \tag{2}$$

where $\theta \equiv \mathbf{u}(t)_{t \geq 0}$, is limited to monitoring decisions only.

We remark that the above problem formulation resembles the multi-armed bandit (MAB) framework. However, unlike standard MABs in which the cost function is decomposable in the individual costs of the bandits, in our problem the overall cost is not decomposable.

3   Numerical Results

In order to validate our approach on real data, for which the Markovian assumption is not perfectly met, we use throughput measurements that were made between 9 AWS (Amazon Web Services) data centres located in several places around the world. In summer 2015, we measured the available throughput between all pairs of data centres every five minutes, by transferring a 10 MB file through the Internet, for a period of four days. We thus collected some 8.3×10^4 measurement data over the 4 days period. Assuming that the available throughput over a path is the minimum of the throughputs of its constituent links, the analysis of these data revealed that the IP route is the maximum throughput route only in 23% of the cases, and that most of the time, the maximum throughput overlay route passes through 1 or 2 intermediate nodes (see [1] for details).

We selected three origin-destination (OD) pairs: Virginia/Ireland, Virginia/Frankfurt and Frankfurt/Tokyo. For the first two pairs, in addition to the IP path, we selected two alternative paths which were sometimes better than the IP path, whereas for the last example there was one alternative path.

For each path, we fitted a Markov model using a clustering method called, Hierarchical Agglomerative Clustering [3]. This method creates a hierarchy between clusters, like a tree. At the beginning, each value of bandwidth is a cluster. The algorithm agglomerate one by one the closest data (in term of a distance metric chosen) together in a new cluster, until it creates one big cluster. On our data, we use the Euclidean distance between the bandwidth values. After that, we decided where to *cut* the tree and obtain a certain number of clusters.

Now that we have our different states, we have to determine the transition probability matrices, P_i. We elaborate this matrix by counting the number of transition between each pair of states. Finally, we search the minimum value τ_{max_i} which satisfy $\max(P_i^{\lim} - P_i^{\tau_{max_i}}) < 10^{-2}$. It appears that, on real data, the τ_{max} per link is lower than 10 and the number of states per link is between 2 and 12.

We evaluate the average utility (see (2)) for four policies: optimal, myopic policy that optimizes the immediate cost only, a receding horizon policy (with a horizon of 3) and a decomposition based heuristic. For a description of the last two policies, we refer the reader to [6].

First, we check that the Markov models we fitted are representative of the real traces. For this, for each OD pair, using the transition matrices, we generate a sample path of throughputs on each of the paths. On these sample paths, we apply the three heuristics (but not the optimal) and compute the average utility for each policy. We then apply the policies on the real traces and compute the average utilities. Table 1 shows the percentage relative error between the average utility computed on a sample path and that on the corresponding real trace. The relative error is less than 2% which indicates a good match. Finally, Table 2 shows the utilities of the four policies for varying monitoring costs. One surprising observation from these examples is that the myopic policy is almost optimal.

Table 1. Percentage relative error between utility computed using Markov model and on real trace.

(a) Virginia/Ireland.

c	1	2	3	4	5
Myopic	1.38	1.14	0.52	0.95	0.01
RH	1.38	0.98	0.50	0.62	0.01
H2	1.73	1.81	1.84	1.90	1.97

(b) Virginia/Frankfurt.

c	1	2	3	4	5
Myopic	1.36	0.91	0.6	1.07	0.03
RH	1.34	1.046	0.62	1.067	0.03
H2	1.63	1.90	1.92	1.97	2.06

Table 2. Utilities for different policies as a function of the monitoring cost.

(a) Virginia/Ireland.

c	OPT	Myopic	RH	H2
1.0	41.98	41.98	41.98	40.45
2.0	39.42	39.40	39.43	38.72
3.0	37.90	37.89	37.90	36.98
4.0	36.99	36.92	36.97	35.25
5.0	36.43	36.43	36.43	33.52

(b) Virginia/Frankfurt.

c	OPT	Myopic	RH	H2
1.0	41.91	41.91	41.91	40.77
2.0	39.45	39.47	39.45	38.84
3.0	38.02	38.03	38.03	36.97
4.0	37.11	37.11	37.11	35.13
5.0	36.38	36.36	36.36	33.30

(c) Frankfurt/Tokyo.

c	OPT	Myopic	RH	H2
1.0	58.44	58.44	58.44	57.83
2.0	57.71	57.67	57.71	56.59
3.0	57.20	57.20	57.20	55.35
4.0	57.20	57.20	57.20	54.11
5.0	57.20	57.20	57.20	53.09

4   Conclusion and Future Work

The results indicate that Markovian models are a good fit for throughput on paths in the Internet. Further, a myopic policy is nearly optimal for minimizing a linear combination of the throughput and monitoring costs.

As future work, we would first like to understand why the myopic policy works well on these examples. It would be interesting to obtain conditions under which this is true. Next, we would like to generalize these models to multi-agent settings in which each node of the overlay can be seen as an agents. These agents can be either cooperative or be non-cooperative. Another possible improvement of the setting would be to allow the routing decision to influence the future evolution of the bandwidth of the path and to get state information from the current routing decision.

Acknowledgment. The work of Marine Ségneré was partially funded by a contract with Direction Générale de l'Armement (DGA), France.

References

1. Brun, O., Hassan, H., Vallet, J.: Scalable, self-healing, and self-optimizing routing overlays. In: IFIP Networking, Vienna, pp. 64–72 (2016)
2. Thraves Caro, C., Doncel, J., Brun, O.: Optimal path discovery problem with homogeneous knowledge. Theory Comput. Syst. **64**(2), 227–250 (2020)
3. James, G., Witten, D., Hastie, T., Tibshirani, R.: An Introduction to Statistical Learning: With Applications in R. Springer, New York (2014). https://doi.org/10.1007/978-1-4614-7138-7
4. Johnsson, A., Melander, B., Björkman, M.: On the analysis of packet-train probing schemes. In: Proceedings of the International Conference on Communication in Computing, Las Vegas, USA, June 2004
5. Melander, B., Bjorkman, M., Gunningberg, P.: A new end-to-end probing and analysis method for estimating bandwidth bottlenecks. In: Global Telecommunications Conference, GLOBECOM 2000, vol. 1, pp. 415–420. IEEE (2000)
6. Mouchet, M.: Scalable monitoring heuristics for improving network latency. In: IEEE/IFIP Network Operations and Management Symposium (IEEE/IFIP NOMS 2020), Budapest, Hungary, April 2020
7. Savage, S., Collins, A., Hoffman, E., Snell, J., Anderson, T.: The end-to-end effects of internet path selection. SIGCOMM Comput. Commun. Rev. **29**(4), 289–299 (1999)
8. Vaton, S., et al.: Joint minimization of monitoring cost and delay in overlay networks: optimal policies with a Markovian approach. J. Netw. Syst. Manag. (JONS) **27**(1), 188–232 (2019)

DNN Based Beam Selection in mmW Heterogeneous Networks

Deepa Jagyasi$^{(\boxtimes)}$ and Marceau Coupechoux

LTCI, Telecom Paris, Institut Polytechnique de Paris, Paris, France
{deepa.jagyasi,marceau.coupechoux}@telecom-paris.fr

Abstract. We consider a heterogeneous cellular network wherein multiple small cell millimeter wave (mmW) base stations (BSs) coexist with legacy sub-6GHz macro BSs. In the mmW band, small cells use multiple narrow beams to ensure sufficient coverage and User Equipments (UEs) have to select the best small cell and the best beam in order to access the network. This process usually based on exhaustive search may introduce unacceptable latency. In order to address this issue, we rely on the sub-6GHz macro BS support and propose a deep neural network (DNN) architecture that utilizes basic components from the Channel State Information (CSI) of sub-6GHz network as input features. The output of the DNN is the mmW BS and beam selection that can provide the best communication performance. In the set of features, we avoid using the UE location, which may not be readily available for every device. We formulate a mmW BS selection and beam selection problem as a classification and regression problem respectively and propose a joint solution using a branched neural network. The numerical comparison with the conventional exhaustive search results shows that the proposed design demonstrate better performance than exhaustive search in terms of latency with at least 85% accuracy.

Keywords: Millimeter wave · Beam selection · Deep neural network · Heterogeneous network · sub-6GHz

1 Introduction

Millimeter Wave (mmW) communication is considered as a promising technique to solve the unprecedented challenge of increasing demand for high data rates in future cellular networks. However, it suffers from limited coverage and in the ultra-dense environment it is significantly prone to blockages such as high density objects like walls, glass, humans, etc. Thus, in-order to provide flexible coverage and minimize the infrastructural cost, it is proposed that mmW networks will be deployed in a multi-tier heterogeneous network, where multiple small cell mmW base stations (BSs) coexist with multiple legacy sub-6GHz macro BSs [9].

Supported by the EXTRANGE4G project with company ETELM funded by DGA. This work has been performed at LINCS laboratory.

S. Lasaulce et al. (Eds.): NetGCooP 2021, CCIS 1354, pp. 172–184, 2021.
https://doi.org/10.1007/978-3-030-87473-5_16

The legacy network operating in sub-6GHz frequencies can handle operations like resource allocation, mobile data offloading, control signalling etc., while the potential mmW BSs can handle massive data traffic [9,13]. In this paper, we propose a solution for optimal resource allocation for a heterogeneous cellular network that enables reliable communication while leveraging the benefits of high data rates from mmW bands.

Beamforming is important in mmW systems in order to overcome the path loss due to shorter wavelength. With the large number of antenna elements associated with mmW transceivers, multiple beams are possible, which can perform directional beamforming and achieve high gain. Thus to ensure high performance, choosing the suitable BS to user equipment (UE) beam-pair from the set of all the possible directional beams is a crucial task. Beam selection has been conventionally addressed using exhaustive search or multi-level selection approach as in [12,15]. However, with these techniques, large number of beams at mmW BSs leads to large beam training overhead and hence unacceptable latency to access the mmW network. Access latency is in turn significantly lower in case of communication at sub-6GHz frequencies. To overcome this challenge, out-of-band spatial information has been used for reducing beam-selection overhead [1]. In recent years, in order to predict the optimal beam and significantly overcome the training overhead, the use of deep learning (DL) and machine learning (ML) tools has proved to be very promising in establishing mmW links [3]. In this paper, we thus propose a deep neural network (DNN)-based mmW BS and beam selection for heterogeneous network by utilizing basic features from the Channel State Information (CSI) available only at sub-6GHz BSs.

DL and ML techniques have been hugely explored for various communication applications which include, channel estimation, design of auto-encoders, spectrum allocation, etc. [7]. In the context of mmW communications, such techniques have been reported for applications such as beam selection, blockage detection, channel estimation, or proactive handover. Various DL and ML techniques to reduce the beam selection overhead in mmW communications use location information, channel information, out-of-band information or measurements from different sensors such as LIDAR, camera, or GPS. Specifically, authors in [6] and [5] have proposed the use of deep convolutional neural networks to perform beam selection task in distributed and centralized architecture respectively. In [11], authors have considered the use of situational knowledge about the environment and location of UEs and proposed the use of ensemble learning-based classification to identify the optimal mmW beam. Later, in [4], authors proposed the applicability of deep learning techniques such as k-nearest neighbours (KNN), support vector classifier (SVC) and multi-layer perceptron by using angle of arrival information to perform the beam-selection task. All these works however assume single-layer networks and ignore the macro-layer of sub-6GHz BSs that will be required for a continuous connectivity. Only two references are dealing with ML/DL-based beam selection in heterogeneous networks [3,14]. Authors in [14] have considered the CSI over sub-6GHz and kernel-based ML algorithms to assist handovers for target vehicle discovery problem and overcome coverage blindness. In [3], authors have proposed the use of sub-6GHz channel and location information for performing the beam-selection and blockage prediction task. However, the solution in [3] is limited to a single BS - single UE communication scenario, where the BS employs co-located sub-6GHz and mmW transceivers. In this paper, we extend the work done in [3] by considering multiple coordinating sub-6GHz and mmW BSs to perform resource allocation for each UE in the network.

Furthermore, location is an important feature that independently can be utilized to perform the task. Most of the previously discussed work on beam selection including [3], considers the availability of the UE location. However, this information may not always be readily available for many cellular devices. Also, location sensors usually has low accuracy and can result in incorrect outputs [8]. Hence, we aim to intentionally eliminate the availability of location information from the set of input features and design the proposed DNN based BS and beam selection framework for a heterogeneous mmW network.

The main contributions of this paper are listed as follows:

1. To guarantee reliable communication and enhanced coverage in mmW communication, we consider the heterogeneous architecture and propose DNN-based BS and beam selection by leveraging basic signal components extracted from the sub-6GHz channel as the input features. We consider multiple coordinated sub-6GHz BSs for optimal mmW resource allocation in order to serve any UE in the network.
2. We propose a branched DNN-structure, which divides the problem into two sub-problems of BS selection and beam selection and is well-adapted for this application.
3. We eliminate the use of location information from the set of input features to perform the considered task. The feature vector considered as input to the network include: the azimuth and elevation angle of arrival (AoA) from the BS, the receive signal power, the signal phase and the propagation delay.

The remainder of this paper is organized as follows. Section 2 describes the network and transceiver model. The proposed problem is formulated in Sect. 3 and then the deep neural network model is discussed in Sect. 4. Section 5 presents the simulation environment and performance evaluation and finally Sect. 6 concludes the paper.

Notations: Throughout this paper, we use bold-faced lowercase letters to denote column vectors and bold-faced uppercase letters to denote matrices. For any matrix $\mathbf{X}$, $\mathbf{X}^T$ denotes the transpose operation.

2   System Model

We consider a heterogeneous cellular network wherein multiple sub-6GHz BSs and mmW BSs operate together in order to serve UEs in the network as shown in the Fig. 1. We assume that there are B_μ sub-6GHz BSs, each equipped with

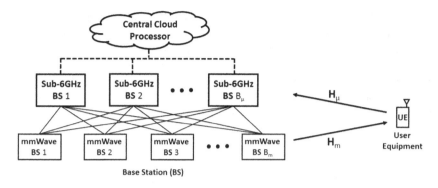

Fig. 1. System model: heterogeneous network architecture with mmW small cells coexisting with sub-6GHz macro BSs. Dashed lines represent the connection of coordinating sub-6GHz BSs with a central cloud processor whereas solid lines represent the connection between any sub-6GHz and mmW BS in a network.

N_μ antenna elements. All the sub-6GHz BSs operate in a coordinated manner for their processing such as channel estimation or precoder design, along with DNN computations being performed at a central cloud processor unit. We assume that there are B_m mmW BSs distributed in the network region that are coordinated with the sub-6GHz BSs to provide high speed data transfer to the UEs in the network. Each of the mmW BS is assumed to be equipped with N_m transceiver antennas. We assume that UEs have a single antenna in both bands[1].

The communication scenario that we study is as follows. A UE is initially connected to a sub-6GHz BS and periodically transmits pilot signals to all macro BSs. Whenever the UE is approaching towards mmW BSs, the coordinated sub-6GHz BSs command the best mmW BS and the best beam that maximizes the achievable rate for this user.

Based on this scenario, the signal received by the macro sub-6GHz BSs at the k-th OFDM sub-carrier, $k = 1, 2, \cdots, K$ can be given by:

$$\mathbf{y}_\mu[k] = \mathbf{h}_\mu[k]d_s + \mathbf{n}_\mu[k], \tag{1}$$

where d_s is the uplink pilot transmitted over the $\mathbf{h}_\mu$ sub-6GHz channel gain matrix and $\mathbf{n}_\mu$ is the additive Gaussian noise vector with zero-mean and covariance matrix $\sigma_\mu^2 \mathbf{I}$ at the sub-6GHz BS antenna arrays. The processing at the sub-6GHz is performed in the baseband domain as the macro BSs are assumed to employ fully-digital architecture.

However, due to the high cost and power consumption of mixed signal RF components at mmW frequencies, mmW transceivers are assumed to employ either fully-analog architecture where the transceiving unit is associated with

---

[1] UEs may be equipped with several antennas but we don't address in this paper the beam alignment problem and we focus on the beam selection at the BS. Once the BS beam is known, the UE may for example perform exhaustive search to select its own beam.

single RF chain or it employs hybrid analog-digital architecture with a number of RF chains less than N_m. In this work, mmW BSs adopt fully-analog beam-forming architecture where, at a given time instant, the signal is transmitted through a single beam which is selected from a finite set $\mathcal{V}$ of M predefined beams, where $\mathcal{V}$ is the codebook. The total transmit power at the mmW BS is P_T. Thus for the downlink transmission, where the mmW BS communicates with the UE, the signal received at the UE can be given as:

$$\mathbf{y}_m[k] = \mathbf{H}_m[k]\mathbf{v}_m[k]d_m + \mathbf{n}_m[k], \tag{2}$$

where $\mathbf{H}_m$ is the mmW channel gain matrix, $\mathbf{v}_m$ is the beamforming vector, d_m is the data transmitted by the mmW BS and $\mathbf{n}_m$ is the additive Gaussian noise at UE with zero-mean and covariance matrix $\sigma_m^2 \mathbf{I}$.

We assume that the mmW channel is modelled as a geometric channel [2] which can be given as:

$$\mathbf{H}_m[k] = \sum_{l=1}^{L} \sqrt{\frac{\rho_l}{K}} e^{j(\kappa_l + \frac{2\pi k}{K}\Gamma_l B_m)} \mathbf{a}(\theta_l, \phi_l) \tag{3}$$

where $\sqrt{\frac{\rho_l}{K}}$ is the path gain for the l-th channel path in the k-th OFDM sub-carrier and κ_l and Γ_l represents the path phase and propagation delay for the l-th channel path respectively. L is the total number of channels paths. The array response vector at the BS is denoted by $\mathbf{a}(\theta_l, \phi_l)$, where θ_l and ϕ_l is the azimuthal and the elevation AoA respectively. The detailed study of the utilized channel model can be obtained in [2]. The sub-6GHz channel is modelled in the same way.

3   Problem Formulation

Given the uplink channel information at sub-6GHz BSs, we aim at designing an optimal mmW BS and beam predictor such that it maximizes the achievable sum-rate for each user in the network. Thus the optimal beamforming vector $\mathbf{v}_m^o$ can be obtained as:

$$\mathbf{v}_m^o = \arg \max_{\mathbf{v}_m \in \mathcal{V}} \sum_{k=1}^{K} \log_2 \left(1 + \gamma |\mathbf{H}_m[k]^T \mathbf{v}_m|^2\right) \tag{4}$$

where $\gamma = P_T/K\sigma_m^2$. To design this optimal predictor we aim to find a map-ping from sub-6GHz channel to mmW BS and beam selection. [3] has shown that, under the assumption that there is a bijective mapping between sub-6GHz channel and user location, there also exists a bijective mapping between sub-6GHz channel and mmW channel. Motivated by this result, we rely on sub-6GHz channel features to deduce the resources in mmW band. We can thus define two mapping functions ζ_{BS}, ζ_b as follows:

$$\zeta_{BS} : \mathbf{f}_\mu \rightarrow \mathcal{P}_{BS} \tag{5}$$

$$\zeta_b : \mathbf{f}_\mu \rightarrow \mathbf{r}_b \tag{6}$$

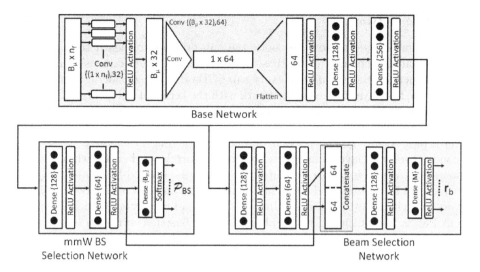

Fig. 2. Deep Neural Network (DNN) model for optimal mmW BS and beam selection.

where $\mathbf{f}_\mu$ is a feature vector of size n_f extracted from the CSI in the sub-6GHz band, $\mathcal{P}_{BS}$ is a probability mass function on the set of mmW BSs and $\mathbf{r}_b$ is a vector of achievable rates for every possible beam out of M at a mmW BS. To find this mapping, we utilize the DNN-based approach which are well-suited for obtaining the non-linear relationships between different data distributions [10]. The ζ_{BS} mapping is formulated as a classification problem, in which each input feature is mapped into a finite set of labels; each label representing candidate mmW BSs, while ζ_b mapping is obtained by solving this sub-problem as a regression task wherein, a real valued achievable rate is obtained for each beam for the selected BS from ζ_{BS} mapping. The proposed DNN based solution is presented in details in Sect. 4.

4 Deep Neural Network Model

In this section, we discuss the DL model adopted to learn the mapping from sub-6GHz channel information to mmW-BS identifier and its beam for a given user. In an environment with multiple mmW BSs and large number of beams, it is important to have flexibility in the network to incorporate new BSs or beams for future requirements. To allow this scalability, the overall beam selection problem can be divided into two sequential sub-problems: optimal mmW BS selection and then optimal beam selection. In order to incorporate the two sub-problems in a single neural network, we consider a branched network which takes feature vectors $\mathbf{f}_\mu$ from sub-6GHz CSI as input and predicts both mmW BS and beams for that user as shown in Fig. 2.

4.1 Base Network

We consider a base network for both the sub-problems to learn the common feature vectors. The input of this base network is a matrix of dimension $B_\mu \times n_f$ which gathers all the features for every sub-6GHz BS. We consider a convolution layer as the first layer of the base network with the kernel of size $1 \times n_f$. This layer acts as a shared weight perceptron layer which is intended to find the correlation within the feature vector of each coordinating sub-6GHz BS. The output of this layer is passed through another convolution layer having kernel size $B_\mu \times 32$. The second convolution layer is intended to learn the correlation between the different macro BSs. We then flatten the output and pass the learned features through a stack of two fully-connected dense layers of size 128 and 256 respectively. All the layers are with Rectified Linear Unit (ReLU) non-linearity activation function as in Fig. 2. The output of the final layer of the base network is branched into two sub-networks that are designed to solve each of the sub-problem of mmW BS and beam selection as discussed in following subsections.

4.2 mmW BS Selection Network

This sub-network is designed to predict the optimal mmW BS in order to serve the desired UE in the communication area. The input to this network are the features learned from the base network. This input vector is further passed through two fully-connected dense layers of size 128 and 64 respectively, for the optimal BS selection specific feature learning. These learned features are then projected onto the B_m feature space using a final dense layer. The output of this layer is then fed to a softmax activation which results in a probability distribution over the number of mmW BSs. The BS with the highest probability is selected as the optimal BS.

4.3 Beam Selection Network

The beam selection sub-network utilizes the learned features from the base network in order to predict the best beam. We incorporate two fully-connected dense layers of size 128 and 64 respectively, each of which is followed by ReLU activation. Moreover as selection of the best beam also depends or gets impacted by the selected BS, we concatenate the feature from the hidden layer of the BS selection network with the output of the previous dense layer from this network as depicted in Fig. 2. These concatenated features provide added information and hence result in better performance. The output of this layer is further passed through a fully-connected dense layer of size 128, to learn the correlation within the concatenated features. Finally, we project these learned features to M dimensional space and pass it though a ReLU activation layer to get the regression output for the achievable sum-rate at each beam. The index with maximum sum-rate value is the selected beam for the selected BS from the mmW BS selection network.

4.4   Discussions

The proposed branched neural network architecture has been obtained after experimenting several DNN configurations. In this section, we discuss these experimented models and provide reasons for adapted changes in the final DNN model. We initially considered a multi-layer sequential DNN with single output vector. We took a concatenated vector of features from all sub-6GHz BSs as input and expected a single vector of achievable rates for each beam at each mmW BS as an output. Though this network architecture is simple and performs the task directly, we observed that this network show large variations for small changes in the environment. Moreover, when the number of mmW BSs increases, the number of output nodes increases dramatically and the system thus requires extensive training to achieve good performance.

To overcome this issue, we adopted a branched network, where we separately selected the optimal mmW BS and then the optimal beam by solving both mappings as a classification problem. Branching the complete problem to two sub-problems helped in the learning of the system and also showcased small variations for small changes in the environment. The consideration of BS selection as a classification problem performed well. It was however much less efficient for beam selection. The reason lies in the fact that due to the large number of narrow beams at mmW BSs, the angular difference between any two adjacent beams is very small, implying that multiple beams can be selected as best beam for certain user locations. We observed that this overlapping beam behaviour could not be solved by classification and the network was unable to converge to a solution.

To tackle this issue, we modified our branched network where this time, we considered the beam selection as a regression problem. To further improve the performance of the overall system, we formed a link between the BS selection branch and the beam selection branch as both of these operations are not mutually independent.

We adopt a soft decision for the BS selection process, i.e., we compute for every BS the selection probability and retain the one with the highest probability. In contrast, a hard decision would have selected a BS with a probability higher than a certain and given threshold. Hard decision has been observed to be training data centered and can guarantee to provide good solutions for features within the bounded range of the training data. However, a hard decision may fail to give good solutions for feature values outside these bounds. A soft decision however, will still provide a solution. Also, when all the BSs are equally probable for selection, a hard decision threshold greater than $1/B_m$ will not provide any solution, while a soft decision will select any one of them. Furthermore, beam selection task is also modelled with soft decision, where the best beam is selected as the one with highest achievable sum-rate. This allows for multiple beams selection (by considering the first highest sum-rates), a characteristic we will use to improve the accuracy of the results, as shown in the next section.

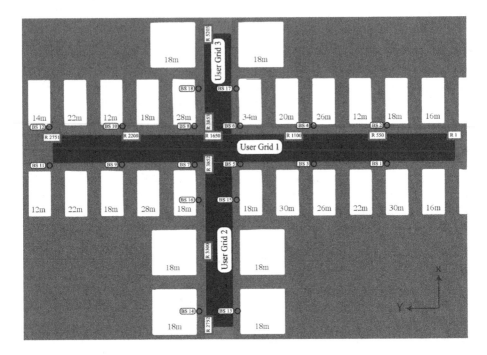

Fig. 3. Simulation environment [2].

5 Simulation Results and Evaluation

In this section, we illustrate the performance of the proposed DNN based BS and beam selection in a heterogeneous mmW networks. We first describe the setting of a simulation environment considered throughout the simulations in Subsect. 5.1 and then discuss the performance results in Subsect. 5.2.

5.1 Simulation Environment

We consider the outdoor simulation environment provided with the available open source DeepMIMO dataset [2]. From the dataset, we consider two different ray tracing scenarios 'O1_3p5' and 'O1_28' operating at 3.5 GHz and 28 GHz frequencies respectively, in order to construct a heterogeneous simulation environment. We consider two sub-6GHz coordinated BSs and eight mmW BSs. The deepMIMO dataset generates the channel at these frequencies. Given the CSI, we extract the basic components and construct the feature vectors from utilizing only the sub-6GHz channel, which acts as the input to our proposed DNN model. Essentially, we consider the azimuthal and elevation AoA, signal power, path loss and signal phase as the extracted features from the sub-6GHz CSI. Intentionally, we don't assume the availability of the UE location, as this information may not be available at the device. The hyperparameters considered for the generation of

Table 1. Dataset parameters for mmW BSs operating at 28 GHz and macro sub-6GHz BSs operating at 3.5 GHz.

Parameters	28 GHz scenario	3.5 GHz scenario
Active BSs	$2,3,4,5,6,7,8,17$	$1,18$
Active users	1651–2200, 3500–5203	1651–2200, 3500–5203
Number of BS antennas	256	16
Antenna spacing ($\times$wavelength)	0.5	0.5
Bandwidth (GHz)	0.5	0.02
Number of OFDM subcarriers	1024	1024
OFDM sampling factor	1	1
OFDM limit	64	64
Number of paths	1	1

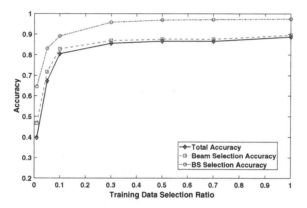

Fig. 4. Performance evaluation of the BS selection, beam selection and total accuracy Vs varying ratio of training data with respect to total training dataset.

the dataset for training and testing are given in Table 1. The outdoor simulation environment we considered is given in Fig. 3. It is an urban environment with the BSs placed along the side of the road. We considered a subset of BSs and users for our experiments, the list of which is given in Table 1. Users are considered to be present on the road and are densely populated for better data generation. Building of varying height, width and material are placed along the road providing blockages and reflections. For both scenarios, we considered 1024 OFDM subcarriers with an OFDM sampling factor of one, where sampling factor is the rate at which we can sample the OFDM subcarriers. Furthermore, the OFDM limit specifies the number of sampled subcarriers to be considered. We set this limit to 64 for both scenarios, which implies that we calculate the channels only at the first 64 sampled subcarriers. Detailed explanation about the simulation environment can be referred in [2].

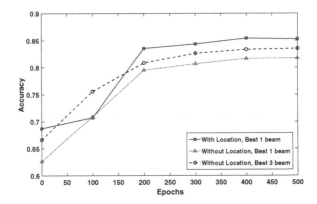

Fig. 5. Accuracy vs number of epochs comparing proposed DNN architecture based beam selection predicting best 1 beam and best 3 beams without considering location information in features and proposed DNN architecture predicting best beam while considering the location information.

5.2   Performance Evaluation

In this subsection, we present the simulation results demonstrating the performance of the proposed scheme, while analyzing the effect of the number of selected beams, the training dataset selection, and the location parameter on accuracy and latency.

In Fig. 4, we evaluate the performance of the proposed DNN-based BS prediction, best beam prediction and overall prediction accuracy, where the total accuracy is obtained by correct prediction of both BS and beam against the varying size of the training dataset. We divide the overall data with a 80:20 ratio where 80% of the total data is used for training whereas the remaining 20% dataset is used for validation/testing purpose. Out of this total available 80% training dataset, we utilise varying training data ratios and observe the performance in terms of accuracy for the proposed system. The system performance illustrates that the network is able to achieve high accuracy for both BS and beam selection tasks. The achievable BS selection accuracy is around 97% whereas the beam can be predicted with 88%. The total accuracy of correctly predicting both the optimal BS and beam is close to 86% when we use complete training dataset. We observe comparable performance with 50% of training data as compared to complete training dataset. This means that we can quickly obtain good results offline and apply the algorithm online and then improve the performance by training over the time.

We compare the performance of the proposed DNN architecture while now considering the UE location as one of the input features. Figure 5 shows this performance as a function of the number of epochs. We compare the performance for the best beam and the best three beams with and without location. As expected, we observe that the location-aided design performs better. However, the performance can be improved by selecting the best b beams, $b = 1...M$, hence

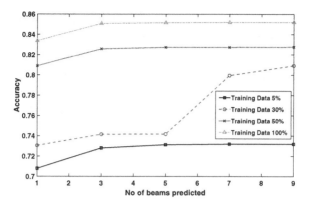

Fig. 6. Accuracy vs number of predicted beams by the proposed DNN beam selection with varying training dataset.

reducing the performance gap between the architecture with or without location parameter. In Fig. 6, we demonstrate the beam selection accuracy with respect to number of beams predicted for the proposed DNN for varying size of the training dataset. As expected, it is observed that the beam selection accuracy increases with the increasing number of predicted beams as well as with the increasing size of the training data. From this figure, we can further observe and analyse the effect of latency for the proposed system. Indeed, an exhaustive search would require to perform M received power measurements (64 in our case), while with our solution, we can achieve 85% accuracy by measuring only the best three beams selected by the network.

6 Conclusion

In this paper, we propose a branched DNN model that jointly performs the mmW BS prediction and beam selection task in a heterogeneous network architecture. We consider that multiple mmW BSs coexist with multiple legacy sub-6GHz BSs to serve the UE in the network area. The sub-6GHz BSs are assumed to function in a coordinated manner and are supported by the central cloud processor. We formulate the mmW BS prediction as a classification problem whereas the optimal beam selection is mapped into a regression problem. For both the tasks, we utilize the channel components available only at the sub-6GHz BSs as a set of input features. As the location information may not be always available or it can be inaccurate due to sensor errors, we intentionally eliminate the use of location as an input feature for the proposed problem. We compare the performance of the proposed DNN based design with conventional exhaustive search and observe the success probability close to 1 for allocating optimal mmW BS and beam while using reduced computational resources. Comparable performance can be achieved with and without user location available provided that the three best beams are considered. At last, we show that much fewer beam power measurements are required compared to exhaustive search, which results in lower latency.

References

1. Ali, A., González-Prelcic, N., Heath, R.W.: Millimeter wave beam-selection using out-of-band spatial information. IEEE Trans. Wireless Commun. **17**(2), 1038–1052 (2018). https://doi.org/10.1109/TWC.2017.2773532
2. Alkhateeb, A.: DeepMIMO: a generic deep learning dataset for millimeter wave and massive MIMO applications. arXiv:1902.06435 (2019)
3. Alrabeiah, M., Alkhateeb, A.: Deep learning for mmWave beam and blockage prediction using sub-6GHz channels. arXiv:1910.02900 (2019)
4. Anton-Haro, C., Mestre, X.: Data-driven beam selection for mmWave communications with machine and deep learning: an angle of arrival-based approach. In: 2019 IEEE International Conference on Communications Workshops (ICC Workshops), pp. 1–6, May 2019. https://doi.org/10.1109/ICCW.2019.8756991
5. Dias, M., Klautau, A., González-Prelcic, N., Heath, R.W.: Position and LIDAR-aided mmWave beam selection using deep learning. In: 2019 IEEE 20th International Workshop on Signal Processing Advances in Wireless Communications (SPAWC), pp. 1–5, July 2019. https://doi.org/10.1109/SPAWC.2019.8815569
6. Klautau, A., González-Prelcic, N., Heath, R.W.: LIDAR data for deep learning-based mmWave beam-selection. IEEE Wirel. Commun. Lett. **8**(3), 909–912 (2019). https://doi.org/10.1109/LWC.2019.2899571
7. Morocho-Cayamcela, M.E., Lee, H., Lim, W.: Machine learning for 5G/B5G mobile and wireless communications: potential, limitations, and future directions. IEEE Access **7**, 137184–137206 (2019). https://doi.org/10.1109/ACCESS.2019.2942390
8. Odijk, D., Kleijer, F.: Can GPS be used for location based services at Schiphol airport, The Netherlands? In: 2008 5th Workshop on Positioning, Navigation and Communication, pp. 143–148, March 2008. https://doi.org/10.1109/WPNC.2008.4510368
9. Sakaguchi, K., et al.: Millimeter-wave evolution for 5G cellular networks. IEICE Trans. Commun. **E98.B**, December 2014. https://doi.org/10.1587/transcom.E98.B.388
10. Schmidhuber, J.: Deep learning in neural networks: an overview. Neural Netw. **61**, 85–117 (2015)
11. Wang, Y., Klautau, A., Ribero, M., Soong, A.C.K., Heath, R.W.: MmWave vehicular beam selection with situational awareness using machine learning. IEEE Access **7**, 87479–87493 (2019). https://doi.org/10.1109/ACCESS.2019.2922064
12. Wei, L., Li, Q., Wu, G.: Initial access techniques for 5G NR: Omni/beam SYNC and RACH designs. In: 2018 International Conference on Computing, Networking and Communications (ICNC), pp. 249–253, March 2018. https://doi.org/10.1109/ICNC.2018.8390325
13. Xiao, M., et al.: Millimeter wave communications for future mobile networks. IEEE J. Sel. Areas Commun. **35**(9), 1909–1935 (2017). https://doi.org/10.1109/JSAC.2017.2719924
14. Yan, L., et al.: Machine learning-based handovers for sub-6 GHz and mmWave integrated vehicular networks. IEEE Trans. Wirel. Commun. **18**(10), 4873–4885 (2019). https://doi.org/10.1109/TWC.2019.2930193
15. Yu, H., Qu, W., Fu, Y., Jiang, C., Zhao, Y.: A novel two-stage beam selection algorithm in mmWave hybrid beamforming system. IEEE Commun. Lett. **23**(6), 1089–1092 (2019). https://doi.org/10.1109/LCOMM.2019.2913385

A Bilevel Model for Centralized Optimization of Charging Stops for EV on Highways

Anthony Woznica[1], Dominique Quadri[1(✉)], Yezekael Hayel[2], and Olivier Beaude[3]

[1] University Paris Saclay LRI, Bat. 650 Univ. Paris Sud,
91405 Orsay Cedex, France
woznica@u-psud.fr, quadri@lri.fr
[2] University of Avignon, 339 Chemin des Meinajaries,
84911 Avignon Cedex 9, France
yezekael.hayel@univ-avignon.fr
[3] EDF Lab Paris Saclay, Palaiseau, France
olivier.beaude@edf.fr
https://www.lri.fr/equipe.php?eq=29
http://univ-avignon.fr/m-yezekael-hayel--3520.kjsp
https://sites.google.com/site/olivierbeaudeshomepage/

Abstract. This paper addresses the multi-period problem of fixing the energy charging price at a set of charging stations deployed to support long journeys of electric vehicles along highways. In order to model the problem, we propose a non-linear bilevel program, in which the leader (a single centralized operator) fixes the charging price over the time horizon to maximize its profit and the follower (the operator of the electric vehicles) chooses the optimal assignment of the electric vehicles to the stations so to minimize a cost function. To solve the resulting model, we suggest to adopt an adaption of the Branch-and-Cut algorithm for Mixed Integer Linear Bilevel Programming proposed by Fischetti et al. (2018). Preliminary computational results are provided to show the interesting performance of the new modeling and algorithmic approach.

Keywords: Electrical vehicles · Bilevel programming · Branch-and-Cut algorithm

1 Introduction

Recently, with the EV driving range[1] (up to 300 km in a single ride) and the deployment of rapid charging stations along highways, the long journey trip

---

[1] Between the two Renault Zoe delivered respectively in 2012 and 2016, the cost of a unit of battery storage (1 kWh) has been divided by two, while the energy density has been doubled by 2. For a same battery size and a same price, Zoe 40 has a doubled driving range.

Supported by PGMO.

S. Lasaulce et al. (Eds.): NetGCooP 2021, CCIS 1354, pp. 185–194, 2021.
https://doi.org/10.1007/978-3-030-87473-5_17

becomes more and more possible [2]. The main problem, tackled in this article, is how to deal with the dimensioning of those rapid charge stations. In fact, the autonomy is not sufficient for a long trip without recharging. Therefore, the problem of managing such stations considering how EV customers react is considered in this work. Such complex system involves a two levels hierarchical structure and a specific mathematical problem has to be considered where the manager of the charging station will be at the upper level and the set of EV at the lower level (the decisions of the EV are managed by a single operator, centralized context). In the past few decades, there have been many advances in the field of Operations Research applied to the transport sector to solve planning or pricing problems [3]. In many cases, these applications do not take into account the feedback effects - also called reactions to the optimum- of the third parties of the organ subject to the problem. Typically, this means optimizing the profit of a service provider without taking into account the beneficiary's reaction, whose decision can potentially vary according to the service provider optimum. Hence, decision problems need to take into account simultaneously two different actors (or agents) who do not have the same objective. The corresponding model has to imply both decision agents interacting sequentially and hierarchically. This type of process can be formulated as a bilevel program [4] where a decision agent, called the leader, integrates explicitly the reaction of another agent, named the follower, when the leader has to make optimal choices. At the upper level, the leader can be for instance faced to a pricing problem as in [5]. At the lower level, on the other hand, the followers can interact in terms of a congestion problem as in [8]. Bilevel models have been proposed recently to address EV problematic. In [9] the authors propose a bilevel optimization framework for designing optimal charging strategies for a fleet of EV. In [10], the authors propose also a bilevel model (here a Stackelberg game), where EV charging station operators compete at the upper level, offering charging prices to attract EV to their station. The present paper proposes an approach for the modeling of the best pricing for charging stations taking into account a set of EV whose decisions are made by a single operator (centralized model). We make this first strong assumption of centralized model because of the following two reasons: (i) the centralized context gives an optimum social bound very interesting in terms of user performances and recharging system; (ii) this bound will help us, for future work, to quantify the price of anarchy obtained in a incitation/information decentralized model. Most of new vehicles, and particularly EV, are basically equipped with GPS driving facilities. Then, it is plausible, and in fact already present in particular EV brands like Tesla, that such GPS system can provide information to drivers on where and how much to charge during the trip. This driving aided system can be computed in a centralized way by an operator of a fleet of EV. Two actors are thus considered, one being the Charging Station Manager (CSM) whose objective is to maximize his profit, the other being the Set of Electric Vehicles (SEV) responsive to the prices assigned to recharging stations and to the waiting times at the charging stations. The SEV is managed by a single operator. The paper is organized as follows. We first state the studied problem in Sect. 2. Section 3 details the bilevel model proposed to formulate both CSM and SEV problems. We establish in Sect. 4 the adaptive exact algorithm implemented to solve our problem. Section 5 is dedicated to numerical experiments. We finally conclude in Sect. 6.

2   Problem Statement

The study of this paper addresses the problem of determining the unit charging price at each charging station (CS) in order to influence EV stops in a multi-dimensional case (several CS and several EV), modeling long travel on highways with EV. The pricing adopted by the Charging Station Manager (CSM) is considered to follow a single objective: to maximize its profit by taking into account the unit charging price and the waiting time at its CS. Inline with the case of highways, we consider we consider a rectilinear road that the EV can use.

In this situation, it is essential to take into account the behavior of the SEV since the unit charging price is implicitly dependent on the distribution of vehicles to charging stations. Therefore, it is primordial to know the SEV's reaction to a pricing decision.

The work here presents a large framework and thus some assumptions limit the complexity of the analysis and allow to focus on implicit CSM/SVE negotiation through an applied Intersection Cuts method [6]. Such method has been recently proposed and is very efficient to solve bilevel problems. The authors suggest the use of basic Branch-and-Bound algorithm first employed for the relaxed bilevel problem (i.e. for a single level problem concerning the upper level where the optimization problem of the lower level is relaxed) called High Point Relaxation (HPR) problem and then an exploration at the node of the search tree is done in order to find a feasible bilevel solution. In addition, the authors suggest for the first time, the use for bilevel program of intersection cuts, initially developed by Balas [1] for Integer Linear Programming. The following hypotheses are then considered: Prices are considered as natural integers and they are also bounded; Time is discrete; Each charging station has only one charging point; EV decisions are managed by a single operator (centralized context).

The main problem for the operator of the SEV is to determine, *a priori*, where each EV has to stop during his long trip (assuming that each EV has necessary the need to stop in order to reach its destination), considering pricing decisions (the unit charging price) proposed by charging stations operator. In addition, such optimization problem depends on the state-of-charge (SoC) of each vehicle. It also involves a bilevel approach. We present in the next section the bilevel model we have proposed to formulate the problem mentioned. Figure 1 gives an overview of the considered problem. Let us consider $S = 6$ Charging Stations (denoted by S on the figure) and 6 EV (denoted by EV on the figure). Each EV can reach several CS according to its battery level, i.e. its SoC. The decision of each EV will be taken according to the charging price and the waiting time before service at each CS. For instance, let us focus on EV_1. EV_1 can reach S_1 or S_2. Then, from S_1 the EV_1 can reach S_3 or from S_2 can reach S_3 or S_4 and so on, until EV_1 reaches the last station S_6. However, each station has its own charging price, making them more or less attractive, and as a consequence, generating more or less demand. This difference of attractiveness induces different queues between the CS, which is another cost for EV.

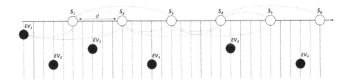

Fig. 1. Possible stopping scenarios for EV_1 where initial SoC for EV 1 is 10 units of energy which is the maximum value.

3 The Proposed Bilevel Model

In this section, the bilevel model is established and notations are given. The problem is concerned with two decision makers: a recharging station manager (RSM) (*the leader*) and a Set of EV (SEV) (*the follower*). The output of the model are the following:

1. *Leader's problem*: The **leader** (RSM) manages a set of charging stations S and a unit charging price p_s has to be assigned to station s on the overall time duration T, in order to maximize his benefit. This price is assumed to be stationary which is more acceptable for customers in general like flat rate pricing schemes in telecom, but the model can be simply enriched by considering dynamic prices without too much complexity.
2. *Follower's problem*: The **follower** (operator of the SEV) has to decide an optimal repartition of the EV over the stations in order to minimize the total charging cost and the waiting time; that is to minimize the cost.

Following notations are considered:

- T is the time interval (time horizon) of the model.
- I is the fleet of EV indexed by i. The EV are initialized to random position (entrance of the road) and SoC within the discrete set $\mathbf{E} := \{0, \ldots, E\}$.
- S is the discrete set of charging stations (CS) indexed by s and geographically positioned such that the distance between two consecutive CS is d distance units uniformly located on the rectilinear considered road.
- All EV are identical and the maximal SoC is denoted by E. The SoC of EV i at time t is denoted by $e_{i,t}$ such that $0 \le e_{i,t} \le E$.
- When EV i stops at a charging station s at time slot t, the EV charging time is denoted by $C_{i,s,t}$ and is equal to $C_{i,s,t} := \alpha_C \times (E - e_{i,t})$ (recovering full battery level), where α_C represents the ratio of the time needed for a full

charge (when at CS) to the one for full discharge (when driving). $\alpha_C \leq 1$, and in the context of highways typical values lie in the range $[1/20, 1/3]^2$.

- Each charging station s has an unit charging price p_s in the discrete ordered set $P = \{p_0, p_1, ..., p_P\}$ with $p_0 = P_{min}$ and $p_P = P_{max}$.

The proposed bilevel model aims to explore the tradeoff between the charging price and waiting time at the charging station. The pricing approach is inspired from [5] which is standard for congestion control. More precisely our model is based on the following decision variables:

- Pricing of charging stations: the decision variables p_s (upper level) are integer representing the pricing applied at station s per unit of charging time,
- Decision variables of the EV: the variables $y_{i,s,t}$ (lower level) are binaries and determine if EV i has to stop at charging station s, knowing that the vehicle will reach it at time t.

Taking ordering indices for the decision variables allows to determine the stops of an EV along his travel and then deduce the arrival time at each station corresponding to the stopping decision.

Let us describe the objective function of the leader (upper level optimization problem), which is dependent on the stopping decision of the SEV (for charging). The benefit of the leader for one particular charging station s is:

$$\sum_i y_{i,s,t} \times C_{i,s,t} \times p_s. \tag{1}$$

This benefit, coming from EV i who stops at time t in station s, depends on the charging time of this EV that is $C_{i,s,t}$. The maximization problem of the leader who manages all CS is then expressed by:

$$max_{p:=(p_1,...,p_S)} \sum_i^I \sum_s^S \sum_t^T y_{i,s,t} \times C_{i,s,t} \times p_s. \tag{2}$$

Concerning the follower (the manager of the SEV), it aims to minimize the total charging cost for EV as well as the waiting times at the charging stations (we make a sum of both follower's objectives). We denote by $\triangle_s$ the total waiting time of all EV who stop at station s, which gives the following objective function:

$$min_y(\sum_i^I \sum_s^S \sum_t^T y_{i,s,t} \times C_{i,s,t} \times p_s + \sum_s^S \triangle_s) \tag{3}$$

where

$$\triangle_s = \sum_i^I \sum_t^T \delta_{i,s,t} \tag{4}$$

---

^{2} This ratio being mainly dependent on the (ultra)fast charging power available currently in the range 50–350 kW in this use-case; the driving part being a "physical constant".

This last expression represents the waiting time of EV i arrived at station s at time t, before starting to get served (i.e. to start charging energy). More precisely, if an EV, for example EV i arrives after EV k at the same station s at time t, and if EV_k is still charging then the waiting time for EV i at station s depends on the remaining charging time of EV k like in a queueing system. In addition, $\ell_{i,s,t}$ is the travel time for EV i given its position at time t, to reach station s such that station s is in front of him but not the next one on the travel, and it is determined by:

$$\ell_{i,s,t} = d_{i,s,t} + \sum_{s'=1}^{s-1} \sum_{t}^{T} C_{i,s',t} y_{i,s',t} + \sum_{s'=1}^{s-1} \sum_{t}^{T} \delta_{i,s',t}, \tag{5}$$

with $d_{i,s,t} = s \times d - pos(i,t)$ is the distance between station s and the current position of EV i at time t denoted by $pos(i,t)$. Indeed, this variable has to take into account the waiting and charging times for EV i in any other charging stations s' where the EV has stopped before reaching station s. If station s is the next one on the travel, the travel time for EV i to reach s is simply:

$$\ell_{i,s,t} = d - pos(i,t).$$

Let us now describe the constraints of our model. We have to guarantee that only one travel can be taken by an EV. The $y_{i,s,t}$ are the decision variables related to the decision that EV i stops at station s at time t and the order corresponds to the t-th visited charging station since the beginning of the travel. This order allows to guarantee that an EV can not turn back on the road. Let us give a simple example to illustrate our constraints. Let us consider for example EV 1. Let us assume that EV 1 stops at the station 1 at time 1 and then at station 3 at time 3 (recall that $d = 1$), consequently $y_{1,1,1} = y_{1,3,3} = 1$, and $y_{1,s,t} = 0$ $\forall s, t$, that is $y_{1,2,4} = 1$ is forbidden. More formally, this constraint can be stated as follows:

$$\sum_{s} y_{i,s,t_0} = 1 \ \forall \ i. \tag{6}$$

That is at t_0 EV_i has to choose one and only one station s. Nevertheless, after t_0 it has the choice to stop or not to a possible station. The constraints can be expressed as follows:

$$(y_{i,s,t} \times \sum_{s'=s+1}^{S} y_{i,s',t+1}) + (\neg y_{i,s,t} \times (\sum_{s'=s+1}^{S} \neg y_{i,s',t+1})/S) = 1. \tag{7}$$

If EV_i stops at the station s at time t then it has to visit one and only one station s' (corresponding to accessible stations from station s) at time $t + 1$ otherwise if EV_i does not visit station s at time t (expressed by $\neg y_{i,s',t+1}$) then EV_i can not stop at any station s' with $\neg y_{i,s',t+1}$. Finally constraints which allow link between waiting time and charging price at CS are described hereafter:

$$p_s \leq \frac{(\Delta_{max} - \Delta_s)}{\Delta_{max}} \times (P_{max} - P_{min}) + P_{min} \ \forall s \in S. \tag{8}$$

Indeed, the constraint below aims to fix a price if and only if the waiting time at station s allows it. It can be noticed that if, for instance, the waiting time at station s is such that $\Delta_s = 0$ then the charging price will be set at maximum value P_{max}. On the contrary, the bigger the waiting time is at station s, the lower the charging price at this station. In addition, note that since Δ_s includes a product of $y_{i,s,t}$ decision variables, those constraints are quadratic. Note that Δ_{max}, which is the maximum waiting time at a CS can be pre-calculated. We are now able to establish the proposed non-linear bilevel problem $(NLBLP)$ as follows:

Note that the proposed bilevel model includes characteristics which make it non-standard:

- the objective functions of the leader (2) and follower (3) are not linear;
- constraints (8) relative to the relation between of the price and waiting time are quadratic.

Consequently, the recent exact solution method developed in [6] has been used by linearization of quadratic constraints in order to consider our non-standard $(NLBLP)$. The algorithm is described in the next section.

4   Solution Algorithm for the NLBLP

Our bilevel program solution method is an adaptation of the Branch-and-Cut algorithm proposed by [6] for our non linear problem. The authors have suggested a general exact solution method dealing with Mixed-Integer Bilevel Program $(MIBLP)$ where both objective functions and constraints for leader and follower are linear. The decision variables of the leader which influence the decision of the follower have to be pure integer and bounded. The other decision variables of the leader can be integer or continuous, and the one of the follower are integers. The general $(MIBLP)$ studied in [6] can be written as follows:

$$\min F(x, y) \tag{9}$$

$$\text{s.t.}$$

$$G(x, y) \leq 0 \tag{10}$$

$$g(x, y) \leq 0 \tag{11}$$

$$x, y \in R^n \tag{12}$$

$$f(x, y) \leq \phi(x) \tag{13}$$

where for a given $x \in R^{n_1}$ the follower value function is:

$$\phi(x) = \min_{y \in R^{n_2}} \{f(x, y) : g(x, y) \leq 0\} \tag{14}$$

When constraint (14) is relaxed, the program is called High Point Relaxation (HPR) which is assumed to be feasible. This is a strong property that our model has to satisfy. The authors suggest a finitely-convergent branch-and-bound algorithm based on the two following assumptions: (i) the variables x and y have finite lower and upper bounds in (HPR) and in the follower $(MILP)$; (ii) the continuous leader variables (if they exist) do not appear in the follower problem.

We apply [6] to our bi-level program. Actually, even if the objective function of the leader is non-linear, it is convex then it can be dealt by CPLEX. In addition, since we deal with (HPR) the problem concerning the non linearity of the objective function of the follower disappears. Concerning the quadratic constraints, a linearization is possible by the use of classical and well-known crossed quadratic terms linearization techniques developed by Glover [7]. Consequently, both assumptions (i) and (ii) established by Fischetti *et al.* are satisfied in our context, we are then allow to use their Branch-and-Cut algorithm.

5   Preliminary Experiments

In this last section, preliminary experiments are conducted in order to illustrate the performance of the Branch-and-Cut algorithm used to solve our nonstandard (NLBLP). MILP problems have been solved using IBM ILOG CPLEX 12.6.3 using callbacks. We have implemented the algorithm in C++ language. Numerical examples were run on a bi-xeon 3.4 Ghz with 4Go of main memory computer. For all instances, maximal SoC is $E = 17$ kWh, $d = 5$ Km for the distance between two consecutive stations, $\alpha_C = 1$ (that is one unit of energy is consumed when traveling and one unit is recharged when charging, per unit of time) and all initial SoC have been randomly determined. We have conducted several tests with a number of EV and CS taking respectively their value in the following discrete sets $\{2; 4; 8; 16; 32; 64\}$ and $\{2; 4; 6; 8; 10\}$.

Table 1 displays: (i) the CPU times (in seconds) required by the exact algorithm (B&C) we have coded, (ii) the CPU times (in seconds) needed by a feasible solution method (HEUR) and (iii) the relative gap in percentage (Gap = (Optimum − Lower Bound)/Lower Bound). The Lower Bound is provided by the heuristic denoted by (HEUR) which corresponds to the first feasible bilevel solution found in the Branch-and-Bound algorithm. Each line of Table 1 is an average on ten random instances. The (B&C) and the heuristic are very fast on small instances (2 EV). Moreover, the lowest relative gap ($\leq 1\%$) is obtained for this size of instances. Such good performance comes from the first step consisting in finding a first feasible solution very quickly, which is a key step here in our approach. When the number of EV increases, both methods become more time consuming and the relative gap between the solutions of both methods also increases. Note that our heuristic can be helpful when the exact solution can not provide the optimum solution like for the instance $(\#EV, \#CS) = (8, 4)$. Obviously, more experiments have to be conducted in future work in order to provide a precise size limit of the instances.

Table 1. Average computational time (in sec.) of the B&C algorithm and our heuristic

Method (# EV, # CS)	B&C	HEUR	Gap (%)
(2,2)	0.12	0.119	0.7
(2,4)	4.418	1.645	0.8
(2,6)	11.1	7.223	1
(2,8)	35.732	13.739	1.2
(2,10)	86.919	21.280	1.5
(4,2)	0.24	0.182	1
(4,4)	23.9	3.431	1.2
(4,6)	2995.201	123.667	1.4
(8,2)	0.715	0.221	3
(8,4)	–	124.931	–
(16,2)	3.233	2.487	10
(32,2)	113.727	98.64	20
(64,2)	1370.292	1114.387	30

6   Conclusion

This work deals with the problem of managing a set of EV into different charging stations, taking into account the charging and the waiting costs. The charging price at each station is settled by a central controller, with the aim to induce an optimal profit. The global problem is a non-linear bi-level problem (NLBLP). The use of a recent tool is possible in our context and significant speedup of the computational time of our heuristic has been obtained without losing the performance. This work then shows on a simple example that managing optimally many EV for long distance travel is possible trough the use of bi-level programs. Many perspectives are possible, for example to take into account a network topology and also to integrate decentralized decision taken by EV individually.

References

1. Balas, E.: Intersection cuts - a new type of cutting planes for integer programming. Oper. Res. **19**(1), 19–39 (1971)
2. Beaude, O., Lasaulce, S., Hennebel, M., Mohand-Kaci, I.: Reducing the impact of electric vehicles charging operations on the distribution network. IEEE Trans. Smart Grid **7**(2), 2666–2678 (2016)
3. Ben-Akiva, M., Lerman, S.: Discrete Choice Analysis: Theory and Application to Travel Demand. MIT Press, Cambridge (1985)
4. Colson, B., Marcotte, P., Savard, G.: An overview of bilevel optimization. Ann. Oper. Res. **153**, 235–256 (2007)
5. Colson, B., Marcotte, P., Savard, G.: A bi-level modelling approach to pricing and fare optimisation in airline industry. J. Revenue Pricing Manag. **2**(1), 23–36 (2003)

6. Fischetti, M., Ljubić, I., Monaci, M., Sinnl, M.: On the use of intersection cuts for bilevel optimization. Math. Program. 77–103 (2017). https://doi.org/10.1007/s10107-017-1189-5

7. Glover, F.: Improved linear integer programming formulations of nonlinear integer problems. Manag. Sci. **22**(4), 455–460 (1975)

8. Gourves, L., Monnot, J., Moretti, S., Thang, N.: Congestion games with capacitated resources. Theory Comput. Syst. **57**(3), 598–616 (2015)

9. Skugor, B., Deur, J.: A bi-level optimisation framework for electric vehicle fleet charging management. Appl. Energy **184**, 1332–1342 (2016)

10. Tan, J., Wang, L.: Real-time charging navigation of electric vehicles to fast charging stations: a hierarchical game approach. IEEE Trans. Smart Grid **8**(2), 846–856 (2017)

Coordinated Scheduling Based on Automatic Neighbor Beam Relation

Marie Masson[1,2,3](✉) 🆔, Zwi Altman[1] 🆔, and Eitan Altman[2,3] 🆔

[1] Orange Labs, Châtillon, France
{marie.masson,zwi.altman}@orange.com
[2] INRIA Sophia Antipolis and LINCS, Nancy, France
[3] Avignon University, Avignon, France
eitan.altman@inria.fr

Abstract. We consider a network with Massive Multiple Input Multiple Output (M-MIMO) base stations using a Grid of Beams (GoB) for data and control channels. 5G allows to establish interference relations between beams of neighboring cells. Such relations can be used to automatically generate a beam relation matrix, denoted as Automatic Neighbor Beam Relation (ANBR) matrix that can be very useful for optimizing different resource allocation processes. This paper shows how the ANBR matrix can be used to coordinate scheduling of neighboring cells with a small amount of information exchange. The coordination is performed by judiciously muting or reducing the bandwidth of certain beams in the process of Multi-User (MU) Proportional Fair (PF) scheduling. Numerical results show how the coordination approach can bring about significant performance gain.

Keywords: Beam relations · ANBR · Massive MIMO ·
Coordination · Multi-user scheduling · Interference management · 5G

1 Introduction

M-MIMO is among the pillars of 5G technology that allows to significantly improve user rates, system capacity and Energy Efficiency (EE) [1]. The concept of GoB has been introduced to allow beamforming of control channels that are used among others to transmit synchronization signals and to broadcast system information to allow initial access and mobility procedures [9]. Beam sweeping is used in conjunction with GoB by switching rapidly the beams one by one in a manner that covers the entire cell surface. Similarly, data channels can be transmitted using the beams of the GoB. In this case the GoB can be seen as a predefined codebook of beams that can be selected by a MU scheduler to serve users.

Beam management is an important feature introduced in 3rd Generation Partnership Project (3GPP) for 5G networks. It is part of the New Radio (NR)

S. Lasaulce et al. (Eds.): NetGCooP 2021, CCIS 1354, pp. 195–207, 2021.
https://doi.org/10.1007/978-3-030-87473-5_18

Automatic Neighbor Relation (ANR) and is considered as a central Self- Organizing Network (SON) function. NR ANR allows to automatically establish different types of relations involving gNodeBs (gNBs) and/or beams [7]: (i) gNB to gNB relations that consist of establishing connectivity over the Xn interface between neighboring gNBs. Such relations are necessary to support mobility, load and traffic sharing or multi-connectivity and were already standardized for 4G networks [4]; (ii) gNBs to Beam relations; and (iii) beam to beam relations, intra- and inter-cell.

The feature of automatically establishing relations between beams is denoted as ANBR. The association of user traffic Quality of Service (QoS), serving and interfering beams, and input from the ANBR enables the exploitation of the beam level spatial resolution to further optimize resource management functions.

This paper investigates the use of ANBR to optimize MU-Multiple Input Multiple Output (MIMO) scheduling. We assume that the ANBR feature is available and provides a binary static matrix, denoted as *ANBR matrix*, with non-zero elements representing beam relations. The way to define relations between beams is not standardized. It can be derived as in classical ANR, e.g. by calculating the average interference users of beam i experience from beam j of a neighboring cell during a long period of time, and comparing it to a predefined threshold. We show how the ANBR matrix can be used to coordinate the MU schedulers of neighboring cells in a manner to minimize collisions between potentially interfering beams, while taking into account the distribution of the traffic.

Traditional approaches for collaboration between adjacent cells with M-MIMO deployment such as Coordinated Multipoint transmission (CoMP) [5] require signaling and computation in both the PHY and the MAC layer and are thus demanding an intensive signal processing and the exchange of important amount of information. They also have strict requirements on the backhaul capacity. In the proposed approach, the coordination is performed at the MAC level and requires little information exchange and processing power.

The paper is organized as follows. Section 2 presents the system model. Section 3 develops a coordinated MU scheduling solution relying on the ANBR. The architecture supporting the solution is also described. Numerical results are presented in Sect. 4 followed by concluding remarks in Sect. 5.

2   System Model

Consider a hexagonal network with tri-sectoral sites, each with M-MIMO system in the Down Link (DL). GoB with beam sweeping is used for initial access (to attach users to the Base Station (BS)) and for synchronization. Data transmissions use the (fixed) beams of the GoB, and the choice of users to be served is made by the MU-scheduler.

Each sector (BS or cell) m is equipped with a M-MIMO antenna with $N = N_x \times N_z$ radiating elements, and is serving N_m mobiles, each with one receiving antenna (the beam generation and antenna modeling is explained in details in

[8]). Figure 1 presents the coverage area provided by GoBs of two adjacent cells for the case of $N_x = N_z = 16$. The color code is used to simplify visualization and has no physical meaning. Fading is removed for clarity. It is noted that the zeros of the beam radiation patterns in the azimuth axis close to the BSs is due to the zeros in the beam radiation patterns and the fact that reflections and fading are omitted. It is recalled that beams for control channels are activated one by one (via beam sweeping) whereas several beams for data channels can be scheduled simultaneously.

We consider two BSs (i.e. macro cells) m and m' which interfere each other, each of which having a GoB denoted by $\mathcal{B}_m$ and $\mathcal{B}_{m'}$ respectively as shown in Fig. 1. We assume that a cell can serve up to K users in a time slot, with at most one user per beam b. The users are selected according to a PF criterion.

Fig. 1. GoBs of two neighboring cells projected on the surface

Assume that the BS serves k users at a given time slot with $k \leq K$ and denote by p_{max} the maximum transmission power of the BS. The power p_m that the BS m transmits to user u, equals $\frac{p_{max}}{k}$. Denote by $C_u^{m,b}$ the useful signal received power of user u from beam b of BS m, by $d_{u,m}$ - the distance between m and u, and by σ - the thermal power. $C_u^{m,b}$ can be written as a function of the channel gain $h_{b_u}^m(u)$:

$$C_u^{m,b} = p_m |h_b^m(u)|^2. \tag{1}$$

$|h_{b_u}^m(u)|^2$ is modeled as the product between the pathloss, the antenna gain $G_{b_u}^m(u)$ of the beam b_u serving user u and measured at the direction of user u, and the fast fading term $Z(u)$. The latter is modeled as a realization (per user) of a Nakagami distribution, and can be parameterized for different propagation environment [2].

$$|h_{b_u}^m(u)|^2 = \frac{c}{d_{u,m}^\gamma} G_{b_u}^m(u) Z(u) \tag{2}$$

where c and γ are constants that depend on the type of environment. The interference generated by a cell m' on u is written as the sum of interferences from its active beams $I_u^{b'}$:

$$I_u^{m'} = \sum_{b' \in \mathcal{B}_{m'}} I_u^{b'} \tag{3}$$

where

$$I_u^{b'} = p_{m'} \sum_{u' \in b', u' \neq u} |h_{b'_{u'}}^{m'}(u)|^2 \tag{4}$$

The Signal to Interference plus Noise Ratio (SINR) of a user u attached to cell m is written as:

$$S_u^m = \frac{p_m |h_{b_u}^m(u)|^2}{\sum\limits_{m'} I_u^{m'} + \sigma^2} \tag{5}$$

A full buffer traffic model is assumed. The baseline scheduler is based on PF without coordination.

3 Coordinated Self-organizing Scheduling

3.1 Architecture Framework

The architecture supporting the ANBR based coordination is shown in (Fig. 2). A centralized ANBR SON function is deployed at the management and orchestration plane as an application (but can be implemented in a distributed manner as well). It provides a static matrix $\mathbf{A}^{m,m'}$ for any two cells m and m', with elements $A_{b,b'}^{m,m'}$. For simplicity of notations we omit the superscripts m and m' from the matrix $\mathbf{A}$ in the rest of the sequel. $A_{b,b'} = 1$ if the beam b of cell m and beam b' of cell m' interfere each other (as mentioned in Sect. 1), and 0 otherwise. The matrix $\mathbf{A}$ is calculated and updated (not often) according to the operator policy and is considered here as constant.

The Distributed Unit (DU) (i.e. Radio Link Control (RLC), Media Access Control (MAC) and Physical (PHY)-High control protocol stack), hosts a new functional block denoted in Fig. 2 as D-ANBR. The D-ANBR. receives and processes measurements from the Radio Units (RUs). It dynamically updates the beam relations' matrix $\mathbf{A}$ and keeps only those relations corresponding to the present traffic distribution (as described in the next subsection). The resulting sparser matrix, $\mathbf{Q}$, is transmitted to the schedulers of BSs m and m' and is used to coordinate them. The time scale for generating the matrix $\mathbf{Q}$ is that of the traffic dynamics, e.g. of the order of a second.

It is noted that the introduction of new control, Radio Resource Management (RRM) or machine learning algorithms is currently studied in standardization fora the ORAN Alliance which is standardizing new interfaces and protocols to support deployment of such algorithms in different network nodes [6].

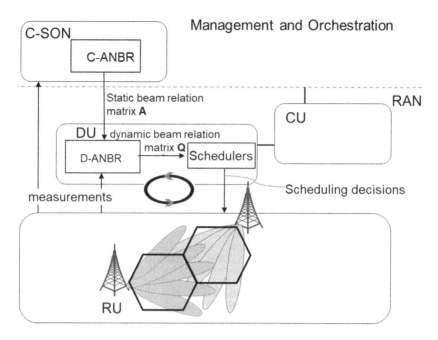

Fig. 2. System architecture

3.2 ANBR-Assisted Coordinated Scheduling

We first describe the generation of the matrix $\mathbf{Q}$ and then explain how it is used to coordinate the MU schedulers of BSs m and m'.

Denote by $\mathcal{U}_{b,m}^{serv}$ the set of users served by beam b of BS m:

$$\mathcal{U}_{b,m}^{serv} = \{u \in m | b = \underset{\bar{b}\in\mathcal{B}_m}{\operatorname{argmax}}\ C_u^{m,\bar{b}}\} \qquad (6)$$

We define two indicators used to generate the matrix $\mathbf{Q}$ using $\mathbf{A}$. The first, $\mathcal{A}_1(b, u')$, indicates whether user u' achieves low Signal Interference Ratio (SIR) that is below a predefined threshold γ_{th} and is thus likely to experience strong interference from beam b:

$$\mathcal{A}_1(b, u') = \begin{cases} 1 & if\ \frac{C_{u'}^{m'}}{I_{u'}^b} < \gamma_{th} \\ 0 & otherwise \end{cases} \qquad (7)$$

For clarity of notation the beam b' serving user u' is not included in (7).

Denote by $\mathcal{U}_b^{int}$ the set of users $u' \in m'$ for which $\mathcal{A}_1(b, u') = 1$, namely the set of users that could benefit the most from reducing the interference from beam b:

$$\mathcal{U}_b^{int} = \{u' \in m' | \mathcal{A}_1(b, u') = 1\}. \qquad (8)$$

The cardinality of $\mathcal{U}_b^{serv}$ and $\mathcal{U}_b^{int}$ are denoted by n_b^{serv} and n_b^{int} respectively,

$$n_b^{int} = |\mathcal{U}_b^{int}| \qquad (9)$$

$$n_b^{serv} = |\mathcal{U}_b^{serv}| \tag{10}$$

The second indicator, $\mathcal{A}_2(b)$, is used to verify whether the ratio between the number of users that beam b interferes and the number of users it serves is above a threshold η_{th}. The rational is that the benefit from muting or limiting the allocated resources to a beam during a Transmission Time Interval (TTI) increases with the amount of users it interferes and decreases with the amount of users it serves.

$$\mathcal{A}_2(b) = \begin{cases} 1 & \text{if } \frac{n_b^{int}}{n_b^{serv}} > \eta_{th} \\ 0 & \text{otherwise.} \end{cases} \tag{11}$$

The matrix element $Q_{b,b'}$ of $\mathbf{Q}$ is defined as follows:

$$Q_{b,b'} = A_{b,b'} \times 1_{\{\mathcal{A}_2(b)=1\}} \times 1_{\{\mathcal{A}_2(b')=1\}} \tag{12}$$

The threshold values of η_{th} and γ_{th} are determined using a simple optimization procedure (see details is Sect. 4).

The rationale for (12) is the following: consider the case where coordination is based on constraining the MU schedulers, namely not to schedule users served by beams b and b' in the same TTI for which $Q_{b,b'} = 1$ (see Algorithm 2). From Eq. (12), the coordination is performed if both users served by b and b' can benefit from coordination. The algorithm for generating the matrix $\mathbf{Q}$ is given by Algorithm 1.

Algorithm 1. Code Generation

Input: $\mathbf{A}$
Init: $\mathbf{Q} = [0]$, matrix of the same dimension as $\mathbf{A}$
for all couples (b, b') for which $A_{b,b'} = 1$ **do**
    $n_b^{int} \leftarrow |\mathcal{U}_b^{int}|$, $n_b^{serv} \leftarrow |\mathcal{U}_b^{serv}|$
    $n_{b'}^{int} \leftarrow |\mathcal{U}_{b'}^{int}|$, $n_{b'}^{serv} \leftarrow |\mathcal{U}_{b'}^{serv}|$
    $Q_{b,b'} = A_{b,b'} \times 1_{\{\mathcal{A}_2(b)\}} \times 1_{\{\mathcal{A}_2(b')\}}$
end for

The MU scheduling uses a known beam selection feature known as *beam skipping* technique that is applied independently in each cell. It allows to reduce intra- and inter-beam interference and to improve users' rates. It is noted that the ANBR based coordinated scheduling is independent of the beam skipping feature and can be applied without it. Each user is attached to the beam of the GoB achieving the best SINR. The user attachment provides certain spatial information that is exploited by the scheduler to mitigate interference: (i) by avoiding scheduling two users attached to the same control beam and (ii) by avoiding scheduling two users attached to adjacent beams of the GoB, both in the same TTI.

We first present a time based ANBR coordination scheme, denoted for sake of brevity as *time-ANBR* scheme. In this scheme, coordination is achieved by muting of certain beams during certain TTIs as explained below. For sake of simplicity, delay has been ignored but can be easily incorporated into the coordination scheme.

Denote by $R_{u,t_{M+1}}^b$ the instantaneous rate of a user $u \in m$ when it is scheduled at t_{M+1}. The average rate at t_{M+1} is calculated using exponential moving average (or Abel average) with a small parameter ϵ [3].

$$\overline{R_{u,t_{M+1}}} = (1 - \epsilon)\overline{R_{u,t_M}} + \epsilon R_{u,t_{M+1}} \tag{13}$$

Denote by $\mathcal{U}_{candidates}$ the set of users that can still be scheduled and by $\mathcal{U}_K$ - the set of users already selected for scheduling, both at t_{M+1}.

Consider next the scheduling algorithm of cell m (or m') (see Algorithm 2). In the initialization phase, $\mathcal{U}_{candidates}$ contains all the users attached to m (or m'). The scheduler ranks the users in $\mathcal{U}_{candidates}$ with respect to a PF criterion, namely $\dfrac{R_{u,t_{M+1}}^b}{R_{u,t_M}^b + d}$.

All beams $b \in \mathcal{B}_m$ for which $Q_{b,b'} = 1$ are muted at an even TTI, whereas the beams $b' \in \mathcal{B}_{m'}$ are muted at an odd TTI. The users of a muted beam are removed from $\mathcal{U}_{candidates}$. The scheduler selects the top-ranked candidate. Then, it removes the selected users attached to the adjacent beams from the candidate list (following the beam skipping scheme). We repeat the above two operations until K candidates are selected or until the set of candidates is empty.

Algorithm 2. ANBR-assisted MU Scheduler

Input : **Q**
Init: $\mathcal{U}_K = \{\}, \mathcal{U}_{candidates} = \{u \in m\}$
if TTI is even **then**
    for all b for which $Q_{b,b'} = 1$ **do**
        Remove $u \in b$ from $\mathcal{U}_{candidates}$
    end for
end if
while $|\mathcal{U}_k| < k$ or $\mathcal{U}_{candidates} \neq \emptyset$ **do**
    $u_{select} \leftarrow \mathrm{argmax}_{u \in \mathcal{U}_{candidates}} \dfrac{R_{u,t_{M+1}}^b}{R_{u,t_M}^b + d}$
    $\mathcal{U}_k \leftarrow \mathcal{U}_k \cup u_{select}$
    Remove users attached to the beam of u_{select} and to the adjacent beams from $\mathcal{U}_{candidates}$
end while

The second coordination scheme is denoted as frequency based ANBR coordination scheme and is denoted for sake of brevity as *frequency-ANBR* scheme. This scheme is similar to the time-ANBR namely instead of consecutively muting beams for which $Q_{b,b'} = 1$, we allocate to these beams half of the available non-overlapping resources. It is noted that in spite of being very simple, both time- and frequency-ANBR coordination schemes achieve high performance with little computational efforts.

Lastly, the generalization of the coordinated scheduling to the case where beams from three cells interfere with each other is straightforward. Sparse three dimensional matrices $\mathbf{A}$ and $\mathbf{Q}$ need to be generated, and non-overlapping resources (e.g. Physical Resource Blocks (PRBs) in the frequency-ANBR) should be allocated to the beams. One should bear in mind that significant co-located interference from several cells should be minimized in the cell-planning phase and not by the scheduler.

4 Numerical Results

4.1 Simulation Scenario

Consider a network comprising 19 sectors (cells): a central sector and two tiers of 18 neighboring sectors (6 sites located on a hexagonal grid) surrounding it. Each base station is equipped with a M-MIMO antenna as described in Sect. 2. The central sector and one of its direct neighbors denoted hereafter as cell 1 and cell 2 (respectively upper left and lower right in Fig. 1), implement the coordinated scheduling. The reference scenario with no coordination serves as a baseline. It implements a PF based MU-scheduler with the beam skipping feature. Both the time- and frequency-based ANBR coordinated scheduling are simulated and compared to the baseline scenario. The simulation parameters are summarized in Table 1.

The traffic distribution of cells 1 and 2 is shown in Fig. 3. A red rectangle surrounds a hotspot zone with high traffic located around the cell edge area of the two cells. Each cell has 35 users in the hotspot area and 10 users in the rest of the cell, drawn according to a uniform distribution in each zone. The users' color code in Fig. 3 is the following: red and blue squares are the users outside of the hotspot and belonging to the cell 1 and 2 respectively, green and yellow are the users of cell 1 and 2 located in the hotspot.

The best values for the thresholds η_{th} and γ_{th} are determined by means of an exhaustive search. We define a uniform grid of 10×10 points (η_{th}, γ_{th}), with η_{th} varying from $1/10$ to 1 and γ_{th} - from 1 to 10. For each point of the grid we compute the Mean User Throughput (MUT) gain using the time-ANBR coordination scheme as depicted in Fig. 4. The gain increases with the decrease in η_{th} while the MUT gain is not sensitive to variations in γ_{th} except for small values. Too small value of γ_{th} results in too few mobiles that can benefit from the coordination between the two cells. Similarly, a small value of η_{th} allows more beam relations to be included in the matrix $\mathbf{Q}$ and more users will participate in the coordinated scheduling. In the rest of the paper, the thresholds' values of $\gamma_{th} = 2.5$ and $\eta_{th} = 1/6$ are set. A gain of 105% in MUT with respect to the baseline is achieved on a plateau of 27 points in the γ_{th} η_{th} plane, indicating little sensitivity of the thresholds (see Fig. 4).

Figure 6 presents the distribution of the served - and interfered users per beam for both cells using equations (10) and (9) respectively.

Table 1. Network and Traffic characteristics

Network parameters	
Number of BSs	2
Number of interfering macros	6 × 3 sectors
Macro-cell layout	Hexagonal trisector
Number of beams b per macro cell	16
Bandwidth	20 MHz
Channel characteristics	
Thermal noise	−174 dBm/Hz
Path Loss (d in km)	$128.1 + 37.6 \log_{10}(d)$ dB
Nakagami-m shape parameter	5
Intersite distance	500 m
Traffic characteristics	
Number of user in the hotspot zone of each cell	35
Number of user outside the hotspot zone in each cell	10
Traffic distribution in hotspot zone	Uniform
Traffic distribution outside hotspot zone	Uniform
Service Type	Full buffer, data

4.2 Performance Analysis

Figures 5 and 7 compare MUT results for the two coordination schemes, the time- and frequency-ANBR and the baseline. The average results per cell are depicted in Fig. 5 and the time evolution of the MUT for both cells is shown in Fig. 7. The improvement brought about by the coordination schemes is very significant, of the order of 100%. One can see that the time-ANBR performs a bit better than the frequency-ANBR based coordination. This is explained by the fact that when a beam is muted, time resources will be used by users served by other beams. In the case of frequency-ANBR, the coordinated beams are allocated non-overlapping frequency resources and hence not all the available resource are used.

In the following results, we consider the frequency-ANBR coordination scheme compared to the baseline. We divide the users into two groups as a function of their locations, namely outside and inside the hotspot area. The users' throughput are presented in the form of horizontal bars, in an increasing order of throughput values in the baseline case (in blue). The same order is kept for the coordinated scheduling case (in red) to ease comparison.

Figures 8 and 10 present the throughputs of the users outside the hotspot zone of cells 1 and 2 respectively. In cell 1 (Fig. 8) certain users see their throughput grows significantly since they benefit from the cells coordination, while other users see their throughputs slightly decreased. In cell 2 (Fig. 10) a non-significant throughput reduction is observed.

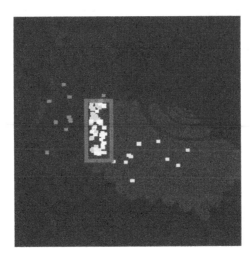

Fig. 3. Traffic map

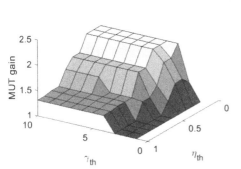

Fig. 4. Comparison of MUT as a function of the thresholds γ_{th} and η_{th}

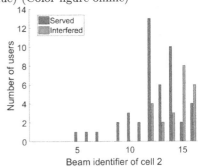

Fig. 5. MUT for frequency(in red)- and time(in yellow)-MUT and baseline(in blue) (Color figure online)

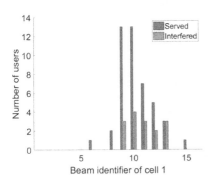

Fig. 6. Number of served and interfered users per beam

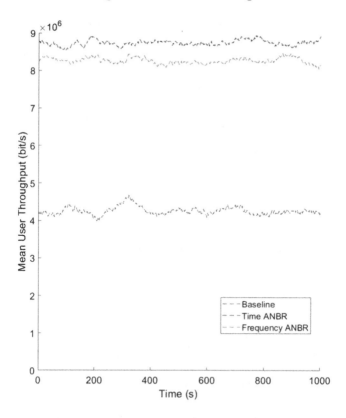

Fig. 7. Time evolution of the MUT of the network

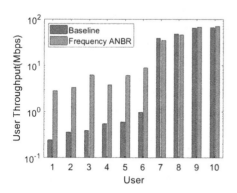

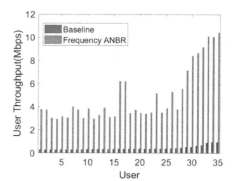

Fig. 8. Throughputs of users outside the hotspot zone in cell 1

Fig. 9. Throughputs of users in the hotspot zone in cell 1

Figures 9 and 11 show the throughputs of the users in the hotspot area of cells 1 and 2 respectively. One can clearly see that most of the users benefit from the coordinated scheduling and see their throughputs significantly increased.

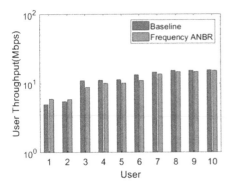 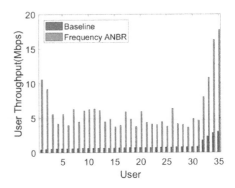

Fig. 10. Throughputs of users outside the hotspot zone in cell 2

Fig. 11. Throughputs of users in the hotspot zone in cell 2

5 Conclusion

This paper has shown how ANBR can be used to coordinate MU scheduling of a pair of neighboring cells with M-MIMO deployment. The strong interference at cell edge motivates the coordination approach. Two solutions have been proposed, a time- and a frequency based coordinated scheduling. The coordination solution exploits the capability to derive beam relations of neighboring cells, which is supported by 5G technology. The dynamic beam relations are updated at the time scale of arrival and departure of users, namely in the order of a second. It makes this approach attractive with respect to traditional techniques such as CoMP which operates at a millisecond time scale and requires high processing capabilities. The coordinated scheduling solution brings about significant throughput gains to users located close to cell edge or in highly interfered area. The ANBR feature has an important potential for other resource allocation and optimization problems such as mobility or load balancing.

References

1. Björnson, et al.: Massive MIMO networks: spectral, energy, and hardware efficiency. Found. Trends® Signal Process. **11**(3–4), 154–655 (2017)
2. Charash, U.: Reception through Nakagami fading multipath channels with random delays. IEEE Trans. Commun. **27**(4), 657–670 (1979)
3. Combes, R., Altman, Z., Altman, E.: Scheduling gain for frequency-selective Rayleigh-Fading channels with application to self-organizing packet scheduling. Perform. Eval. **68**(8), 690–709 (2011)
4. Dahlen, A., et al.: Evaluations of LTE automatic neighbor relations. In: 2011 IEEE 73rd Vehicular Technology Conference (VTC Spring), pp. 1–5 (2011)
5. Irmer, R., et al.: Coordinated multipoint: concepts, performance, and field trial results. IEEE Commun. Mag. **49**(2), 102–111 (2011)

6. O-RAN Alliance: O-RAN: towards an open and smart ran. White Paper (2018)
7. Ramachandra, P., et al.: Automatic neighbor relations (ANR) in 3GPP NR. In: IEEE Wireless Communications and Networking Conference Workshops (WCNCW), pp. 125–130 (2018)
8. Tall, A., et al.: Multilevel beamforming for high data rate communication in 5G networks. arXiv preprint arXiv:1504.00280 (2015)
9. Vook, F.W., et al.: MIMO and beamforming solutions for 5G technology. In: 2014 IEEE MTT-S International Microwave Symposium (IMS2014), pp. 1–4 (2014)

Advance in Game Theory

Financial Replicator Dynamics: Emergence of Systemic-Risk-Averting Strategies

Indrajit Saha[(✉)] and Veeraruna Kavitha

IEOR, IIT Bombay, Mumbai, India
{indrojit,vkavitha}@iitb.ac.in

Abstract. We consider a random financial network with a large number of agents. The agents connect through credit instruments borrowed from each other or through direct lending, and these create the liabilities. The settlement of the debts of various agents at the end of the contract period can be expressed as solutions of random fixed point equations. Our first step is to derive these solutions (asymptotically), using a recent result on random fixed point equations. We consider a large population in which the agents adapt one of the two available strategies, risky or risk-free investments, with an aim to maximize their expected returns (or surplus). We aim to study the emerging strategies when different types of replicator dynamics capture inter-agent interactions. We theoretically reduced the analysis of the complex system to that of an appropriate ordinary differential equation (ODE). We proved that the equilibrium strategies converge almost surely to that of an attractor of the ODE. We also derived the conditions under which a mixed evolutionary stable strategy (ESS) emerges; in these scenarios the replicator dynamics converges to an equilibrium at which the expected returns of both the populations are equal. Further the average dynamics (choices based on large observation sample) always averts systemic risk events (events with large fraction of defaults). We verified through Monte Carlo simulations that the equilibrium suggested by the ODE method indeed represents the limit of the dynamics.

Keywords: Evolutionary stable strategy (ESS) · Replicator dynamics · Ordinary differential equation · Random graph · Systemic risk · Financial network

1 Introduction

We consider a financial network with large number of agents. These agents are interconnected to each other through financial commitments (e.g., borrowing - lending etc.). In addition they make investments in either risk-free (risk neutral) or risky derivatives. In such a system the agents not only face random economic shocks (received via significantly smaller returns of their risky investments),

© Springer Nature Switzerland AG 2021
S. Lasaulce et al. (Eds.): NetGCooP 2021, CCIS 1354, pp. 211–228, 2021.
https://doi.org/10.1007/978-3-030-87473-5_19

they are also affected by the percolation of the shocks faced by their neighbours (creditors), neighbours of their neighbours etc. In the recent years from 2007–2008 onwards, there is a surge of activity to study the financial and systemic level risks caused by such a percolation of shocks [1,3–5]. Systemic risk is the study of the risks related to financial networks, when individual or entity level shocks can trigger severe instability at system level that can collapse the entire economy (e.g., [3–5]). In this set of papers, the author study the kind of topology (or graph structure) that is more stable towards the percolation of shocks in financial network, where stability is measured in terms of the total number of defaults in the network.

In contrast to many existing studies in literature related to systemic risk, we consider heterogeneous agents and we consider evolutionary framework. In our consideration, there are two groups of agents existing simultaneously in the network; one group of agents invest in risk-free instruments, while the other group considers risky investments. The second group borrows money from the other members of the network to gather more funds towards the risky investments (with much higher expected returns). These investments are subjected to large (but rare) economic shocks, which can potentially percolate throughout the network and can even affect the 'risk-free' agents; the extent of percolation depends upon relative sizes of the two groups. We consider that new agents join such a network after each round of investment; they choose their investment type (risky or risk-free) based on their observations of the returns (the surplus of the agents after paying back their liabilities) of a random sample of agents that invested in previous round. The relative sizes of the two groups changes, the network structure changes, which influences the (economic shock-influenced) returns of the agents in the next round, which in turn influences the decision of the new agents for the round after. Thus the system evolves after each round. We study this evolution process using the well known evolutionary game theoretic tools.

In a financial network perspective, this type of work is new to the best of our knowledge. We found few papers that consider evolutionary approach in other aspects related to finance; in [7], the authors study the financial safety net (a series of the arrangement of the firms to maintain financial stability), and analyze the evolution of the bank strategies (to take insurance or not); recently in [6] authors consider an evolutionary game theoretic model with three types of players, i) momentum traders ii) contrarian traders iii) fundamentalists and studied the evolution of the relative populations. As already mentioned, these papers relate to very different aspects in comparison with our work.

Evolutionary Stable Strategies. Traditionally evolutionary game models have been studied in the literature to study animal behaviour. The key ingredients of the evolutionary game models are a) a large number of players, b) the dynamics and c) the pay-off function (e.g., see the pioneering work [10]). Replicator dynamics deals with evolution of strategies, reward based learning in dynamic evolutionary games. Typically it is shown that these dynamics converge to a stable equilibrium point called Evolutionary Stable Strategy (ESS), which can be seen as a refinement of a strict Nash Equilibrium [10]; a strategy prevailing in a large population is called evolutionary stable if any small fraction of mutants playing a different strategy get wiped out eventually. Formally, in a 2-player symmetric game, a pure strategy $\hat{s}$ is said to be evolutionary stable if

1. $(\hat{s}, \hat{s})$ is a Nash Equilibrium; i.e., $u(\hat{s}, \hat{s}) \geq u(s', \hat{s})\ \forall s'$ and
2. If $(\hat{s}, \hat{s})$ is not a strict NE (i.e., $\exists$ some $s' \neq \hat{s}$ such that $u(\hat{s}, \hat{s}) = u(s', \hat{s})$), then $u(\hat{s}, s') > u(s', s')$.

We study the possible emergence of evolutionary stable strategies, when people choose either a risky or a risk-free strategy; the main difference being that the returns of either group are influenced by the percolation of shocks. The returns of the portfolios depend further upon the percolation of shocks due to layered structure of financial connections, and not just on the returns of the investments, i.e., not just on economic shocks. Our main conclusions are two fold; a) when agents consider large sample of data for observation and learning, the replicator dynamics can settle to a mixed ESS, at which the expected returns of the two the groups are balanced; b) in many other scenarios, through theoretical as well as simulation based study, we observed that the replicator dynamics converges to one of the two strategies, i.e., to a pure ESS (after completely wiping out the other group).

The analysis of these complex networks (in each round) necessitated the study of random fixed point equations (defined sample path-wise in large dimensional spaces), which represent the clearing vectors of all the agents ([1,3–5] etc.). The study is made possible because of the recent result in [1], which provided an asymptotically accurate one dimensional equivalent solution.

2   Large Population Finance Network

We consider random graphs, where the edges represent the financial connection between the two nodes. Any two nodes are connected with probability $p_{ss} > 0$ independent of the others, but the weights on the edges depend on (the number of) neighbors. This graph represents a large financial network where borrowing and lending are represented by the edges and the weights over them. The modeller may not have access to the exact connections of the network, but random graph model is a good approach to analyse such a complex system. In particular we consider the graphs that satisfy the assumptions of [1].

The agents are repeatedly investing in some financial projects. In each round of investment, the agents borrow/lend to/from some random subset of the agents of the network. Some of them may invest the remaining in a risk-free investment (which has a constant rate of interest r_s). While the others invest the rest of their money in risky investments which have random returns; we consider a binomial model in which returns are high (rate u) with high probability δ and can have large shocks (rate d), but with small probability $(1 - \delta)$; it is clear that $d < r_s < u$. We thus have two types of agents, we call the group that invests in risk-free projects as 'risk-free' group (G_1), the rest are being referred to as 'risky' group (G_2).

New agents join the network in each round of investment. They choose their
investment type, either risk-free or risky, for the first time based on the previous
experience of the network and continue the same choice for all future rounds of
investment. The new agents learn from network experience (returns of agents of
the previous round of investments) and choose a suitable investment type, that
can potentially give them good returns. The new agents either learn from the
experience of a random sample (returns of two random agents) of the network or
learn from a large number of agents. In the former case, their choice of investment
type depends upon the returns of the random sample in the previous round.
While in the latter case the decision can also depend on the average utility of
each group of the agents, obtained after observing large number of samples.

Two Strategies: As mentioned before, there are two strategies available in
the financial market. Risk-free agents of G_1 use strategy 1; these agents lend
some amount of their initial wealth to other agents (of G_2) that are willing to
borrow, while the rest is invested in a government security, for example, bonds,
government project etc. Risky agents of G_2 are adapting strategy 2, wherein they
borrow funds from the other agents and invest in risky security, for example,
derivative markets, stocks, corporate loans etc. These agents also lend to other
agents of G_2. Let ϵ_t be the fraction of the agents in G_1 group and let $n(t)$ be
the total number of agents in round t. Thus the total number of agents (during
round t) in group 1 equals $n_1(t) := |G_1| = n(t)\epsilon_t$ and $n_2(t) := |G_2| = n(t)(1-\epsilon_t)$.

We consider that one new agent is added in each round[1], and thus size of
the graph/network is increasing. The agents are homogeneous, i.e., they reserve
the same wealth $w > 0$ for investments (at the initial investment period) of each
round. *Each round is composed of two time periods, the agents invest during the
initial investment period* and they obtain their returns after some given time gap.
The two time period model is borrowed from [1,3,4] etc. The new agents make
their choice for the next (and the future) round(s), based on their observations
of these returns of the previous round.

Initial Investment Phases: During the initial investment phases (of any round
t), any agent $i \in G_1$ lends to any agent $j \in G_2$ with probability p_{ss} and it lends
(same) amount[2] $w/(n(t)p_{ss})$ to each of the approachers based on the number that
approached it for loan; let I_{ij} be the indicator of this lending event. Note that for
large $n(t)$, the number of approachers of G_2 approximately equals $n(t)(1-\epsilon_t)p_{ss}$,
and, thus any agent of G_1 lends approximately $w(1 - \epsilon_t)$ fraction to agents of
G_2. The agents of G_1 invest the rest $w\epsilon_t$ in risk-free investment (returns with
fixed rate of interest r_s).

Let $\tilde{w}$ be the accumulated wealth[3] of any agent of G_2 out of which a positive
fraction α is invested towards the other banks of G_2 and $(1 - \alpha)$ portion is

---

[1] This approach can easily be generalized to several other types of dynamics and we
briefly discuss a few of them towards the end.
[2] This normalization, (after choosing the required parameters, like w, appropriately)
is done to derive simpler final expressions.
[3] These amounts could be random and different from agent to agent, but with large
networks (by law of large numbers) one can approximate these to be constants.

invested in risky security. Thus the accumulated wealth of a typical G_2 agent is governed by the following equation,

$$\tilde{w} = \underbrace{w + w\epsilon}_{\text{Initial wealth + Borrowed from } G_1} + \underbrace{\tilde{w}\alpha}_{\text{Lend/borrow } G_2} \quad \text{and thus } \tilde{w} = \frac{w(1+\epsilon)}{(1-\alpha)}.$$

(1)

Thus the total investment towards the risky venture equals $\tilde{w}(1-\alpha) = w(1+\epsilon)$. The G_2 agents have to settle their liabilities at the end of the return/contract period (in each round) and this would depend upon their returns from the risky investments. Thus the total liability of any agent of G_2 is $y = (w\epsilon + \tilde{w}\alpha)(1+r_b)$, where r_b is the borrowing rate[4]; by simplifying

$$y = \frac{w(\epsilon + \alpha)(1 + r_b)}{(1-\alpha)}.$$

Similarly, any agent of G_2 lends the following amount to each of its approachers (of G_2):

$$\frac{\alpha\tilde{w}}{n(t)(1 - \epsilon_t)p_{ss}} = \frac{\alpha w(1+\epsilon)}{n(t)(1 - \epsilon_t)p_{ss}(1-\alpha)}.$$

(2)

Return and Settling Phases, Clearing Vectors: We fix the round t and avoid notation t for simpler notations. The agents of G_2 have to clear their liabilities during this phase in every round. Recall the agents of G_2 invested $w(1+\epsilon)$ amount in risky-investments and the corresponding random returns (after economic shocks) are:

$$K_i = \begin{cases} w(1+\epsilon)(1+u) =: k_u, & \text{w.p. (with probability) } \delta \\ w(1+\epsilon)(1+d) =: k_d, & \text{otherwise} \end{cases}$$

(3)

This is the well known binomial model, in which the upward moment occurs with probability δ and downward moment with $(1-\delta)$. The agents have to return y (after the interest rate r_b) amount to their creditors, however may not be able to manage the same because of the above economic shocks. In case of default, the agents return the maximum possible; let X_i be the amount cleared by the ith agent of group G_2. Here we consider a standard bankruptcy rule, limited liability and pro-rata basis repayment of the debt contract (see [3,4]), where the amounts returned are proportional to their liability ratios. Thus node j of G_2 pays back $X_i L_{ji}/y$ towards node i, where L_{ji} the amount borrowed (liability) during initial investment phases equals (see the details of previous subsection and Eq. (2)):

$$L_{ji} = \begin{cases} I_{ji}\frac{w}{np_{ss}}, & \text{if } i \in G_1 \\ I_{ji}\frac{\alpha w(1+\epsilon)}{np_{ss}(1-\alpha)(1-\epsilon)}, & \text{if } i \in G_2. \end{cases}$$

(4)

---

[4] For simplicity of explanation, we are considering constant terms to represent all these quantities, in reality they would be i.i.d. quantities which are further independent of other rounds and the asymptotic analysis would go through as in [1].

Thus the maximum amount cleared by any agent $j \in \mathcal{G}_2$, X_j, is given by the following fixed point equation in terms of the clearing vector $\{X_i\}_{i \in \mathcal{G}_2}$ composed of clearing values of all the agents (see [3,4] etc.):

$$X_i = \min \left\{ \left(K_i + \sum_{j \in G_2} X_j \frac{L_{ji}}{y} - v \right)^+, y \right\},$$ (5)

with the following details: the term K_i is the return of the risky investment, the term $\sum_{j \in G_2} X_j L_{ji}/y$ equals the claims form the other agents (those borrowed from agent i) and v denotes the taxes to pay. In other words, agent i will pay back the (maximum possible) amount $K_i + \sum_{j \in G_2} X_j \frac{L_{ji}}{y} - v$ in case of a default, and in the other event, will exactly pay back the liability amount y.

Surplus of any agent is defined as the amount obtained from various investments, after clearing all the liabilities. This represents the utility of the agent in the given round. The surplus of the agent $i \in G_2$:

$$R_i^2 = \left(K_i + \sum_{j \in G_2} X_j \frac{L_{ji}}{y} - v - y \right)^+,$$ (6)

while that of agent $i \in G_1$ is given by:

$$R_i^1 = \left(w\epsilon(1 + r_s) + \sum_{j \in G_2} X_j \frac{L_{ji}}{y} - v \right)^+.$$ (7)

In the above, the first term is the return from the risk free investment. The second term equals the returns or claims form G_2 agents (whom they lent) and v denotes the amount of taxes.

3   Asymptotic Approximation of the Large Networks

We thus have dynamic graphs whose size increases with each round. In this section, we obtain appropriate asymptotic analysis of these graphs/systems, with an aim to derive the pay-off of each group after each round. Towards this, we derive the (approximate) closed form expression of the Eqs. (6) and (7), which are nothing but the per-agent returns after the settlement of the liabilities.

The returns of the agents depend upon how other agents settle their liabilities to their connections/creditors. Thus our first step is to derive the solution of the clearing vector fixed point Eqs. (5). Observe that the clearing vector $\{X_j\}_{j \in G_2}$ is the solution of the vector-valued random fixed point Eqs. (5) in n-dimensional space (where n is the size of the network), defined sample-path wise.

Clearing Vectors Using Results of [1]: Our financial framework can be analysed using the results of [1], as the details of the model match[5] the assumptions of

---

[5] Observe that $\alpha(1 + \epsilon)/(\alpha + \epsilon) < 1$.

the paper. By [1, Theorem 1], the aggregate claims converge almost surely to constant values (as the network size increases to infinity):

$$\text{(claims of agents of } G_1\text{)}, \quad \sum_{j \in G_2} X_j \frac{L_{ji}}{y} \to \frac{(1-\alpha)(1-\epsilon)}{\alpha + \epsilon} \bar{x}^\infty \quad \text{a.s., and}$$

$$\text{(claims of agents of } G_2\text{)}, \quad \sum_{j \in G_2} X_j \frac{L_{ji}}{y} \to \frac{\alpha(1+\epsilon)}{(\alpha + \epsilon)} \bar{x}^\infty \quad \text{a.s.,}$$

where the common expected clearing value $\bar{x}^\infty$ satisfies the following fixed point equation in one-dimension:

$$\bar{x}^\infty = E \left[\min \left\{ \left(K_i + \frac{\alpha(1+\epsilon)}{\alpha + \epsilon} \bar{x}^\infty - v \right)^+, y \right\} \right]. \tag{8}$$

Further by the same Theorem, the clearing vectors converge almost surely to (asymptotically independent) random vectors:

$$X_i \to \min \left\{ \left(K_i + \frac{\alpha(1+\epsilon)}{\alpha + \epsilon} \bar{x}^\infty - v \right)^+, y \right\}, \quad \text{for each } i \in G_2. \tag{9}$$

By virtue of the above results, the random returns given by Eqs. (6) and (7), converge almost surely:

$$R_i^1 \to \left(w\epsilon(1 + r_s) + \frac{(1-\alpha)(1-\epsilon)}{(\alpha + \epsilon)} \bar{x}^\infty - v \right)^+, \quad \text{for each } i \in G_1 \tag{10}$$

$$R_i^2 \to \left(K_i + \frac{\alpha(1+\epsilon)}{\alpha + \epsilon} \bar{x}^\infty - v - y \right)^+, \quad \text{for each } i \in G_2. \tag{11}$$

Probability of Default is defined as the fraction of agents of G_2 that failed to pay back their full liability, i.e., $P_d := P(X_i < y)$. For large networks (when the initial network size n_0 itself is sufficiently large), one can use the above approximate expressions and using the same we obtain the default probabilities and the aggregate clearing vectors in the following (proof in Appendix).

Lemma 1. *The asymptotic average clearing vector and the default probability of G_2 is given by:*

$$(\bar{x}^\infty, P_d) = \begin{cases} (y, & 0) & \text{if } c_\epsilon > \frac{y-\underline{w}}{y} \\ \left(\frac{\delta y + (1-\delta)\overline{w}}{1-(1-\delta)c_\epsilon}, & 1-\delta \right) & \text{if } \frac{y-\overline{w}}{y-(1-\delta)(\overline{w}-\underline{w})} < c_\epsilon < \frac{y-\underline{w}}{y} \\ \left(\frac{k_d(1-\delta) + k_u \delta - v}{1-c_\epsilon}, & 1 \right) & \text{if } c_\epsilon < \frac{y-\overline{w}}{y-(1-\delta)(\overline{w}-\underline{w})} \end{cases} \tag{12}$$

where, $c_\epsilon = \frac{\alpha + \alpha\epsilon}{\alpha + \epsilon}$, $E[W] = \delta k_u + (1-\delta)k_d - v$, $\underline{w} = k_d - v$ *and* $\overline{w} = k_u - v$. ∎

Expected Surplus: By virtue of the Theorem developed in [1, Theorem 1] we have a significantly simplified limit system, whose performance is derived in the above Lemma. We observe that this approximation is sufficiently close (numerical simulations illustrate good approximations), and assume the following as the pay-offs of each group after each round of the investments:

$$\phi_1(\epsilon) := E(R_i^1) = \left(w\epsilon(1 + r_s) + \frac{(1-\alpha)(1-\epsilon)}{\alpha + \epsilon} \bar{x}^\infty - v \right)^+, \text{ for any agent of } G_1$$

$$\phi_2(\epsilon) := E(R_i^2) = E\left(K_i + \frac{(1+\epsilon)\alpha}{\alpha + \epsilon} \bar{x}^\infty - v - y \right)^+, \tag{13}$$

$$= \left(k_u + \frac{\alpha(1+\epsilon)}{\alpha + \epsilon} \bar{x}^\infty - v - y \right)^+ \delta + \left(k_d + \frac{\alpha(1+\epsilon)}{\alpha + \epsilon} \bar{x}^\infty - v - y \right)^+ (1 - \delta),$$

for any agent of G_2. Observe here that the aggregate limits are almost sure constants, hence the expected surplus of all the agents of the same group are equal, while the random returns of the same group are i.i.d. (independent and identically distributed).

4   Analysis of Replicator Dynamics

In every round of investments, we have a new network that represents the liability structure of all the agents of that round formed by the investment choices of the agents, and, in the previous two sections we computed the (asymptotically approximate) expected returns/utilities of each agent of the network. As already mentioned in Sect. 2, new agents join the network in each round, and choose their strategies depending upon their observations of these expected returns of the previous round.

These kind of dynamics is well described in literature by name replicator dynamics (e.g.,[2,6,9] etc.). The main purpose of such a study is to derive asymptotic analysis and answer some or all of the following questions: will the dynamics converge, i.e., would the relative fractions of various populations settle as the number of rounds increase? will some of the strategies disappear eventually? if more than one population type survives what would be the asymptotic fractions? etc. These kind of analysis are common in other types of networks (e.g., wireless networks (e.g., [9]), biological networks [2]), but are relatively less studied in the context of financial networks (e.g., [6]). We are interested in knowing the asymptotic outcome of these kind of dynamics (if there exists one) and study the influence of various network parameters on the outcome. We begin with precise description of the two types of dynamics considered in this paper.

4.1   Average Dynamics

The new agent contacts two random (uniformly sampled) agents of the previous round. If both the contacted agents belong to the same group, the new agent adapts the strategy of that group. When it contacts agents from both the groups it investigates more before making a choice; the new agent observes significant portion of the network, in that, it obtains a good estimate of the average utility of agents belonging to both the groups. It adapts the strategy of the group with maximum (estimated) average utility.

Say it observes the average of each group with an error that is normally distributed with mean equal to the expected return of the group and variance proportional to the size of the group, i.e., it observes (here $\mathcal{N}(0,\sigma^2)$ is a zero mean Gaussian random variable with variance σ^2)

$$\hat{\phi}_i(\epsilon) = \phi_i(\epsilon) + \mathcal{N}_i \text{ with } \mathcal{N}_1 \sim \mathcal{N}\left(0, \frac{1}{\bar{c}\epsilon}\right) \text{ and } \mathcal{N}_2 \sim \mathcal{N}\left(0, \frac{1}{\bar{c}(1-\epsilon)}\right),$$

for some $\bar{c}$ large. Observe by this modeling that: *the expected values of the observations are given by $(\phi_1(\epsilon), \phi_2(\epsilon))$ and are determined by the relative proportions of the two populations, while the variance of any group reduces as its proportion increases to 1 and increases as the proportion reduces to zero.* We also assume that the estimation errors $\{\mathcal{N}_1, \mathcal{N}_2\}$ (conditioned on the relative fraction, ϵ) corresponding to the two groups are independent. Then the probability that the new agent chooses strategy 1 is given by

$$Prob(\hat{\phi}_1(\epsilon) - \hat{\phi}_2(\epsilon) > 0) = Prob(\mathcal{N}_2 - \mathcal{N}_1 \le \phi_1(\epsilon) - \phi_2(\epsilon)),$$

which by (conditional) independence of Gaussian random variables equals[6]

$$g(\epsilon) := \int_{-\infty}^{(\phi_1(\epsilon)-\phi_2(\epsilon))\sqrt{\bar{c}\epsilon(1-\epsilon)}} e^{-x^2/2} \frac{dx}{\sqrt{2\pi}}. \tag{14}$$

Let $(n_1(t), n_2(t))$ respectively represent the sizes of G_1 and G_2 population after round t and note that $\epsilon_t = \frac{n_1(t)}{n_1(t)+n_2(t)}$. Then the system dynamics is given by the following ($g(\cdot)$ given by (14)):

$$(n_1(t+1), n_2(t+1)) = \begin{cases} (n_1(t)+1, n_2(t)) & \text{w.p. } \epsilon_t^2 + 2\epsilon_t(1-\epsilon_t)g(\epsilon_t) \\ (n_1(t), n_2(t)+1) & \text{w.p. } (1-\epsilon_t)^2 + 2\epsilon_t(1-\epsilon_t)(1-g(\epsilon_t)). \end{cases} \tag{15}$$

It is clear that (with ϵ_0 and n_0 representing the initial quantities),

$$\epsilon_{t+1} = \frac{n_1(t+1)}{t+n_0+1} = \frac{(t+n_0)\epsilon_t + Y_{t+1}}{t+n_0+1} = \epsilon_t + \frac{1}{t+n_0+1}(Y_{t+1}-\epsilon_t) \text{ where}$$

$$Y_{t+1} = \begin{cases} 1 & \text{wp } \epsilon_t^2 + 2\epsilon_t(1-\epsilon_t)g(\epsilon_t) \\ 0 & \text{wp } (1-\epsilon_t)^2 + 2\epsilon_t(1-\epsilon_t)(1-g(\epsilon_t)), \text{ for all } t \ge 1. \end{cases}$$

One can rewrite the update equations as

$$\epsilon_{t+1} = \epsilon_t + \frac{1}{t+n_0+1}(h(\epsilon_t) + M_{t+1}), \text{ with, } M_{t+1} := Y_{t+1} - \epsilon_t - h(\epsilon_t), \text{ where,}$$

$$h(\epsilon) := E[Y_{t+1} - \epsilon_t | \epsilon_t = \epsilon] = \epsilon(1-\epsilon)(2g(\epsilon)-1) \text{ for any } 0 \le \epsilon \le 1.$$

---
[6] Because $\frac{1}{\epsilon} + \frac{1}{1-\epsilon} = \frac{1}{\epsilon(1-\epsilon)}$.

and observe that (with $\mathcal{F}_t$ the natural filtration of the process till t)

$$E[M_{t+1}|\mathcal{F}_t] = E[M_{t+1}|\epsilon_t] = 0 \text{ and } E[M_{t+1}^2|\mathcal{F}_t] \leq C \text{ for some constant } C < \infty.$$

Further observe that $0 \leq \epsilon_t \leq 1$ for all t and all sample paths.

Thus our algorithm satisfies assumptions[7] **A.1** to **A.4** of [8] and hence we have using [8, Theorem 2] that

Theorem 1. *The sequence $\{\epsilon_t\}$ generated by average dynamics (15) converges almost surely (a.s.) to a (possibly sample path dependent) compact connected internally chain transitive invariant set of ODE:*

$$\dot{\epsilon}_t = h(\epsilon_t). \tag{16}$$

∎

The dynamics start with initial condition $\epsilon_0 \in (0,1)$ and clearly would remain inside the interval $[0,1]$, i.e., $\epsilon_t \in [0,1]$ for all t (and almost surely). Thus we consider the invariant sets of ODE (16) within this interval for some interesting case studies in the following (Proof in Appendix).

Corollary 1. *Define $\bar{r}_r := u\delta + d(1-\delta)$. And assume $w(1+d) > v$ and observe that $u > r_b \geq r_s > d$. Assume $\epsilon_0 \in (0,1)$. Given the rest of the parameters of the problem, there exists a $\bar{\delta} < 1$ (depends upon the instance of the problem) such that the following statements are valid for all $\delta \geq \bar{\delta}$:*

(a) *If $\bar{r}_r > r_b > r_s$ then $\phi_2(\epsilon) > \phi_1(\epsilon)$ for all ϵ, and $\epsilon_t \to 0$ almost surely.*
(b) *If $\phi_1(\epsilon) > \phi_2(\epsilon)$ for all ϵ then $\epsilon_t \to 1$ almost surely.*
(c) *When $r_b > \bar{r}_r > r_s$, and case (b) is negated there exists a unique zero ϵ^* of the equation $\phi_1(\epsilon) - \phi_2(\epsilon) = 0$ and*

$$\epsilon_t \to \epsilon^* \text{ almost surely; further for } \delta \approx 1, \ \epsilon^* \approx \frac{r_b - \bar{r}_r}{\bar{r}_r - r_s}.$$

∎

From (13) and Lemma 1, it is easy to verify that all the limit points are evolutionary stable strategies (ESS). Thus the replicator dynamics either settles to a pure strategy ESS or mixed ESS (in part (c) of the corollary), depending upon the parameters of the network; after a large number of rounds, either the fraction of agents following one of the strategies converges to one or zero or the system reaches a mixed ESS which balances the expected returns of the two groups.

In many scenarios, the expected rate of return of the risky investments is much higher than the rate of interest related to lending/borrowing, i.e., $\bar{r}_r > r_b$. Further the assumptions of the corollary are satisfied by more or less all the scenarios (due to standard no-arbitrage assumptions) and because the shocks are usually rare (i.e., δ is close to 1). Hence by the above corollary, in majority

---

[7] The assumptions require that the process is defined for the entire real line. One can easily achieve this by letting $h(\epsilon) = 0$ for all $\epsilon \notin [0,1]$, which still ensures required Lipschitz continuity and by extending $M_{t+1} = 0$ for all $\epsilon_t \notin [0,1]$.

of scenarios, the average dynamics converges to a pure strategy with all 'risky' agents (i.e., $\epsilon_t \to 0$). The group G_1 gets wiped out and almost all agents invest in risky ventures, as the expected rate of returns is more even in spite of large economic shocks. One can observe a converse or *a mixed ESS when the magnitude of the shocks is large (d too small) or when the shocks are too often to make* $\bar{r}_r < r_b$.

4.2   Random Dynamics

When the new agent contacts two random agents of different groups, its choice depends directly upon the returns of the two contacted agents. The rest of the details remain the same as in average dynamics. In other words, the new agents observe less, their investment choice is solely based on the (previous round) returns of the two contacted agents. In this case the dynamics are governed by the following (see (10)–(11)):

$$(n_1(t+1), n_2(t+1)) = \begin{cases} (n_1(t)+1, & n_2(t)) & \text{wp } \epsilon_t^2 \\ (n_1(t), & n_2(t)+1) & \text{wp } (1-\epsilon_t)^2 \\ (n_1(t)+G(\epsilon_t), & n_2(t)+(1-G(\epsilon_t))) & \text{else, with} \end{cases}$$

$$G(\epsilon_t) = 1_{\{R^1 \geq R^2\}} \tag{17}$$

$$= 1_{\left\{ \left(we(1+r_s) + \frac{(1-\alpha)(1-\epsilon)}{(\alpha+\epsilon)} \bar{x}^\infty - v \right)^+ \geq \left(K_i + \frac{\alpha(1+\epsilon)}{\alpha+\epsilon} \bar{x}^\infty - v - y \right)^+ \right\}},$$

where $\bar{x}^\infty = \bar{x}^\infty(\epsilon_t)$ is given by Lemma 1. Here we assume people prefer risk-free strategy under equality, one can easily consider the other variants. Once again this can be rewritten as

$$\epsilon_{t+1} = \epsilon_t + \frac{Z_{t+1} - \epsilon_t}{t + n_0 + 1} \text{ with } Z_{t+1} = \begin{cases} 1 & \text{wp } \epsilon_t^2 \\ 0 & \text{wp } (1-\epsilon_t)^2 \\ G(\epsilon_t) & \text{else.} \end{cases} \tag{18}$$

As in previous section the above algorithm satisfies assumptions[8] **A**.1 to **A**.4 of [8] and once again using [8, Theorem 2], we have:

Theorem 2. *The sequence $\{\epsilon_t\}$ generated by average dynamics (17) converges almost surely (a.s.) to a (possibly sample path dependent) compact connected internally chain transitive invariant set of ODE:*

$$\dot{\epsilon}(t) = h_R(\epsilon(t)), \ h_R(\epsilon) := E_\epsilon \left[Z_{t+1} - \epsilon_t | \epsilon_t = \epsilon \right] = \epsilon(1-\epsilon)(2E[G(\epsilon)] - 1). \tag{19}$$

∎

One can derive the analysis of this dynamics in a similar way as in average dynamics, however there is an important difference between the two dynamics; we can never have random dynamics converges to an intermediate attractor, like

---

[8] In the current paper, we consider scenarios in which $h_R(\cdot)$ is Lipschitz continuous, basically under the conditions of Corollary 2.

the attractor in part (c) of Corollary 1 (unique ϵ^* satisfying $\phi_1 = \phi_2$). This is because $E_\epsilon[G] = P(R^1(\epsilon) > R^2(\epsilon))$ equals 0, $1 - \delta$ or 1 and never $1/2$ (unless $\delta = 1/2$, which is not a realistic case). Nevertheless, we consider the invariant sets (corresponding to pure ESS) within $[0,1]$ for some cases (Proof in Appendix):

Corollary 2. *Assume $\epsilon_0 \in (0,1)$. Given the rest of the parameters of the problem, there exists a $1/2 < \delta < 1$ (depends upon the instance of the problem) such that the following statements are valid:*

(a) If $E_\epsilon[G] = 0$ for all ϵ or $1 - \delta$ for all ϵ, then $\epsilon_t \to 0$ almost surely.
(b) If $E_\epsilon[G] = 1$ for all ϵ, then $\epsilon_t \to 1$ almost surely.
(c) When $w(1 + d) > v$ and $u > r_b \geq r_s > d$, there exists a $\bar{\delta} < 1$ such that for all $\delta \geq \bar{\delta}$, the default probability $P_d \leq (1 - \delta)$ and $E[G] = 1 - \delta$ and this is true for all ϵ. Hence by part (a), $\epsilon_t \to 0$ almost surely. ∎

Remarks: *Thus from part (c), under the conditions of Corollary 1, the random dynamics always converges to all 'risky' agents (pure ESS), while the average dynamics, as given by Corollary 1, either converges to pure or mixed ESS further based on system parameters (mainly various rates of return).*
From this partial analysis (corollaries are for large enough δ) it appears that one can never have mixed ESS with random dynamics, and this is a big contrast to the average dynamics; when agents observe sparsely the network eventually settles to one of the two strategies, and if they observe more samples there is a possibility of emergence of mixed ESS that balances the two returns. We observe similar things, even for δ as small as 0.8 in numerical simulations (Table 4). We are keen to understand this aspect in more details as a part of the future work.

To summarize we have a financial network which grows with new additions, in which the new agents adapt one of the two available strategies based on the returns of the agents that they observed/interacted with. Our asymptotic analysis of [1] was instrumental in deriving these results. This is just an initial study of the topic. One can think of other varieties of dynamics, some of which could be a part of our future work. The existing agents may change their strategies depending upon their returns and observations. The agents might leave the network if they have reduced returns repeatedly. The network may adjust itself without new additions etc.

5   Numerical Observations

We performed numerical simulations to validate our theory. We included Monte-Carlo (MC) simulation based dynamics in which the clearing vectors are also computed by directly solving the fixed point equations, for any given sample path of shocks. Our theoretical observation well matches the MC based limits.

In Tables 1, 2 and 3 we tabulated the limits of the average dynamics for various scenarios, and the observations match the results of Corollary 1. The configuration used for Table 1 is: $n_0 = 2000, \epsilon_0 = 0.75, r_s = 0.18, r_b = 0.19, w = 100, v = 46, \alpha = 0.1$, while that used for Table 2 is: $n_0 = 2000, \epsilon_0 = 0.5, r_s = 0.17, r_b = 0.19, w = 100, v = 40, \alpha = 0.1$. For both these tables risky expected

Table 1. When the shocks are too large along with larger taxes ($v = 46$), the average dynamics converges to a configuration with all 'risk-free agents'!

u	d	δ	ϕ_1	ϕ_2	ϵ^*
0.2	−0.05	0.8	72	0	1
0.2	−0.1	0.8	72	0	1
0.2	−0.15	0.8	72	0	1
0.2	−0.2	0.8	72	0	1
0.2	−0.25	0.8	72	0	1

Table 2. Average dynamics converges to mixed ESS, at which both populations survive with $\phi_1 = \phi_2$.

u	d	δ	ϕ_1	ϕ_2	ϵ^*
0.2	−0.1	0.95	78.33	78.27	0.3326
0.2	−0.11	0.95	78.24	78.31	0.3791
0.2	−0.12	0.95	78.14	78.14	0.4288
0.2	−0.13	0.95	78.04	78.04	0.4820
0.2	−0.14	0.95	77.92	77.92	0.5385

Table 3. Average Dynamics converges to all 'risky-agents'; Configuration: $n_0 = 2000, \epsilon_0 = .5, r_s = 0.10, r_b = 0.12, w = 100, v = 30, \alpha = 0.5$

u	d	δ	ϕ_1	ϕ_2	ϵ^*
0.15	−0.1	0.9	0	82.12	0
0.16	−0.1	0.9	0	83.24	0
0.17	−0.1	0.9	0	84.29	0
0.18	−0.1	0.9	0	85.19	0

Table 4. Average and Random dynamics, Comparison of MC results with theory Configuration: $n_0 = 2000, u = 0.2, r_s = 0.17, r_b = 0.19, w = 100, \alpha = 0.1$

Config	ϵ^* (Theory)		ϵ^* (Monte Carlo)	
(d, δ, v)	Avg	Rndm	Avg	Rndm
0.10, 0.95, 40	0	0	.0016	0.0011
−0.10, 0.95, 40	0.33	0	.3214	0.0004
−0.15, 0.95, 40	0.6	0	.5988	0.0014
0.10, 0.80, 46	1	0	.9896	0.0065

rate of returns $\bar{r}_r$ is smaller than r_b and the dynamics converges either to 'all risky' agents configuration or to a mixed ESS. In Table 3, the risky expected rate of returns $\bar{r}_r = .1250$ which is greater than r_b and r_s, thus the dynamics converges to all risky-agents, as indicated by Corollary 1.

In Table 4 we considered random dynamics as well as average dynamics. In addition, we provided the Monte-Carlo based estimates. There is a good match between the MC estimates and the theory. Further we have the following observations: a) random dynamics always converge to a configuration with all 'risky' agents, as given by Corollary 2; b) when $\bar{r}_r > r_b$, the average dynamics also converges to $\epsilon^* = 0$ as suggested by Corollary 1; and c) when $\bar{r}_r < r_b$, the average dynamics converges to mixed ESS or to a configuration with all 'risk-free' agents, again as given by Corollary 1.

As the 'risk increases', i.e., as the amount of taxes increase and or as the expected rate of return of risky investments $\bar{r}_r$ decreases, one can observe that the average dynamics converges to all 'risk-free' agents (last row of Table 4) thus averting systemic risk event (when there are large number of defaults, P_d). While the random dynamics fails to do the same. As predicted by theory (the configurations satisfying part (b) of Corollary 2), random dynamics might also succeed in averting the systemic risk event, when the expected number of defaults is one for all $\epsilon > 0$. It is trivial to verify that the configuration with $w(1 + u) < v$, is one such example. Thus, average dynamics is much more robust towards averting systemic risk events.

6   Conclusions

We consider a financial network with a large number of agents. The agents are interconnected via liability graphs. There are two types of agents, one group lends to others and invests the rest in risk-free projects, while the second group borrows/lends and invests the rest in risky ventures. Our study is focused on analysing the emergence of these groups, when the new agents adapt their strategies for the next investment round based on the returns of the previous round. We considered two types of dynamics; in average dynamics the new agents observe large sample of data before deciding their strategy, and in random dynamics the decision is based on a small random sample.

We have the following important observations: a) when the expected rate of return of the risky investments is higher (either when the shocks are rare or when the shocks are not too large) than the risk-free rate, then 'risk-free' group wipes out eventually, almost all agents go for risky ventures; this is true for both types of dynamics; b) when the expected rate of risky investments is smaller, a mixed ESS can emerge with average dynamics while the random dynamics always converges to all risky agents; at mixed ESS the expected returns of both the groups are equal; more interestingly, when the risky-expected rate is too small, the average dynamics converges to a configuration with all risk-free agents.

In other words, in scenarios with possibility of a systemic risk event, i.e., when there is a possibility of the complete-system collapse (all agents default), the average dynamics manages to wipe out completely the risky agents; the random dynamics can fail to do the same. Thus when agents make their choices rationally and after observing sufficient sample of the returns of the previous round of investments, there is a possibility to avoid systemic risk events. These are some initial results and we would like to investigate further in future to make more affirmative statements in this direction.

Appendix

Proof of Lemma 1: We consider the following:

<u>Case 1:</u> First consider the case when downward shock can be absorbed, in this case the clearing vector $\bar{x}^\infty = y\delta + y(1 - \delta) = y$, default probability is $P_d = 0$. The region is true if the following condition is meet i.e., if

$$k_d - v + yc_\epsilon > y \implies c_\epsilon > \frac{y - w}{y}$$

<u>Case 2:</u> Consider the case with banks receive shock will default and the corresponding average clearing vector $\bar{x}^\infty = y\delta + (\underline{w} + c_\epsilon \bar{x}^\infty)(1 - \delta)$ which simplifies to:

$$\bar{x}^\infty = \frac{y\delta + \underline{w}(1 - \delta)}{1 - c_\epsilon(1 - \delta).}$$

This region lasts if the following conditions hold to be true

$$k_d - v + c_\epsilon \bar{x}^\infty < y, \text{ and } k_u - v + c_\epsilon \bar{x}^{2\infty} > y.$$

Substituting $\bar{x}^\infty = \frac{y\delta + \underline{w}(1-\delta)}{1 - c_\epsilon(1-\delta)}$ we have,

$$\frac{y - \overline{w}}{y - (1 - \delta)(\overline{w} - \underline{w})} < c_\epsilon < \frac{y - \underline{w}}{y}.$$

In this regime the default probability is $P_d = (1 - \delta)$. Case 3 can be proved similarly, more details are in [11]. ∎

Proof of Corollary 1: First consider the system with $\delta = 1$, i.e., system without shocks. From Lemma 1, $P_d \leq (1 - \delta)$ for all ϵ because (with $\delta = 1$)

$$y\left(c_\epsilon - \frac{y - \overline{w}}{y}\right) = w(1+\epsilon)(1+u) - v - w\epsilon(1+r_b) = w(1+u) - v + w\epsilon(u - r_b) \geq w(1+u) - v,$$

for all ϵ (the lower bound independent of ϵ). Under these assumptions, there exists $\bar{\delta} < 1$ by continuity of the involved functions such that

$$y\left(c_\epsilon - \frac{y - \overline{w}}{y - (1 - \delta)(\overline{w} - \underline{w})}\right) > 0 \text{ for all } \delta \geq \bar{\delta} \text{ and for all } \epsilon.$$

Thus from Lemma 1 $\bar{x}^\infty = y$ or $\bar{x}^\infty = \frac{\delta y + (1-\delta)\underline{w}}{1 - (1-\delta)c_\epsilon}$ for all such $\delta \geq \bar{\delta}$. We would repeat a similar trick again, so assume initially $\bar{x}^\infty = y$ for all ϵ and consider $\delta \geq \bar{\delta}$. With this assumption we will have:

$$R^1(\epsilon) = \left(w\epsilon(1 + r_s) + \frac{(1 - \alpha)(1 - \epsilon)}{(\alpha + \epsilon)}y - v\right)^+ \tag{20}$$
$$= (w\epsilon(1 + r_s) + w(1 - \epsilon)(1 + r_b) - v)^+$$
$$= (w(1 + r_b) - v + w\epsilon(r_s - r_b)), \text{ under the given hypothesis, and}$$

$$R^2(\epsilon) = \left(K_i + \frac{\alpha(1 + \epsilon)}{\alpha + \epsilon}y - v - y\right)^+ = (K_i - w\epsilon(1 + r_b) - v)^+ \tag{21}$$
$$= \begin{cases} R_u^2 & \text{w.p. } \delta & \text{where } R_u^2 := w(1 + u) - v + w\epsilon(u - r_b) \\ (R_d^2)^+ & \text{w.p. } 1 - \delta & \text{where } R_d^2 := w(1 + d) - v + w\epsilon(d - r_b). \end{cases}$$

Note that $R_u^2 \geq w(1 + u) - v > 0$ (for any ϵ) under the given hypothesis.

Proof of part (a): When $\bar{r}_r > r_b$, from (13), it is clear that (inequality only when R_d^2 is negative)

$$\phi_2(\epsilon) - \phi_1(\epsilon) \geq R_u^2 \delta + R_d^2(1 - \delta) - \phi_1(\epsilon) = w(\bar{r}_r - r_b) + w\epsilon(\bar{r}_r - r_s) > 0.$$

Thus in this case $\phi_2 > \phi_1$ for all ϵ and hence

$$g(\epsilon) < 1/2 \text{ and } 2g(\epsilon) - 1 < 0 \text{ for all } 0 < \epsilon < 1.$$

Therefore with Lyaponuv function $V_0(\epsilon) = \epsilon/(1-\epsilon)$ on defined on neighbourhood $[0,1)$ of 0 (in relative topology on $[0,1]$) we observe that

$$\frac{dV_0}{d\epsilon} h(\epsilon) = \frac{\epsilon}{1-\epsilon}(2g(\epsilon) - 1) < 0 \text{ for all } 0 < \epsilon < 1 \text{ and equals } 0 \text{ for } \epsilon = 0.$$

Further $V_0(\epsilon) \to \infty$ as $\epsilon \to 1$, the boundary point of $[0,1)$. Thus $\epsilon^* = 0$ is the asymptotically stable attractor of ODE (16) (see [8, Appendix, pp.148]) and hence the result follows by Theorem 1.

For all $\delta \geq \bar{\delta}$, from Lemma 1, we have the following

$$\sup_\epsilon |y - \bar{x}^\infty| = \sup_\epsilon (1-\delta) \left| \frac{y - c_\epsilon - w}{1 - (1-\delta)c_\epsilon} \right| < \frac{1-\delta}{\delta} \eta \qquad (22)$$

for some $\eta > 0$, which decreases to 0 as $\delta \to 1$. (The last inequality is due to $c_\epsilon < 1$ and then taking supremum over ϵ). By continuity of the above upper bound with respect to δ and the subsequent functions considered in the above parts of the proof, there exists a $\bar{\delta} < 1$ (further big if required) such that all the above arguments are true for all $\delta > \bar{\delta}$.

Proof of part (b): The proof follows in similar way, now using Lyaponuv function $V_1(\epsilon) = (1-\epsilon)/\epsilon$ on neighbourhood $(0,1]$ of 1, and by observing that $g(\epsilon) > 1/2$ for all $\epsilon < 1$ and hence

$$\frac{dV_1}{d\epsilon} h(\epsilon) = -\frac{1-\epsilon}{\epsilon}(2g(\epsilon) - 1) < 0 \text{ for all } 0 < \epsilon < 1 \text{ and equals } 0 \text{ for } \epsilon = 1.$$

Proof of part (c): It is clear that $\phi_1(\epsilon) = R^1(\epsilon)$ decreases linearly as ϵ increases:

$$\phi_1(\epsilon) = w(1 + r_b) - v + w\epsilon(r_s - r_b).$$

For ϵ in the neighbourhood of 0, $\phi_2(\epsilon) > 0$ and is decreasing linearly with slope $\bar{r}_r - r_b$, because $R_d^2(0) = w(1+d) - v > 0$ and thus for such ϵ

$$\phi_2(\epsilon) = w(1 + \bar{r}_r) - v + w\epsilon(\bar{r}_r - r_b).$$

From (21), $R_d^2(\epsilon)$ is decreasing with increase in ϵ. There is a possibility of an $\bar{\epsilon}$ that satisfies $R_d^2(\bar{\epsilon}) = 0$, in which case ϕ_2 increases linearly with slope $\delta w(u - r_b)$, i.e.,

$$\phi_2(\epsilon) = \delta \left[w(1 + u) - v + w\epsilon(u - r_b) \right] \text{ for all } \epsilon \geq \bar{\epsilon}.$$

When $\bar{r}_r < r_b$ we have,

$$\phi_1(0) = w(1 + r_b) - v > w(1 + \bar{r}_r) - v = \phi_2(0).$$

By hypothesis $\phi_1(\epsilon) < \phi_2(\epsilon)$ for some ϵ, hence by intermediate value theorem there exists at least one ϵ^* that satisfies $\phi_1(\epsilon^*) = \phi_2(\epsilon^*)$. Further the zero is unique because ϕ_2 is either linear or piece-wise linear (with different slops), while ϕ_1 is linear.

Consider Lyaponuv function $V_*(\epsilon) := (\epsilon - \epsilon^*)^2/(\epsilon(1 - \epsilon))$ on neighbourhood $(0, 1)$ of ϵ^*, note $V_*(\epsilon) \to \infty$ as $\epsilon \to 0$ or $\epsilon \to 1$ and observe by (piecewise) linearity of the functions we will have

$$\phi_1(\epsilon) > \phi_2(\epsilon) \text{ and thus } (2\,g(\epsilon) - 1) > 0 \text{ for all } 0 < \epsilon < \epsilon^* \text{ and}$$
$$\phi_2(\epsilon) > \phi_1(\epsilon) \text{ and thus } (2\,g(\epsilon) - 1) < 0 \text{ for all } 1 > \epsilon > \epsilon^*.$$

Thus we have[9],

$$\frac{dV_*}{d\epsilon} = 2\frac{\epsilon - \epsilon^*}{\epsilon(1 - \epsilon)} + \frac{(\epsilon - \epsilon^*)^2(2\epsilon - 1)}{\epsilon^2(1 - \epsilon)^2} \text{ and hence}$$

$$\frac{dV_*}{d\epsilon}h(\epsilon) = (\epsilon - \epsilon^*)\left(2 + \frac{(\epsilon - \epsilon^*)(2\epsilon - 1)}{\epsilon(1 - \epsilon)}\right)(2\,g(\epsilon) - 1) < 0 \text{ for all } \epsilon \notin \{0, 1, \epsilon^*\}.$$

Thus ϵ^* is the asymptotically stable attractor of ODE (16) and hence the result follows by Theorem 1. The result can be extended for $\delta < 1$ as in case (a) and the rest of the details follow by direct verification (at $\delta = 1$), i.e., by showing that $\phi_1(\epsilon^*) = \phi_2(\epsilon^*)$ at $\delta = 1$ and the equality is satisfied approximately in the neighbourhood of $\delta = 1$. ∎

Proof of Corollary 2: For part (a), $h_R(\epsilon) = -c_G\epsilon(1 - \epsilon)$, where the constant $c_G = 1$ (or respectively $c_G = 2\delta - 1$). Using Lyanponuv function of part (a) of Corollary 1, the proof follows in exactly the same lines.
For part (b), $h_R(\epsilon) = \epsilon(1 - \epsilon)$, and proof follows as in part (b) of Corollary 1.
For part (c), first observe (using equations (20)–(21) of proof of Corollary 1)

$$R_u^2(\epsilon) - R^1(\epsilon) \geq w(1 + u) + w\epsilon(u - r_s) - y + \bar{x}^\infty\left(\frac{2\alpha + \epsilon - 1}{\alpha + \epsilon}\right)$$
$$= w(1 + u) + w\epsilon(u - r_s) + (\bar{x}^\infty - y) - \bar{x}^\infty\left(\frac{1 - \alpha}{\alpha + \epsilon}\right)$$
$$= w(u - r_b) + w\epsilon(u - r_s) + (\bar{x}^\infty - y)\left(1 - \frac{1 - \alpha}{\alpha + \epsilon}\right) > 0.$$

The last inequality is trivially true for $\delta = 1$ (and so $\bar{x}^\infty = y$) for the given hypothesis, and then by continuity as in proof of Corollary 1, one can consider

---

[9] When $\epsilon < 1/2$ and $\epsilon < \epsilon^*$ then clearly $\frac{(\epsilon - \epsilon^*)(2\epsilon - 1)}{\epsilon(1 - \epsilon)} > 0$. When $\epsilon > 1/2$ we have $(2\epsilon - 1)/\epsilon < 1/2$ and with $\epsilon < \epsilon^*$ we have $\epsilon^* - \epsilon < 1 - \epsilon$ and thus

$$2 + \frac{(\epsilon - \epsilon^*)(2\epsilon - 1)}{\epsilon(1 - \epsilon)} \geq 3/2 > 0 \text{ for all } \epsilon < \epsilon^*.$$

In a similar way $\epsilon > \epsilon^*$, then we will have that the above term is again positive.

$\bar{\delta} < 1$ such that for all $\delta \geq \bar{\delta}$, the term $(\bar{x}^\infty - y)(1 - \frac{1-\alpha}{\alpha+\epsilon})$ (uniformly over ϵ) can be made arbitrarily small. When $P_d = 0$, i.e., $\bar{x}^\infty = y$ for some ϵ, then $R_d^2(\epsilon) - R^1(\epsilon) = w(d - r_b) + w\epsilon(d - r_s) < 0$ for all such ϵ. When $P_d \neq 0$, then $R_d^2 = 0 \leq R^1$. Thus in either case $R_d^2(\epsilon) \leq R^1(\epsilon)$ for all ϵ.

By virtue of the above arguments we have $P_d \leq (1 - \delta)$ and $E[G] = 1 - \delta$ and this is true for all ϵ, for all $\delta \geq \bar{\delta}$. The rest of the proof follows from part(a). ∎

References

1. Kavitha, V., Saha, I., Juneja, S.: Random fixed points, limits and systemic risk. In: 2018 IEEE Conference on Decision and Control (CDC), pp. 5813–5819. IEEE (2018)
2. Miekisz, J.: Evolutionary game theory and population dynamics. In: Capasso, V., Lachowicz, M. (eds.) Multiscale Problems in the Life Sciences. LNM, vol. 1940, pp. 269–316. Springer, Heidelberg (2008). https://doi.org/10.1007/978-3-540-78362-6_5
3. Eisenberg, L., Noe, T.H.: Systemic risk in financial systems. Manag. Sci. **47**, 236–249 (2001)
4. Acemoglu, D., Ozdaglar, A., Tahbaz-Salehi, A.: Systemic risk and stability in financial networks. Am. Econ. Rev. **105**, 564–608 (2015)
5. Allen, F., Gale, D.: Financial contagion. J. Polit. Econ. **108**, 1–33 (2000)
6. Li, H., Chensheng, W., Yuan, M.: An evolutionary game model of financial markets with heterogeneous players. Procedia Comput. Sci. **17**, 958–964 (2013)
7. Yang, K., Yue, K., Wu, H., Li, J., Liu, W.: Evolutionary analysis and computing of the financial safety net. In: Sombattheera, C., Stolzenburg, F., Lin, F., Nayak, A. (eds.) MIWAI 2016. LNCS (LNAI), vol. 10053, pp. 255–267. Springer, Cham (2016). https://doi.org/10.1007/978-3-319-49397-8_22
8. Borkar, V.S.: Stochastic Approximation: A Dynamical Systems Viewpoint, vol. 48. Springer, Heidelberg (2009)
9. Tembine, H., Altman, E., El-Azouzi, R., Hayel, Y.: Evolutionary games in wireless networks. IEEE Trans. Syst. Man Cybern. Part B (Cybern.) **40**(3), 634–646 (2009)
10. Smith, J.M., Price, G.R.: The logic of animal conflict. Nature **246**(5427), 15–18 (1973)
11. Saha, I., Kavitha, V.: Financial replicator dynamics: emergence of systemic-risk-averting strategies. arXiv preprint arXiv:2003.00886 (2020)

Impact of Private Observation in the Bayesian Persuasion Game

Rony Bou Rouphael$^{(\boxtimes)}$ and Maël Le Treust

ETIS UMR 8051, CY Université, ENSEA, CNRS, 6, avenue du Ponceau,
95014 Cergy-Pontoise Cedex, France
{rony.bou-rouphael,mael.le-treust}@ensea.fr

Abstract. In the Bayesian persuasion setting, the sender aims at persuading the decision maker, so called the decoder, to choose a certain action among a set of possible actions. This paper considers two Bayesian persuasion games: one that involves the observation of a private signal by the decoder in addition to the signal transmitted by the encoder, and another version where no private signal is accessible by the decoder. Our goal is to examine the impact of this private signal on the encoder's optimal utility. In order to do so, we investigate an example involving a binary state, a binary private signal and a binary receiver's actions set. We identify the optimal splitting of the decoder's beliefs satisfying the information constraint imposed by the restricted communication channel, and we compute the encoder's optimal utility value, with and without private signal. Varying the parameters such as the prior belief, the precision of the private signal and the channel capacity, we aim at determining which of the two settings is more favorable to the encoder.

Keywords: Bayesian persuasion · Strategic communication · Side information

1 Introduction

In [5], Kamenica-Gentzkow investigate a persuasion game in which the sender observes the realization of a state variable and commits to some signalling mechanism, then the receiver chooses the best-reply action corresponding to its posterior belief. Communication in persuasion games may be constrained by a limited channel's capacity and messages distorted by some source of noise, as in [9]. Moreover, the receiver may privately observe a signal correlated to the state, as in the source coding problem of Slepian-Wolf and Wyner-Ziv, in [12] and [13]. In such settings, the persuasion problem is hard to solve even for simple models. Tools from information theory, involving entropy and mutual information,

M. Le Treust—Gratefully acknowledges financial support from INS2I CNRS, DIM-RFSI, SRV ENSEA, UFR-ST UCP, The Paris Seine Initiative and IEA Cergy-Pontoise. This research has been conducted as part of the project Labex MME-DII (ANR11-LBX-0023-01).

© Springer Nature Switzerland AG 2021
S. Lasaulce et al. (Eds.): NetGCooP 2021, CCIS 1354, pp. 229–238, 2021.
https://doi.org/10.1007/978-3-030-87473-5_20

provided a solution for certain scenarios of repeated persuasion problems. The optimal solution to the noisy persuasion problem relies on a specific concavification involving an auxiliary utility function for the sender that accounts for the private observation of the receiver as in [8,9] and [10].

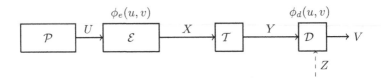

Fig. 1. Bayesian persuasion game with noisy channel $T(y|x)$, with or without decoder's side information Z. The utility functions of the encoder $\mathcal{E}$ and decoder $\mathcal{D}$ are denoted by $\phi_e(u,v)$ and $\phi_d(u,v)$.

1.1   State of the Art

Channel coding and communication problems originally introduced in [11] have been studied in several settings, particularly with the side information setting as in [1] where a hierarchical communication game is considered to treat information disclosure problems originated in economics and involving different objectives for the encoder and the decoder. In [2], Alonso-Câmara provide necessary and sufficient conditions under which a sender benefits from persuading decoders with distinct prior beliefs. The computational aspects of the persuasion game are considered in [4], where the impact of the channel's capacity on the optimal utility is investigated. Persuasion of a privately informed receiver was also investigated in [6], in which the optimal persuasion mechanisms are characterized. In [7], Laclau-Renou depicted the constraints imposed on the sender when multiple receivers have multiple beliefs.

1.2   Contributions

In this paper, we consider a persuasion game with binary source/state and binary decoder's actions and we investigate the effect of the decoder's side observation on the encoder's optimal utility. We compute numerically the two values of the persuasion problems, with and without decoder's side information, depending on two keys parameters, 1) the channel capacity and 2) the precision of the decoder's side information. Depending on these two parameters, the decoder's side information may increase or decrease the encoder's utility.

The paper is organized as follows. The notations are defined in Sect. 2. In Sect. 3, we formulate the two concavification problems. In Sect. 4, we introduce the example of a binary source and state, and we formulate the optimal solutions for the case with no private observation in Sect. 4.1, and for the case where private observation is available at the decoder Sect. 4.2. In Sect. 5, we provide the results of our numerical simulations.

2   Notations

This paper considers a communication model that is illustrated in Fig. 1. Let $\mathcal{E}$ denote the encoder and $\mathcal{D}$ denote the decoder. Notations U, Z, X, Y, and V denote the random variables of information source $u \in \mathcal{U}$, side information $z \in \mathcal{Z}$, channel inputs $x \in \mathcal{X}$, channel outputs $y \in \mathcal{Y}$, and decoder's actions $v \in \mathcal{V}$ respectively. Calligraphic fonts $\mathcal{U}$, $\mathcal{Z}$, $\mathcal{X}$, $\mathcal{Y}$, and $\mathcal{V}$, denote the alphabets and lowercase letters u, z, x, y, and v denote the signal realizations. Notation $\mathcal{P}_U$ stands for the probability distribution of the state U of the game. The private observation Z of the receiver is correlated to U according to the conditional probability distribution $\mathcal{P}_{Z|U}$. We will denote the beliefs of the decoder by $p \in \Delta(\mathcal{U})$ whereas $p(u)$ belongs to $[0, 1]$ for each $u \in \mathcal{U}$. The i.i.d. memoryless channel distribution will be denoted by $\mathcal{T}_{Y|X}$. We denote by $\Delta(\mathcal{X})$ the probability simplex, i.e. the set of probability distributions over $\mathcal{X}$. We denote by $\mathcal{Q}_X$ the probability distribution over $\mathcal{X}$, i.e. the posterior beliefs of the decoder. The joint probability distribution $\mathcal{Q}_{XV} \in \Delta(\mathcal{X} \times \mathcal{V})$ decomposes as follows, $\mathcal{Q}_{XV} = \mathcal{Q}_X \times \mathcal{Q}_{V|X}$. The channel's capacity will be denoted by C. Notations $H(U)$, $H(U|Z)$ and $I(X;Y)$ refer to Shannon's entropy, conditional entropy and mutual information respectively [3, pp. 12], and are given below.

$$H(U) = \sum_{u \in \mathcal{U}} p(u) \log_2 \frac{1}{p(u)}, \qquad H(U|Z) = \sum_{z \in \mathcal{Z}} \sum_{u \in \mathcal{U}} p(z, u) \log_2 \frac{1}{p(u)}, \quad (1)$$

$$I(X;Y) = \sum_{x \in \mathcal{X}} \sum_{y \in \mathcal{Y}} p(x, y) \log_2 \frac{p(x, y)}{p(x) p(y)}, \qquad C = \max_{\mathcal{P}(x)} I(X;Y). \quad (2)$$

3   Concavification Problems

Given a capacity value $C \geq 0$, we consider the two concavification problems below, stated in [10, Def. III.1] and [9, Def. 2.4].

$$\Gamma_0 = \sup_{(\lambda_w, p_w)_{w \in \mathcal{W}}} \left\{ \sum_{w \in \mathcal{W}} \lambda_w \cdot \Phi_e(p_w) \ s.t. \ \sum_{w \in \mathcal{W}} \lambda_w \cdot p_w = \mathcal{P}_U, \right.$$

$$\left. \sum_w \lambda_w \cdot H(p_w) \geq H(U) - C, |\mathcal{W}| = \min(|\mathcal{U}| + 1, |\mathcal{V}|) \right\}, \quad (3)$$

$$\Gamma = \sup_{(\lambda_w, p_w)_{w \in \mathcal{W}}} \left\{ \sum_{w \in \mathcal{W}} \lambda_w \cdot \Psi_e(p_w) \ s.t. \ \sum_{w \in \mathcal{W}} \lambda_w \cdot p_w = \mathcal{P}_U, \right.$$

$$\left. \sum_w \lambda_w \cdot h(p_w) \geq H(U|Z) - C, |\mathcal{W}| = \min(|\mathcal{U}| + 1, |\mathcal{V}|^{|\mathcal{Z}|}) \right\}, \quad (4)$$

where

$$\Phi_e(p) = \mathbb{E}_p \left[\phi_e(U, v^\star(p)) \right], \qquad H(p) = H(U), \quad (5)$$

$$v^\star(p) = \underset{v \in \arg\max \mathbb{E}_p\left[\phi_d(U,v)\right]}{\arg\min} \mathbb{E}_p\left[\phi_e(U,v)\right], \tag{6}$$

and

$$\forall z \in \mathcal{Z}, \quad q_z \in \Delta(\mathcal{U}), \quad q_z(u) = \frac{p(u) \cdot \mathcal{P}(z|u)}{\sum_{u'} p(u') \cdot \mathcal{P}(z|u')}, \qquad \forall u \in \mathcal{U}, \tag{7}$$

$$\Psi_e(p) = \sum_{u,z} p(u) \cdot \mathcal{P}(z|u) \cdot \Phi_e(q_z), \tag{8}$$

$$h(p) = \sum_{u,z} p(u) \cdot \mathcal{P}(z|u) \cdot \log_2 \frac{\sum_{u'} p(u') \cdot \mathcal{P}(z|u')}{p(u) \cdot \mathcal{P}(z|u)}. \tag{9}$$

The notation $v^\star(p) \in V$ stands for the decoder's best reply action with respect to its posterior belief $p \in \Delta(\mathcal{U})$. If several actions maximize the utility of the decoder, we assume that it chooses the one that minimizes the encoder's utility. Thus, the encoder's expected utility $\Phi_e(p)$ is evaluated with respect to the decoder's belief $p \in \Delta(\mathcal{U})$. In the presence of side information $z \in \mathcal{Z}$, the decoder's belief is denoted by $q_z \in \Delta(\mathcal{U})$. As a consequence, the encoder's expected utility $\Psi_e(p)$ is a convex combination between the utilities $\Phi_e(q_z)$ evaluated at different possible beliefs $(q_z)_{z\in\mathcal{Z}}$. The supremum in (4) and (3) are taken over the set of splittings $(\lambda_w, p_w)_{w\in\mathcal{W}}$ of the prior probability distribution $\mathcal{P}_U \in \Delta(\mathcal{U})$, that satisfy the cardinality bound, either $|\mathcal{W}| = \min(|\mathcal{U}|+1, |\mathcal{V}|)$ or $|\mathcal{W}| = \min(|\mathcal{U}|+1, |\mathcal{V}|^{|\mathcal{Z}|})$.

Formulas (3) and (4) are solutions to the persuasion game with noisy channel. The value Γ corresponds to the persuasion problem in which the decoder has a private observation Z correlated with the state U according to the conditional probability distribution $P_{Z|U}$, whereas the value Γ_0 corresponds to the persuasion problem in which the decoder has no access to a side information, or equivalently, has a private observation Z that is independent from the state U. When removing the entropy-based constraints in Γ_0, the concavification problem boils down to the optimal solution provided by Kamenica-Gentzkow in [5].

4   Example with Binary Source and State

In this section, we will illustrate a particular scenario of a strategic communication involving binary source and state and decoder's action. Let $\mathcal{U} = \{u_0, u_1\}$ the state space, $\mathcal{V} = \{v_0, v_1\}$ the action space, and $p_0 = \mathrm{P}(U = u_1) \in [0,1]$ the decoder's prior belief. We consider a binary symmetric noisy channel where $\mathcal{X} = \{x_0, x_1\}$ denotes the set of channel inputs, $\mathcal{Y} = \{y_0, y_1\}$ denotes the set of channel outputs. The channel's capacity for noise level $\epsilon \in [0, \frac{1}{2}]$ is given by $C = 1 - H_b(\epsilon)$ where $H_b(p)$ denotes the binary entropy. Utility functions of both encoder and decoder are given in Tables 1 and 2.

Table 1. Encoder's utility

	v_0	v_1
u_0	0	1
u_1	0	1

Table 2. Decoder's utility

	v_0	v_1
u_0	9	0
u_1	4	10

As shown in the decoder's expected utility graph Fig. 2a, the red lines represent the decoder's best reply action. Therefore, the action of the decoder will only change from v_0 to v_1 depending on the utility threshold γ. In this example, we consider the prior $p_0 = 0.4$ and the utility threshold $\gamma = 0.6$.

4.1 Persuasion Without Side Information (Equation for Γ_0)

The optimal number of posterior beliefs when no side information is available at the decoder is two [9, lemma 6.1]. These posterior beliefs of the decoder need to satisfy the splitting condition and information constraint

$$\lambda q_1 + (1 - \lambda)q_2 = p_0 \iff \lambda = \frac{p_0 - q_2}{q_1 - q_2} \iff 1 - \lambda = \frac{q_1 - p_0}{q_1 - q_2}, \tag{10}$$

$$\lambda H_b(q_1) + (1 - \lambda)H_b(q_2) \geq H_b(p_0) - C. \tag{11}$$

Assuming the information constraint is binding at the optimal, we get

$$\lambda H_b(q_1) + (1 - \lambda)H_b(q_2) = H_b(p_0) - C \tag{12}$$

$$\iff H_b(q_1) = \frac{p_0 H_b(q_2) - q_2(H_b(p_0) - C)}{(p_0 - q_2)} + q_1 \frac{(-H_b(q_2) + H_b(p_0) - C)}{(p_0 - q_2)} \tag{13}$$

The encoder's expected utility function Φ_e depicted in Fig. 2b is defined over $[0,1]$ by $\Phi_e(q) = \mathbb{1}_{q \in [\gamma, 1]}$. For each $q_2 \in [p_0, 1]$, we denote by $q_1(q_2)$ the unique solution of (13) for a given pair (p_0, C). We assume that the decoder's threshold $\gamma > p_0$, hence at the optimum $q_2 = \gamma$, thus

$$\Gamma_0 = \sup_{q_2 \in [0,1]} \left(\lambda \Phi_e(q_1(q_2)) + (1 - \lambda)\Phi_e(q_2) \right) = \frac{q_1(\gamma) - p_0}{q_1(\gamma) - \gamma}. \tag{14}$$

Figure 2b shows an unrestricted communication without decoder's side information. The green dotted line is the concavification of the encoder's expected utility function represented in the red lines. The optimal utility value corresponds to the evaluation of this concavification at the prior belief p_0.

4.2 Persuasion with Side Information (Equation for Γ)

When side information $\mathcal{Z} = \{z_0, z_1\}$ is observed by the decoder, then [9, Lemma 6.3] ensures that the optimal number of posterior beliefs is three. The posterior distributions (q_1, q_2, q_3) from observing the message delivered by the encoder, must satisfy the information constraint given by

$$\lambda_1 \cdot h(q_1) + \lambda_2 \cdot h(q_2) + \lambda_3 \cdot h(q_3) \geq H(U|Z) - \max_{P(x)} I(X;Y) \tag{15}$$

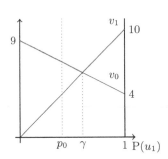
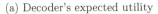

(a) Decoder's expected utility

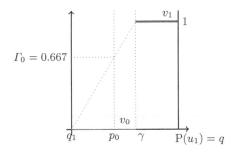

(b) Encoder's expected utility and optimal pay-off with $p_0 = 0.4$, $\delta = 0.5$ and $\gamma = 0.6$.

Fig. 2. Encoder and Decoder's Expected Utilities

Thus $(\lambda_1, \lambda_2, \lambda_3)$ can be computed from the above information constraint, the splitting lemma $\lambda_1 q_1 + \lambda_2 q_2 + \lambda_3 q_3 = p_0$ and the fact that $\lambda_1 + \lambda_2 + \lambda_3 = 1$. We assume that the information constraint is binding. From [10, Eq. (57)-(59)], we have

$$\lambda_1 = \frac{IC \cdot (q_2 - q_3) + h(q_2) \cdot (q_3 - p_0) + h(q_3) \cdot (p_0 - q_2)}{h(q_1) \cdot (q_2 - q_3) + h(q_2) \cdot (q_3 - q_1) + h(q_3) \cdot (q_1 - q_2)}, \tag{16}$$

$$\lambda_2 = \frac{IC \cdot (q_3 - q_1) + h(q_3) \cdot (q_1 - p_0) + h(q_1) \cdot (p_0 - q_3)}{h(q_1) \cdot (q_2 - q_3) + h(q_2) \cdot (q_3 - q_1) + h(q_3) \cdot (q_1 - q_2)}, \tag{17}$$

$$\lambda_3 = \frac{IC \cdot (q_1 - q_2) + h(q_1) \cdot (q_2 - p_0) + h(q_2) \cdot (p_0 - q_1)}{h(q_1) \cdot (q_2 - q_3) + h(q_2) \cdot (q_3 - q_1) + h(q_3) \cdot (q_1 - q_2)}. \tag{18}$$

Given a *interim* belief parameter $q \in [0, 1]$, the decoder's side information might be z_0 or z_1, thus inducing the two following posterior beliefs

$$p_1(q) = \frac{q.\delta}{(1 - q).(1 - \delta) + q.\delta}, \qquad p_2(q) = \frac{q.(1 - \delta)}{(1 - q).\delta + q.(1 - \delta)}. \tag{19}$$

The decoder's threshold γ induces the two corresponding threshold ν_1 and ν_2 for the *interim* belief parameter $q \in [0, 1]$

$$\nu_1 = \frac{\gamma.(1 - \delta)}{\delta.(1 - \gamma) + \gamma(1 - \delta)}, \qquad \nu_2 = \frac{\gamma.\delta}{\gamma.\delta + (1 - \delta).(1 - \gamma)}. \tag{20}$$

Thus the encoder's utility function $\Psi_e(q)$ represented by the red lines in Fig. 3 and the conditional entropy $h(q)$ reformulate as

$$\Psi_e(q) = 0 \cdot \mathbb{1}_{\{q \in]0, \nu_2]\}} + ((1 - q) \cdot \delta + q \cdot (1 - \delta)) \cdot \mathbb{1}_{\{q \in]\nu_2, \nu_1]\}} + 1 \cdot \mathbb{1}_{\{q \in]\nu_1, 1]\}}, \tag{21}$$

$$h(q) = ((1 - q) \cdot (1 - \delta) + q \cdot \delta) \cdot H_b(p_1(q)) + ((1 - q) \cdot \delta + q \cdot (1 - \delta)) H_b(p_2(q). \tag{22}$$

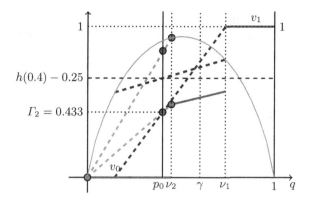

Fig. 3. Splitting over 2 posteriors ($q_1 = 0$; $q_2 = 0.4468$) with $C = 0.25$, $p_0 = 0.4$, $\delta = 0.35$, $\gamma = 0.6$.

The encoder's optimal utility value is given by

$$\Gamma = \sup_{\substack{q_1 \in [0, \nu_2], q_2 \in [\nu_2, \nu_1], \\ q_3 \in [\nu_1, 1]}} \left(\lambda_1 \cdot \Psi_e(q_1) + \lambda_2 \cdot \Psi_e(q_2) + \lambda_3 \cdot \Psi_e(q_3) \right) \tag{23}$$

$$= \sup_{\substack{q_1 \in [0, \nu_2], q_2 \in [\nu_2, \nu_1], \\ q_3 \in [\nu_1, 1]}} \left(\frac{(h(p_0) - C)\left((q_3 - q_1) \cdot \left(q_2 \cdot (1 - 2\delta) + \delta\right) + (q_1 - q_2)\right)}{h(q_1) \cdot (q_2 - q_3) + h(q_2) \cdot (q_3 - q_1) + h(q_3) \cdot (q_1 - q_2)} \right.$$

$$+ \frac{\left(h(q_3) \cdot (q_1 - p_0) + h(q_1) \cdot (p_0 - q_3)\right) \cdot \left(q_2 \cdot (1 - 2\delta) + \delta\right)}{h(q_1) \cdot (q_2 - q_3) + h(q_2) \cdot (q_3 - q_1) + h(q_3) \cdot (q_1 - q_2)}$$

$$+ \left. \frac{h(q_1) \cdot (q_2 - p_0) + h(q_2) \cdot (p_0 - q_1)}{h(q_1) \cdot (q_2 - q_3) + h(q_2) \cdot (q_3 - q_1) + h(q_3) \cdot (q_1 - q_2)} \right) \tag{24}$$

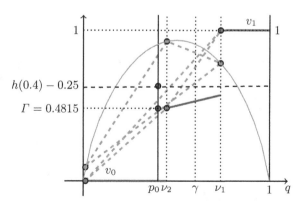

Fig. 4. Optimal Splittings over 3 posteriors ($q_1 = 0.012$; $q_2 = 0.4468$; $q_3 = 0.7358$) with $C = 0.25$, $p_0 = 0.4$, $\delta = 0.35$, $\gamma = 0.6$.

In some cases, the optimal splitting has only two posterior instead of three. Fig. 3 and Fig. 4 represent the optimal utility of the encoder depending on belief parameter q over a constrained communication channel with capacity $C = 0.25$ and with decoder's private observation $\delta = 0.35$. Splitting over three posteriors instead of two, improves the encoder's optimal payoff.

5   Numerical Simulations

In this section we investigate the impact of the private observation on the encoder's optimal utility. Numerical simulations over (C, δ) regions are performed for both concavification problems Γ_0 and Γ, revealing the encoder's optimal payoff values with and without decoder's private observation.

5.1   Encoder's Optimal Payoff Values

The optimal splitting of the prior over 3 posterior beliefs results in the encoder's optimal payoff values shown in Fig. 5 with respect to the (C, δ) regions.

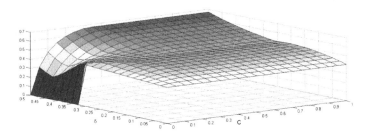

Fig. 5. Encoder's optimal payoff evaluated with three posteriors w.r.t. δ and C for $p_0 = 0.4$ and $\gamma = 0.6$.

As the channel's capacity increases, the encoder's utility is improved without decoder's side information. This is due to the fact that more capacity allows the transmission of more information and hence information can be optimally disclosed. However; with low capacity, the decoder's side observation can enhance the utility of the encoder until the encoder has no capacity at all, it becomes optimal to have private information up to some threshold $\delta^\star$ evaluated in Proposition 1 below.

5.2   Impact of the Decoder's Private Signal

Proposition 1. *Let $C = 0$.*

- *If $p_0 < \gamma$ and $\delta \in [0, \frac{p_0 \cdot (\gamma-1)}{p_0 \cdot (-1+2\gamma)-\gamma}] \cup [\frac{\gamma \cdot (1-p_0)}{p_0 \cdot (1-2\gamma)+\gamma}, 1]$, then $\Gamma > \Gamma_0$.*

- *If $p_0 \geq \gamma$ then $\Gamma_0 \geq \Gamma$.*

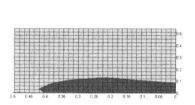
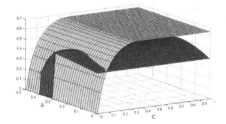

Fig. 6. (δ, C) regions for encoder's optimal utility with (blue) and without (green) decoder's private observation for $p_0 = 0.4$ and $\gamma = 0.6$. (Color figure online)

5.3   Impact of the Number of Posteriors

The encoder could potentially achieve a greater payoff by splitting the prior over three posterior beliefs instead of splitting over two posteriors only (Fig. 7).

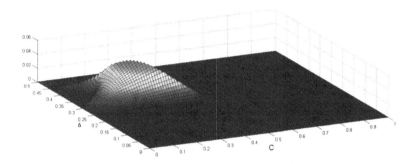

Fig. 7. Difference between optimal utility values obtained by splitting with three posteriors and two posteriors.

References

1. Akyol, E., Langbort, C., Başar, T.: Information-theoretic approach to strategic communication as a hierarchical game. Proc. IEEE **105**(2), 205–218 (2016)
2. Alonso, R., Câmara, O.: Bayesian persuasion with heterogeneous priors. J. Econ. Theory **165**(C), 672–706 (2016)
3. Cover, T.M., Thomas, J.A.: Elements of Information Theory. Wiley, Hoboken (1991)
4. Dughmi, S., Kempe, D., Qiang, R.: Persuasion with limited communication. In: Proceedings of the 2016 ACM Conference on Economics and Computation. EC 2016, pp. 663–680. Association for Computing Machinery, New York (2016). https://doi.org/10.1145/2940716.2940781
5. Kamenica, E., Gentzkow, M.: Bayesian persuasion. Am. Econ. Rev. **101**, 2590–2615 (2011)

6. Kolotilin, A., Mylovanov, T., Zapechelnyuk, A., Li, M.: Persuasion of a privately informed receiver. Discussion Papers 2016–21, School of Economics, The University of New South Wales (2016)
7. Laclau, M., Renou, L.: Public Persuasion (2016). https://hal-pse.archives-ouvertes.fr/hal-01285218
8. Le Treust, M., Tomala, T.: Information-theoretic limits of strategic communication. Preliminary Draft (2018). https://arxiv.org/abs/1807.05147
9. Le Treust, M., Tomala, T.: Persuasion with limited communication capacity. J. Econ. Theory **184**, 104940 (2019)
10. Le Treust, M., Tomala, T.: Strategic communication with side information at the decoder. arXiv preprint arXiv:1911.04950 (2019)
11. Shannon, C.E.: A mathematical theory of communication. Bell Syst. Tech. J. **27**(3), 379–423 (1948)
12. Tuncel, E.: Slepian-wolf coding over broadcast channels. IEEE Trans. Inf. Theory **52**(4), 1469–1482 (2006)
13. Wyner, A.D., Ziv, J.: The rate-distortion function for source coding with side information at the decoder. IEEE Trans. Inf. Theory **22**(1), 1–11 (1976)

Social Networks

Optimal Campaign Strategy for Social Media Marketing with a Contrarian Population

Vineeth S. Varma[1]([✉])([iD]), Bikash Adhikari[1], Irinel-Constantin Morărescu[1], and Elena Panteley[2]

[1] Université de Lorraine, CRAN, UMR 7039 and the CNRS, CRAN, UMR 7039, Nancy, France
vineeth.satheeskumar-varma@univ-lorraine.fr
[2] Laboratoire des signaux et systèmes, CNRS, 3 rue Joliot Curie, 91192 Gif-sur-Yvette, France

Abstract. We address formally the problem of opinion dynamics when the social network composed of conformists and contrarians are not only influenced by their neighbors, but also by an external influential entity referred to as a marketer. The population of contrarians tries to have an opinion that is the opposite of the opinion held by the conformists. The influential entity tries to sway the overall opinion as close as possible to the desired opinion by using a specific influence budget. The main technical issue addressed is finding how the marketer should allocate its budget among the agents such that the agents' opinion will be as close as possible to the desired opinion while taking into account the behavior of the contrarian population. Our main results show that the marketer has to prioritize certain agents over others based on their initial condition, their influence power in the social graph and the population class they belong to. Numerical examples illustrate the analysis.

1 Introduction

During the last decade, social networks gained increasing importance in our daily life. Consequently, more and more companies are using digital social networks to promote specific goods and/or ideas. This motivated the scientific community to give more attention to the analysis of opinion dynamics in social networks. This is a challenging task since human behavior is very different from one individual to another and the interactions in the network can change over time. Various mathematical models [1,4,6,7,9,10] have been proposed in order to capture different features of this complex dynamics. Empirical models based on in vitro and in vivo experiments have also been developed [5,13,16].

Some mathematical model target consensus as collective asymptotic behavior of the network [7,9] while some others lead to the network clustering [1,10,14]. In

---

Supported by CEFIPRA under the project 6001-1/2019.

order to enforce consensus some recent studies propose the control of one or few agents (see [3,8]). Besides these methods of controlling opinion dynamics towards consensus, we also find few attempts to control the discrete-time dynamics of opinions such that, as many as possible reach a certain set after a finite number of influences [11]. In very recent work ([15]) the authors developed a formal method for the optimal space-time allocation of a budget allowing an influencer to bring the consensus value of the network as close as possible to a desired value.

It is noteworthy that most of the proposed mathematical models consider that the opinions of two interacting individuals are approaching one to another. Although this behavior seems to be sociologically accepted, one can also provide real-life situations in which interactions are antagonistic. In this case, the distance between the opinions of two interacting individuals will increase (see [1,12,17]). To the best of our knowledge, there exists no attempt to control this kind of opinion dynamics, and only analysis of the asymptotic behavior is reported.

In this paper, we consider the challenging problem that requires to minimize the distance between the average of opinions and a desired value using a given control/marketing budget over a social network split into two groups. Basically, this social network with the contrarian population represents a model for real cases such as supporters of competing teams, parties, etc. On top of this assumption on the network structure, we also assume that the maximal marketing influence cannot instantaneously make the opinion of one individual to be equal with the desired value.

To provide a mathematical model we consider that the opinion dynamics is fast enough such that we can assume it evolves in continuous time and we want to design a marketing strategy that minimizes the distance between the average of the opinions and the desired value after a campaign with certain budget constraints. This results in a linear-impulsive closed-loop dynamics in which the jumps are controlled by the influencer. Our main result shows that the optimal control strategy consists of influencing as much as possible the most central/popular (see [2] for a formal definition of centrality) individuals of the network.

It is worth highlighting that in this study we do not control the state of the influencing entity which is assumed to be constant. Instead, we control the influence weight that the marketer can have on different individuals of the social network. It was shown in [15] that this approach allows highlighting the effectiveness of targeted marketing with respect to broadcasting strategies when budget constraints are present.

Notation. Let $\mathbb{R} := (-\infty, \infty)$, $\mathbb{R}_{\geq 0} := [0, \infty)$, and $\mathbb{Z}_{\geq 0} := \{0, 1, 2, \ldots\}$. We use E for the expectation taken over the relevant stochastic variables and $\mathbf{1}(\cdot)$ for the indicator function, taking the value 1 when the condition is satisfied and 0 otherwise.

2   Problem Statement

We consider a social network populated by agents belonging to the set $\mathcal{V} := \{1, 2, \ldots, N\}$ with the connections given by $a_{i,j}$ indicating the influence of agent j on agent i. All agents belong to the class of conformists denoted by the set $\mathcal{V}^+$ or to the class of contrarians denoted by $\mathcal{V}^-$. We use $c_n \in \{1, -1\}$ to denote the agent class with $c_n = 1$ for all $n \in \mathcal{V}^+$ and $c_n = -1$ for all $n \in \mathcal{V}^-$. These two sets are non-intersecting and may be interpreted as two hostile camps as considered in [17].

The opinion of all agents belongs to the interval $[-1, 1]$ and we denote the collection of all opinions by $x \in [-1, 1]^N$. The internal opinion dynamics of the network is given by

$$\dot{x}_i = \sum_{j \in \mathcal{V}} a_{i,j} x_j - |a_{i,j}| x_i \tag{1}$$

for all $i \in \mathcal{V}$. The above dynamics is similar to a standard consensus model as in [6] but $a_{i,j}$ may be positive or negative depending on the interaction type.

In this work, we focus on antagonistic interaction between agents that belong to different classes and standard consensus interactions among agents of the same class, i.e., contrarians or conformist. Basically, this implies that $a_{i,j} = c_i c_j |a_{i,j}|$. As a result of this extra term, agents i and j will have a consensus only if i, j belong to the same class. If they belong to different classes, agent i will try to have an opinion in opposition to agent j. We can write the overall dynamics for all agents as

$$\dot{x} = -Lx \tag{2}$$

where L is a Laplacian-like matrix with

$$L_{i,j} = \begin{cases} \sum_{k \in \mathcal{V}} |a_{i,k}| & \text{for all } i = j \in \mathcal{V}, \\ -a_{i,j} & \text{for all } i \neq j \in \mathcal{V} \end{cases} \tag{3}$$

The influence of the external entity (marketer) with a desired opinion d is limited to a campaign occurring at $t = 0$. We model this influence through an impulsive control/jump at $t = 0$. Due to the contrarian population, the marketer may also desire to bring some agents closer to the opinion $-d$ but has must select the best agents due to budget constraints. We, therefore, model the impulsive control as follows

$$x_n(0^+) = (1 - |u_n|) x_n(0) + u_n d \tag{4}$$

for all $n \in \mathcal{V}$ with $u_n \in [-\bar{u}, \bar{u}]$, with $\bar{u}$ denoting the control saturation and $\sum_{n \in \mathcal{V}} |u_n| \leq B$ is the budget constraint.

The main objective of this paper is to provide an optimal control u such that the distance between the average opinion of all agents and the desired opinion $d \in \{-1, 1\}$, given by $\left| \frac{1_N^T x}{N} - d \right|$ is minimized asymptotically. Given $x_0 \in [-1, 1]^N$, the optimization problem can be stated formally as follows

$$\begin{aligned}
&\text{Minimize}_u\, J(u) := \lim_{t\to\infty}\left|\tfrac{\mathbf{1}_N^\top x(t)}{N} - d\right| \\
&\text{where } \dot{x}(t) = -Lx(t),\, x(0) = x_0\, \forall t > 0, \\
&\text{under (4) with } u_n \in [-\bar{u}, \bar{u}]\, \forall n \in \mathcal{V} \\
&\text{and } \sum_{n\in\mathcal{V}} |u_n| \le B.
\end{aligned} \tag{5}$$

3 Analysis

To begin our analysis, we first characterize the asymptotic opinions of the agents under the dynamics (2). For this purpose, we rely on results established in [17].

Proposition 1. *Given the opinions $x(0^+)$ formed after the campaign, we have*

$$\lim_{t\to\infty} x(t) = cv^\top x(0^+) \tag{6}$$

where v is the left eigen-vector of the matrix L associated with eigenvalue 0 with $\sum_{n\in\mathcal{V}} |v_n| = 1$ and $sign(v_n) = sign(c_n)$.

Proof. First, we use Lemma 1 in [17] to conclude that if (2) establishes modular consensus (consensus in absolute value of each x_n, then

$$\lim_{t\to\infty} x(t) = \rho v^\top x(0^+) \tag{7}$$

where $\rho_n \in \{1, -1\}$ for all $n \in \mathcal{V}$. By construction in (3), we have that $a_{i,j}$ is positive for $i \neq j$ only when they both belong to the same set $\mathcal{V}^+$ or $\mathcal{V}^-$ and is negative otherwise. This implies that $\mathcal{V}^+$ or $\mathcal{V}^-$ form hostile camps as defined in [17], allowing us to apply Lemma 2 from [17]. This states that modular consensus is established for quasi-strongly connected graphs with $\rho_i = -\rho_j$ for all i, j in opposite camps. Since we pick $sign(v_n) = sign(c_n)$, we have that $\rho_n = c_n$. □

Proposition 1 allows us to study the impact of the control on the cost function in 5. We provide the optimal control u which will minimize this cost in the following. We define a sorting index

$$\gamma_n := |v_n||c_n d - x_n(0)| \tag{8}$$

Theorem 1. *If $card(\mathcal{V}^+) = card(\mathcal{V}^-)$, the final cost is invariant to the control and $J(u) = 1$ for all u. If $card(\mathcal{V}^+) > card(\mathcal{V}^-)$ and the desired opinion be d, then the optimal control u^* is given by*

$$u^*_{o(n)} = \begin{cases} c_{o(n)}\bar{u} & \text{if} & n \le \left\lfloor\frac{B}{\bar{u}}\right\rfloor \\ Bc_{o(n)} - c_{o(n)}\bar{u}\left\lfloor\frac{B}{\bar{u}}\right\rfloor & \text{if} & n = \left\lfloor\frac{B}{\bar{u}}\right\rfloor + 1 \\ 0 & \text{otherwise} \end{cases} \tag{9}$$

where $o : \mathcal{V} \to \mathcal{V}$ is a bijection such that $\gamma_{o(1)} \ge \gamma_{o(2)} \ge \cdots \ge \gamma_{o(N)}$. When $card(\mathcal{V}^+) < card(\mathcal{V}^-)$, setting $-d$ as the desired opinion and applying u^* minimizes the cost $J(u)$.

Proof: First, we rewrite the minimization (5) problem as follows

$$\text{Minimize}_u \lim_{t\to\infty} \left| \frac{1_N^\top x(t)}{N} - d \right|$$
$$\text{where } \dot{x}(t) = -Lx(t),\ \forall t > 0,$$
$$x_n(0^+) = (1 - u_n^+ - u_n^-)x_n(0) + u_n^+ d - u_n^- d,\ \forall n \in \mathcal{V} \qquad (10)$$
$$u_n^+, u_n^- \in [0, \bar{u}],\ \text{such that } u_n^+ u_n^- = 0,\quad \forall n \in \mathcal{V},$$
$$\sum_{n\in\mathcal{V}} u_n^+ + u_n^- \leq B.$$

in order to separate the positive and negative control action. Note that we have that

$$\lim_{t\to\infty} x(t) = c v^\top x(0^+) \qquad (11)$$

since the dynamics in $(0,\infty)$ is given by $\dot{x} = -Lx$ and using [1].

Denoting $x^* := v^\top x(0^+)$, which is a scalar and belongs to the interval $[-1, 1]$, we have that the final cost is given by

$$J = \left| \frac{\sum_{n\in\mathcal{V}} c_n}{N} x^* - d \right| = \left| \frac{card(\mathcal{V}^+) - card(\mathcal{V}^-)}{N} x^* - d \right|. \qquad (12)$$

We use $G := \left| \frac{card(\mathcal{V}^+) - card(\mathcal{V}^-)}{N} \right|$. Therefore,

1. if $card(\mathcal{V}^+) = card(\mathcal{V}^-)$, we have $J = d$ for any x^*,
2. if $card(\mathcal{V}^+) > card(\mathcal{V}^-)$, we have $J = |Gx^* - d|$,
3. and if $card(\mathcal{V}^+) < card(\mathcal{V}^-)$, $J = |Gx^* + d|$.

This implies that minimizing the final cost J is equivalent of minimizing $(x^* - d)^2$ for case 2 and $(x^* + d)^2$ for case 3. We have

$$\frac{\partial((d-x^*)^2)}{\partial u_n^+} = -2(d - x^*)v_n, \qquad \frac{\partial((d-x^*)^2)}{\partial u_n^-} = 2(d - x^*)v_n,$$
$$\frac{\partial((d+x^*)^2)}{\partial u_n^+} = 2(d + x^*)v_n, \qquad \frac{\partial((d+x^*)^2)}{\partial u_n^-} = -2(d + x^*)v_n. \qquad (13)$$

In the two cases, $2(d-x^*)$ and $2(d+x^*)$ are respectively positive functions which hold true for all n. Applying Lemma 1 (provided in the Appendix A) for each case, we have the result by setting $u_n = u_n^+ - u_n^-$. ∎

Theorem 1 provides the control strategy to be implemented in order to minimize the cost $J(u)$, which implies minimizing the distance between the average final opinion and the desired opinion. When the number of contrarians and conformists are the same, the opinions are polarised around 0 in a symmetric fashion. Therefore, the final average opinion is always 0, leading to a fixed cost of $J(u) = |d|$.

Nominally, we assume that the set of contrarians is a minority, i.e., $card(\mathcal{V}^+) > card(\mathcal{V}^-)$. In this case, we select the most influential agents based on the index γ which depends on the vector centrality v as well as the distance to the desired opinion. However, since the contrarians oppose the conformists, the optimal strategy is to push the contrarians closer to $-d$ and the conformists closer to d. Since the conformists are a majority, the minimal cost is when all conformists are at d and all contrarians at $-d$. The control in Theorem 1 precisely achieves this, i.e. it brings the conformists as close as possible to d and the contrarians as close as possible to $-d$, under the budget constraints.

Note that when $card(\mathcal{V}^-) = 0$, i.e. there are no antagonistic relations, the optimal control u^* in Theorem 1 matches the result in [15] (which considers $a_{i,j} \geq 0$ for all i, j) as expected. When the population of contrarians are larger, i.e. $card(\mathcal{V}^+) < card(\mathcal{V}^-)$, the minimal cost occurs when all conformists are at $-d$ and all contrarians at d. This is achieved by simply setting the new desired opinion to be $-d$ and applying the control in Theorem 1.

4   Numerical Illustration

To demonstrate and compare the numerical results the following budget allocation strategies are adopted:

Strategy 1: Optimal Budget Allocation, where the budget is allocated to agents according to Theorem 1.

Strategy 2: Uniform Budget Allocation, where budget is allocated uniformly to all the agents with negative control for contrarians, i.e. $u_i = c_i \frac{B}{N}$.

Strategy 3: Positive Budget Allocation, where budget is allocated uniformly and an identical control is applied to all the agents, i.e. $u_i = \frac{B}{N}$.

Strategy 3 corresponds to traditional advertising campaigns through Television or radio broadcasts as the same advertising action is applied to all agents regardless of their individual characteristics. The network structure we consider is a strongly connected directed graph with 10 nodes (N=10) and the number of conformists is greater than that of the contrarians ($|V|^+ > |V|^-$). The strongly connected graph is represented in Fig. 1. The initial opinions are uniformly chosen between $[-1, 1]$ and represented in Table 1.

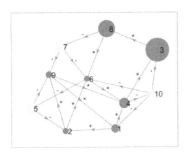

Fig. 1. Red and cyan nodes represent conformists and contrarians, respectively with the size scaled based on vector centrality v. (Color figure online)

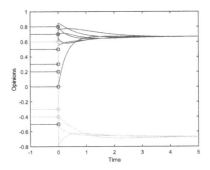

Fig. 2. Opinion dynamics with the control applied impulsively at $t = 0$, red and cyan plots represent conformists and contrarians. (Color figure online)

In Fig. 1, the red and cyan nodes represent the conformist and the contrarian agents, respectively and the size of the nodes represents the agent's centrality. The centralities of all the agents are also represented in Table 1. As stated in Sect. 2, positive and negative values of c represents conformists and contrarians respectively. Consider that the budget $B = 4$, the maximum control that can be allocated to each agent $\bar{u} = 0.7$, and opinions before the campaign as given in Table 1. When no control action is applied, we see that $J(0) = 0.8877$. Using strategy 1, budget allocation is performed based on the influence power of an agent γ (represented in Table 1) while satisfying the budget constraints. The budget of $[0.7, 0.7, 0.7, 0.7, 0.7, 0.5]$ is allocated to the agents $[3\text{--}5, 8\text{--}10]$ and the resultant cost $J(u^*) = 0.7331$ (Fig. 2). To better understand the advantage of the designed marketing strategy (strategy 1), the results obtained are compared with the ones obtained from strategy 2 and 3. Using strategy 2, the total available budget $B = 4$ is allocated uniformly among N agents and the resultant cost is 0.7728. Next, with strategy 3, the positive budget is uniformly allocated to all the agents and the cost obtained is 0.8646.

Table 1. Data

Agents	1	2	3	4	5	6	7	8	9	10
c	1	1	1	1	-1	1	-1	1	1	-1
Initial Opinion	0.8	0.7	0.5	0.3	-0.3	0	-0.4	0.2	-0.5	0.6
Centrality	0.075	0.054	0.224	0.095	-0.111	0.052	-0.108	0.160	0.057	-0.064
Gamma (γ)	0.015	0.016	0.112	0.067	0.078	0.052	0.065	0.128	0.085	0.101

Finally, we consider a large-scale network with 300 nodes, formed by 279 conformists and 21 contrarians. The graph is constructed in the following manner, among the 300 agents, 25 nodes are taken to be central and influence 30 to 60 agents, while the remaining 275 have 1 to 30 agents they influence. The initial opinion of the agents are generated by the formula $x_i(0) = -1 + \frac{2i}{N}$.

As seen in Fig. 3, the cost with no control (or when budget $B = 0$) is the same for all strategies i.e., $J(0) = 1$. It is clear from the plot that the cost is smaller using Strategy 1 (based on Theorem 1) for any budget between 0 and $N\bar{u} = 210$ compared to the other strategies. Note that when $B = N\bar{u}$, there is enough budget to allocate the maximum control to all agents, making Strategy 1 and 2 equivalent. However, for strategy 3, due to positive budget allocation even to contrarians, the cost is significantly higher ($J = 0.48$) even when $B = N\bar{u} = 210$.

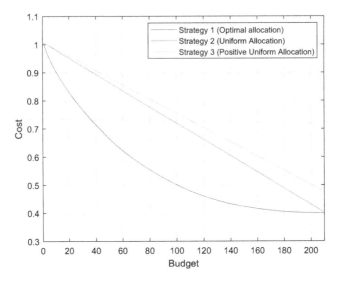

Fig. 3. Cost vs Budget.

5   Conclusion

In this work, we have shown how an optimal campaign strategy can be designed for the control of opinion dynamics over a social network. The main novelty of this work, with respect to previous works on control of opinion dynamics is that we consider the presence of contrarians in the network, which have an antagonistic relationship with the other agents. The external entity or the marketer wants to bring the average opinion of all agents as close as possible to a desired opinion d. Interestingly, we see that the optimal strategy involves bringing the conformists closer to d and the contrarians closer to $-d$, and allocating the available budget among the best agents sorted according to a centrality measure and the distance of their initial opinions to d or $-d$. The future directions to this work would involve considering multiple-campaigns or a continuous control as well as the presence of a competing marketer, resulting in a non-cooperative game.

A   Appendix

Lemma 1. *Given an optimization problem (OP) under the following standard form*

$$\underset{y \in \mathbb{R}^N}{minimize}\ C(y^+, y^-)$$
$$subject\ to\ y_i^s - \bar{y} \leq 0,\ \forall i \in \{1, ..., N\}, s \in \{+, -\}$$
$$-y_i^s \leq 0,\ \forall i \in \{1, ..., N\}, s \in \{+, -\} \tag{14}$$
$$-B + \sum_{i=1}^{N} y_i^+ + y_i^- \leq 0$$

where $N \in \mathbb{N}, N \geq 1$, $\bar{y} < 1$, $B \geq 0$ and $C(y)$ is a decreasing convex function in y_i such that the following condition holds.
For all $i \in \{1, \ldots, N\}$, $\exists g(y) \geq 0$ such that

$$\frac{\partial C(y)}{\partial y_i^+} = -k_i g(y) \text{ and } \frac{\partial C(y)}{\partial y_i^-} = k_i g(y)$$

for some $k_i \in \mathbb{R}$.
Then an optimal solution y^* to this OP is given by water-filling as follows

$$y_{o(i)}^{+*} = \begin{cases} \bar{y} & if & i \leq \left\lfloor \frac{B}{\bar{y}} \right\rfloor & and \; k_{o(i)} > 0 \\ B - \bar{y} \left\lfloor \frac{B}{\bar{y}} \right\rfloor & if & i = \left\lfloor \frac{B}{\bar{y}} \right\rfloor + 1 \; and \; k_{o(i)} > 0 \\ 0 & otherwise & \end{cases} \quad (15)$$

and

$$y_{o(i)}^{-*} = \begin{cases} \bar{y} & if & i \leq \left\lfloor \frac{B}{\bar{y}} \right\rfloor & and \; k_{o(i)} < 0 \\ B - \bar{y} \left\lfloor \frac{B}{\bar{y}} \right\rfloor & if & i = \left\lfloor \frac{B}{\bar{y}} \right\rfloor + 1 \; and \; k_{o(i)} < 0 \\ 0 & otherwise & \end{cases} \quad (16)$$

where $o : \{1, \ldots, N\} \mapsto \{1, \ldots, N\}$ represents an ordering function which can be any bijection for Case 2 and, one satisfying $|k_{o(1)}| \geq |k_{o(2)}| \geq \cdots \geq |k_{o(N)}|$ for Case 1.

Proof: Note that all the constraint functions of the considered OP are affine, which corresponds to sufficient conditions for applying KKT conditions. Since the OP is convex, KKT conditions are necessary and sufficient for optimality. By denoting the Lagrangian by

$$\ell(y, \lambda^+, \bar{\lambda}^+, \hat{\lambda}, \lambda^-, \bar{\lambda}^-) = C(y) + \left(\sum_{i=1}^{N} \sum_{s \in \{+,-\}} \bar{\lambda}_i^s (y_i^s - \bar{y}) - \lambda_i^s y_i^s \right) \\ + \hat{\lambda}^+ \left(-B + \sum_{i=1}^{N} y_i^+ + y_i^- \right). \quad (17)$$

Let us assume that $\frac{\partial C(y)}{\partial y_i} = k_i g(y)$, in this case the first necessary and sufficient condition for optimality can be simplified to write

$$k_i g(y^*) = \lambda_i^{+*} - \bar{\lambda}_i^{+*} + \hat{\lambda}^* + \lambda_i^{-*} - \bar{\lambda}_i^{-*} \quad (18)$$

which must hold for all $i \in \{1, \ldots, N\}$. The primal feasibility conditions write

$$0 \leq y_i^{s*} \leq \bar{y} \; \forall i \in \{1, \ldots, N\}, s \in \{+, -\},$$

and

$$\sum_{i=1}^{N} y_i^{+*} + y_i^{-*} \leq B. \quad (19)$$

All the KKT multipliers must satisfy the dual feasibility conditions: $\lambda_i^{s\star} \geq 0$, $\bar{\lambda}_i^{s\star} \geq 0$, $\widehat{\lambda}^{s\star} \geq 0$ for all $i \in \{1,\ldots,N\}, s \in \{+,-\}$. At last, the complementary slackness conditions are given by

$$\bar{\lambda}_i^{s\star}(y_i^{s\star} - \bar{y}) = 0,$$
$$\lambda_i^{s\star} y_i^\star = 0,$$
$$\widehat{\lambda}^\star\left[\left(\sum_{i=1}^N y_i^{+\star} + y_i^{-\star}\right) - B\right] = 0.$$

Since $g(y^\star)$ is identical for all $i \in \{1,\ldots,N\}$ and $s \in \{+,-\}$ and is non-negative, we must have $\lambda_i^{s\star}, \bar{\lambda}_i^{s\star}$ and $\widehat{\lambda}^\star$ chosen so that (18) holds. We get $y_i^+ y_i^- = 0, \forall i \in \{1,\ldots,N\}$.

Take $y^\star$ from (16). Set $\lambda_j^{s\star} = \bar{\lambda}_j^{s\star} = 0$ for $j = \left(\mathcal{O}\left\lceil\frac{\beta_k}{\bar{u}}\right\rceil\right)$ and $s = +$ if $k_j > 0$ and $s = -$ if $k_j < 0$ as it is the only component with a non-saturated solution. For any n, $y_n^{+\star} = 0$ if $k_n < 0$ and $y_n^{-\star} = 0$ if $k_n > 0$ can be imposed as the left hand side of (18) will be negative and positive correspondingly and letting $\lambda_n^s \neq 0$ will be possible.

For any i such that $\mathcal{O}(i) < j$, we have $k_i \geq k_j$ and this can be satisfied by setting $y_i^{\text{sign}(k_i)\star} = \bar{y}$ and having $\bar{\lambda}_i^{\text{sign}(k_i)\star} > 0$ and $\lambda_i^s = 0$. On the other hand, for any i such that $\mathcal{O}(i) > j$, we set $y_i^\star = 0$ and the KKT conditions are satisfied if $\bar{\lambda}_i^s = 0$ and $\lambda_i^{s\star} > 0$. The solution from (16) can also be verified to satisfy (19) and therefore, we have it satisfying all the KKT conditions.

■

References

1. Altafini, C.: Consensus problems on networks with antagonistic interactions. IEEE Trans. Autom. Control **58**(4), 935–946 (2013)
2. Bonacich, P., Lloyd, P.: Eigenvector-like measures of centrality for asymmetric relations. Soc. Netw. **23**, 191–201 (2001)
3. Caponigro, M., Piccoli, B., Rossi, F., Trélat, E.: Sparse feedback stabilization of multi-agent dynamics. In: Proceedings of the 55th IEEE Conference on Decision and Control, pp. 4278–4283 (2016)
4. Chowdhury, N.R., Morărescu, I.C., Martin, S., Srikant, S.: Continuous opinions and discrete actions in social networks: a multi-agent system approach. In: Proceedings 55th IEEE Conference on Decision and Control (2016)
5. Davis, J.H.: Group decision making and quantitative judgments: a consensus model. In: Understanding Group Behavior: Consensual Action by Small Groups, vol. 1, pp. 35–60. Lawrence Erlbaum Associates (1996)
6. Deffuant, G., Neau, D., Amblard, F., Weisbuch, G.: Mixing beliefs among interacting agents. Adv. Complex Syst. **3**, 87–98 (2000)
7. DeGroot, M.H.: Reaching a consensus. J. Am. Stat. Assoc. **69**(345), 118–121 (1974)
8. Dietrich, F., Martin, S., Jungers, M.: Control via leadership of opinion dynamics with state and time-dependent interactions. IEEE Trans. Autom. Control **63**, 1200–1207 (2017). https://doi.org/10.1109/TAC.2017.2742139

9. Friedkin, N.E., Johnsen, E.C.: Social influence and opinions. J. Math. Sociol. **15**, 193–206 (1990)
10. Hegselmann, R., Krause, U.: Opinion dynamics and bounded confidence models, analysis, and simulation. J. Artif. Soc. Soc. Simul. **5**(3) (2002)
11. Hegselmann, R., Kurz, S., Niemann, C., Rambau, J.: Optimal opinion control: the campaign problem. J. Artif. Soc. Soc. Simul. **18**(3) (2015)
12. Hendrickx, J.M.: A lifting approach to models of opinion dynamics with antagonisms. In: IEEE Conference on Decision and Control. IEEE (2014)
13. Kerckhove, C.V., Martin, S., Gend, P., Rentfrow, P.J., Hendrickx, J.M., Blondel, V.D.: Modelling influence and opinion evolution in online collective behaviour. PLoS ONE **11**(6), 1–25 (2016)
14. Morărescu, I.C., Girard, A.: Opinion dynamics with decaying confidence: application to community detection in graphs. IEEE Trans. Autom. Control Autom. Control **56**(8), 1862–1873 (2011)
15. Morărescu, I.C., Varma, V.S., Buşoniu, L., Lasaulce, S.: Space-time budget allocation policy design for marketing over social networks. Nonlinear Analysis and Hybrid Systems (under review)
16. Ohtsubo, Y., Masuchi, A., Nakanishi, D.: Majority influence process in group judgment: test of the social judgment scheme model in a group polarization context. Group Process. Intergroup Relat. **5**(3), 249–261 (2002)
17. Proskurnikov, A.V., Matveev, A.S., Cao, M.: Opinion dynamics in social networks with hostile camps: consensus vs. polarization. IEEE Trans. Autom. Control **61**(6), 1524–1536 (2015)

A Dynamic Game Formulation for Control of Opinion Dynamics over Social Networks

Jomphop Veetaseveera[(✉)], Vineeth S. Varma[ORCID],
and Irinel-Constantin Morărescu

Université de Lorraine, CRAN, UMR 7039 and the CNRS, CRAN, UMR 7039,
Nancy, France
jomphop.veetaseveera@univ-lorraine.fr

Abstract. This paper considers the case where the opinion of agents in a social network is influenced not only by the other agents, but also by two marketers in competition. The main contributions of this work is to propose a dynamical game formulation of the problem and to conduct the corresponding equilibrium analysis. Due to the competition between the marketers, the opinions never reach consensus but are spread between the desired opinions of the two marketers. Our analysis provides practical insights to know how a marketer should exploit its knowledge about the social network to design the control of opinions using results from optimal control theory. Numerical examples illustrate the analysis.

1 Introduction

A duopoly is a standard problem in economics, politics, and marketing that considers the competition between two (dominant) players over a market, for example, see [10]. Illustrative examples of real-life duopolies are Airbus/Boeing in the market of large commercial airplanes, Republican/Democratic parties in the American politics.

Traditional research on competitive games between marketers assumes a homogeneous population of consumers [6,7]. Unlike these works, we propose a marketing resource allocation based on the influence power that each individual has over the (physical or digital) social network. Basically, we consider that the advertising is done in two steps: the first is done by the marketer that allocate her resources to sway some individuals/agents on her opinion and the second is done by the agents of the social network who influence each other. Consequently, each marketer has to target appropriate influential agents in the network in order to optimize her revenue. Since the focus of the paper is on the resource allocation of the marketer, the second step is modeled by a simple opinion dynamics model introduced in [4].

In this paper we consider the challenging problem that requires to minimize the distance between average of opinions and a desired value using a given control/marketing budget over a social network split in two groups. Basically, this

© Springer Nature Switzerland AG 2021
S. Lasaulce et al. (Eds.): NetGCooP 2021, CCIS 1354, pp. 252–260, 2021.
https://doi.org/10.1007/978-3-030-87473-5_22

social network with contrarian population represents a model for real cases such as supporters of competing teams, parties, etc. On top of this assumption on the network structure we also assume that the maximal marketing influence cannot instantaneously make the opinion of one individual to be equal with the desired value.

In the literature on *viral marketing*, the idea that members of a social network influence each other's purchasing decisions have been studied, with the goal being to select the best set of people to market to such that the overall profit is maximized by propagation of influence through the network [5]. This problem has since received much attention, including both empirical and theoretical results [1], but these results often consider a single entity influencing the network.

In this paper, we consider two competing marketers who want to use their marketing budget in order to sway on their side as many individuals of the network as possible. Thus, the natural framework to exploit is that of game theory and a reasonable solution concept (for arguments see *e.g.*, [8]) for analyzing such a competition situation is the Nash equilibrium (NE). In [9], the authors consider multiple influential entities competing to control the opinion of consumers under a game theoretical setting. However, this work assumes an undirected graph and a voter model for opinion dynamics resulting in strategies that are independent of the node centrality (i.e., agent influence power). On the other hand, the recently published work [11] considers a similar competition with opinion dynamics over a directed graph but with no budget constraints and by considering the average agents' opinion instead of the final one; these two differences change the problem significantly.

Notation. Let $\mathbb{R} := (-\infty, \infty)$ and $\mathbb{R}_{\geq 0} := [0, \infty)$. We use I_n for the identity matrix, $\mathbf{1}_n$ for the column vector of 1 and $\mathbf{0}_n$ for the column vector of 0, $n \in \mathbb{N}$. In the sequel, the symbol $||.||$ corresponds to the norm2 and $x^\top$ stands for the transpose of x, $x \in \mathbb{R}^n$.

2   Problem Statement

We consider a social network populated by agents belonging to the set $\mathcal{V} := \{1, 2, \ldots, n\}$. The parameters $a_{i,j}$ characterize the influence of agent j on agent i, $i \neq j$ and $i, j \in \mathcal{V}$. The opinions of agent i at time step k is denoted by $x_i(k)$ and evolves according to

$$x_i(k+1) = x_i(k) + \delta \left(\sum_{j \in \mathcal{V}} a_{i,j}(x_j(k) - x_i(k)) \right), \tag{1}$$

due to their interaction with other agents in the social network. In our problem, we consider the case where these agents are also influenced by the marketing/advertising of two marketers. The desired opinion of *Marketer m* is $d_m \in \mathbb{R}$, $m \in \{1, 2\}$. As a result of these interactions, the opinion dynamics for $x(k)$, the collective vector of all opinions is given by

$$x(k+1) = Dx(k) + B_1 u_1(k) + B_2 u_2(k), \tag{2}$$

where $D = I_n - \delta L$ is the row stochastic matrix defining the internal dynamics of the network with $L \in \mathbb{R}^{n \times n}$ being the Laplacian matrix associated with adjacency $a_{i,j}$, B_m denotes the manner in which the *Marketers* influence the agents and u_m is the action of *Marketers* m on the agents. We concentrate on two influence models:

1. Uniform broadcasting (UB) with $B_m = (1,1,\ldots,1)^\top$ which implies that all agents in the network receive the same control. This influence model corresponds to traditional advertising/marketing done on television or radio where the control is applied uniformly on all agents.
2. Targeted advertising (TA) with $B_m = I_n$ which implies that the advertising control can be designed for each individual in the network. This model corresponds to modern social media marketing as done by companies like Facebook or Google.

Let $\mathbf{u}_m$ denote the sequence of control actions applied by *Marketer* m. We define the infinite horizon cost for *Marketer* m as

$$J_m(\mathbf{u}_1, \mathbf{u}_2) := \sum_{k=0}^{\infty} \alpha^k \left(||x(k) - d_m \mathbf{1}_n||^2 + u_m(k)^T R_m u_m(k) \right), \qquad (3)$$

where $\alpha \in (0,1)$ is a discount factor which is often related to inflation rates/interest rates in economic literature. We assume that the revenue generated by the firm associated to *Marketer* m at time k depends on the market share captured by the firm. As the distance between the opinions of agents in the social network and the desired opinion d_m decreases, the revenue increases. Alternately, we say that the loss incurred by not capturing the entire market is characterized by $\sum_{i \in \mathcal{V}} ||x_i(k) - d_m||^2$. On the other hand, advertising to agent i incurs a cost which we take to be $u_m^T R_m u_m$. The term R_m is like the price for advertisements.

3   Analysis

Unlike standard control theory problems, in our framework, we have two competing marketers who attempt to bring $x_i(k)$ to a desired opinion $d_m \in \mathbb{R}$ with $d_1 \neq d_2$. If $d_1 = d_2$, the problem can be seen as a distributed optimal control problem. However, when $d_1 \neq d_2$, each marketer has its own objective resulting in a non-cooperative game. Due to the cost function depending on $x(k)$ and its dynamics, we have a dynamic game [2].

The problem we consider corresponds to a difference-game as introduced in [2]. However, since these games are hard to analyse in the most general case, we make the following assumption.

Assumption 1. *We assume that each marketer applies a static state feedback strategy,*

$$u_m(k) = G_m(d_m \mathbf{1}_n - x(k)) - M_m, \qquad (4)$$

where $G_m \in \mathbb{R}^{n \times n}$ is a fixed feedback gain and $M_m \in \mathbb{R}^n$ is an offset applied to balance the opposing marketer resulting in the strategy given by $K_m := (G_1 \ M_1) \in \mathbb{R}^{n \times n+1}$.

We can redefine the cost J_m in terms of the static feedback strategy K_m as

$$J_m(K_1, K_2) := \sum_{k=0}^{\infty} \alpha^k \left(||x(k) - d_m \mathbf{1}_n||^2 + u_m(k)^T R_m u_m(k) \right), \qquad (5)$$

where $x(k)$ follows (2) and $u_m(k)$ is given by (4).

Under Assumption 1, we can reformulate the dynamic game with actions $\mathbf{u}_m$ as a static one-shot game with the action space of each player corresponding to being the control gain and offset. Formally, we define the game $\mathcal{G}$ in strategic form as follows

$$\mathcal{G} := (\{1, 2\}, \{\mathcal{K}_1, \mathcal{K}_2\}, \{J_1, J_2\}), \qquad (6)$$

where:

1. $\{1, 2\}$ is the set of players or *Marketers*;
2. $\mathcal{K}_m$ is the set of pure actions for player m, specifically, we have $K_m = (G_1 \ M_1) \in \mathcal{K}_m = \mathbb{R}^{n \times (n+1)}$;
3. $J_m(K_1, K_2)$ as defined in (5) is the cost function for player m;

A natural solution concept for a game is that of the Nash equilibrium. For convenience, we use $-m$ to denote the other player, i.e. $3 - m$ for all $m \in \{1, 2\}$. A pure Nash equilibrium (NE) is defined for the game $\mathcal{G}$ as follows.

Definition 1. *We say that (K_1^*, K_2^*) form a NE of the game $\mathcal{G}$ if and only if*

$$J_m(K_m, K_{-m}^*) \geq J_m(K_m^*, K_{-m}^*), \qquad (7)$$

for all $K_m \in \mathcal{K}_m$ and for all $m \in \{1, 2\}$.

The NE is a suitable notion to study player interactions in a non-cooperative game when all players behave rationally and are capable of computing the best decisions to make for a given opponent strategy. However, this assumption on player behavior may not always hold and some players may behave differently or play simpler/naive strategies.

As a first step, we consider the case where player 2 is naive, i.e. it plays a given strategy K_2. In the following, we compute the best response player m can do for a given strategy player K_{-m} played by $-m$.

3.1 Best Response to a Given Opponent Strategy

Let $\beta_m(K_{-m})$ denote the best response function defined as

$$\beta_m(K_{-m}) = \arg\min_{K_m} J_m(K_m, K_{-m}). \qquad (8)$$

In general, the best response function is set valued as the arg min is not unique. However, the following proposition provides a unique best response and the method to find this value.

Proposition 1. *The best response function is unique and can be evaluated as*

$$\beta_m(K_{-m}) = \alpha(\tilde{B}_m^\top P_m \tilde{B}_m + R_m)^{-1}\tilde{B}_m^\top P_m \tilde{D}_m, \tag{9}$$

where $\tilde{D}_m = \begin{pmatrix} D - B_{-m}G_{-m} & B_{-m}(G_{-m}\mathbf{1}_n(d_{-m} - d_m) - M_{-m}) \\ \mathbf{0}_n^\top & 1 \end{pmatrix}$, $\tilde{B}_m = \begin{pmatrix} B_m \\ \mathbf{0}_n^\top \end{pmatrix}$ *and* P_m *is the solution to the Algebraic Riccati equation*

$$P_m = \tilde{D}_m^\top(\alpha P_m - \alpha^2 P_m \tilde{B}_m(\alpha \tilde{B}_m^\top P_m \tilde{B}_m + R_m)^{-1}\tilde{B}_m^\top P_m)\tilde{D}_m + Q_m, \tag{10}$$

where $Q_m = \begin{pmatrix} I_n & \mathbf{0}_n. \\ \mathbf{0}_n^\top & 1 \end{pmatrix}$

Proof. Here, we search for u_1^* minimizing the cost $J_1(u_1, u_2)$ supposing u_2 is known. The reasoning is still the same to find u_2^* for u_1 known. Under Assumption 1, we have

$$x(k+1) = Dx(k) + B_1 u_1(k) + B_2 G_2(d_2 \mathbf{1}_n - x(k)) - B_2 M_2. \tag{11}$$

Let $e_1(k) = x(k) - d_1 \mathbf{1}_n$ be the error for the desired opinion d_1. The error dynamics is

$$e_1(k+1) = (D - B_2 G_2)e_1(k) + B_1 u_1(k) + B_2 G_2 \mathbf{1}_n(d_2 - d_1) - B_2 M_2, \tag{12}$$

where $B_2 G_2 \mathbf{1}_n(d_2 - d_1) - B_2 M_2$ is a constant affine term. We modify (12) by including the affine term in the state variable and we use an algorithm from the Sect. 4.2 of Vol. II, 4th Ed. of [3], to solve (10). The modified system is

$$\begin{pmatrix} e_1(k+1) \\ 1 \end{pmatrix} = \tilde{D}_1 \begin{pmatrix} e_1(k) \\ 1 \end{pmatrix} + \tilde{B}_1 u_1, \tag{13}$$

where $\tilde{D}_1 = \begin{pmatrix} D - B_2 G_2 & B_2 G_2 \mathbf{1}_n(d_2 - d_1) - B_2 M_2 \\ \mathbf{0}_n^\top & 1 \end{pmatrix}$ and $\tilde{B}_1 = \begin{pmatrix} B_1 \\ \mathbf{0}_n^\top \end{pmatrix}$.

Finally, the optimal control u_1^* depending on u_2 is

$$u_1^*(x) = -\alpha(\alpha \tilde{B}_1^\top P_1 \tilde{B}_1 + R_1)^{-1}\tilde{B}_1^\top P_1 \tilde{D}_1 \begin{pmatrix} e_1 \\ 1 \end{pmatrix}$$

$$= -(G_1 \; M_1)\begin{pmatrix} e_1 \\ 1 \end{pmatrix} = -G_1(x - d_1) - M_1, \tag{14}$$

which is consistent with Assumption 1. Thus, the the best response is $\beta_m(K_{-m}) = (G_1 \; M_1)$. The uniqueness of the solution comes from the uniqueness of the Riccati solution.

3.2   Nash Equilibrium

We propose the following iterative asynchronous best response algorithm to find a Nash equilibrium of the game $\mathcal{G}$. While the convergence of the iterative best response dynamics, i.e., Algorithm 1 is not guaranteed for all game classes, our numerical tests show that this property holds as illustrated in the next section.

Algorithm 1. Sequential Gain computation

Data : $\alpha = 0.999$; $\epsilon = 10^{-6}$;
Initialization : $K_1^0, K_2^0, k = 0$;
   while $\|K_1^{k+1} - K_1^k\| > \epsilon$ **or** $\|K_2^{k+1} - K_2^k\| > \epsilon$ **do**
      $K_1^{k+1} = \beta_1(K_2^k)$;
      $K_2^{k+1} = \beta_2(K_1^{k+1})$;
      $k = k + 1$;
   end while
   Result : (K_1^k, K_2^k) is a NE of the game $\mathcal{G}$.

---

4 Numerical Illustration

In this section, we will study the performance of targeted advertising (TA) when compared to uniform broadcasting (UB). For all our numerical tests, we will consider the graph presented in Fig. 1. We take $d_1 = 2$ and $d_2 = -2$. When player m is implementing TA, we take $R_m \in \{I_n, 2I_n\}$, and $R_m \in \{n, 2n\}$ while implementing UB. We consider the following initial conditions

$$x_A = (1, 2, -3, 0, 6, -5, 4, 3, -2, 4)^\top \text{ and } x_B = (4, -2, -2, -3, 2, 0, 2, -1, 1, 0)^\top.$$

As a first step, we consider the situation when both marketers apply the NE strategy, but when *Marketer 1* applies TA and *Marketer 2* UB. Since for UB, the control u_2 is applied to all n agents, we take $R_2 = n$ and $R_1 = I_n$ to look at a symmetric situation in terms of the *Marketer* revenue and cost. If we take a graph where all agents have the same centrality, we notice that the opinions of all agents converge to 0 which lies at the middle of d_1 and d_2 due to the symmetry.

However, if we consider the graph as in Fig. 1 with $x(0) = x_A$, the opinions evolve as seen in Fig. 2 with the average final opinion being closer to d_1, the desired opinion of *Marketer 1*. This demonstrates the advantage of TA and Table 1 further illustrates how *Marketer 1* prioritizes advertising the agents with a higher centrality (vector centrality associated to the Laplacian matrix).

Table 1. Centrality of each agent

Agent	1	2	3	4	5	6	7	8	9	10
Centrality	0.476	0.068	0.006	0.011	0.231	0.003	0.003	0.006	0.162	0.034
$x_i(t \mapsto \infty)$	2.484	0.959	0.550	0.418	1.692	0.024	-0.864	0.206	1.188	-0.144

In Fig. 3 and Fig. 4, we plot the evolution of average opinion when both *Marketers* apply TA or UB respectively. Interestingly, consensus is reached when TA is applied only when $R_1 = R_2$ and in this case all the opinions converge

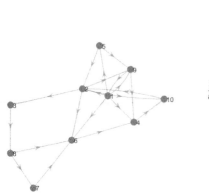

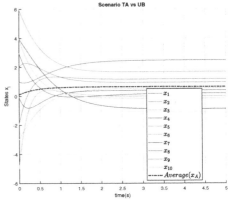

Fig. 1. Directed graph of 10 agents.

Fig. 2. Opinion dynamics with $R_1 = I_n$ and $R_2 = n$.

to 0 which lies halfway between d_1 and d_2. However, taking $R_2 = 2R_2$ results in the opinions converging to

$$\lim_{t \to \infty} x(t) = (2.48; 0.96; 0.55; 0.42; 1.69; 0.02; -0.86; 0.2055; 1.19; -0.14)^T.$$

We notice that for all R_m, UB vs UB results in consensus. The corresponding costs are provided in Table 2.

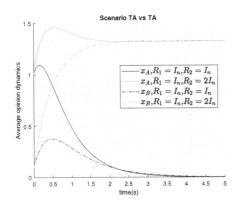

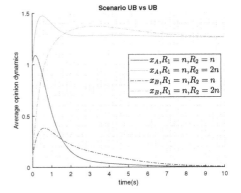

Fig. 3. Opinion dynamics in the scenario TA

Fig. 4. Opinion dynamics in the scenario UB

Table 2. Costs depending on the scenario and the initial condition

Scenario	R_1	R_2	$Cost$ with K^*			
			$J_1(x_A)$	$J_2(x_A)$	$J_1(x_B)$	$J_2(x_B)$
TA	I_n	I_n	86 453	102 759	87 233	98 272
	I_n	$2I_n$	30 731	127 644	31 428	122 783
UB	n	n	77 755	97 619	78 200	97 093
	n	$2n$	29 549	122 263	30 163	120 843
TA vs UB	I_n	n	54 367	120 470	56 075	115 919

5 Conclusion

In this work, we have studied the behavior of two competing firms/marketers that control the opinions of a set of consumers interacting over a social network. We consider a linear interaction model and quadratic costs for each firm related to the revenue earned and the amount they spent in order to control the network (via advertisements or other marketing strategies). As a first step, we find the optimal static state feedback control which must be applied by a marketer assuming that the other marketer applies a given strategy. Next, we provide an iterative algorithm, which we observe to converge to the Nash equilibrium after extensive numerical tests. We provide numerical examples which illustrate the advantage of viral marketing techniques (targeted advertising based on the social graph) when compared to traditional advertising strategies (uniform broadcast). Future work will focus on theoretically studying the existence and uniqueness of the Nash equilibrium of the game we have studied.

References

1. Arthur, D., Motwani, R., Sharma, A., Xu, Y.: Pricing strategies for viral marketing on social networks. In: Leonardi, S. (ed.) WINE 2009. LNCS, vol. 5929, pp. 101–112. Springer, Heidelberg (2009). https://doi.org/10.1007/978-3-642-10841-9_11
2. Basar, T., Olsder, G.J.: Dynamic Noncooperative Game Theory, vol. 23. SIAM (1999)
3. Bertsekas, D.P., Bertsekas, D.P., Bertsekas, D.P., Bertsekas, D.P.: Dynamic Programming and Optimal Control, vol. 1. Athena Scientific, Belmont (1995)
4. DeGroot, M.H.: Reaching a consensus. J. Am. Stat. Assoc. **69**(345), 118–121 (1974)
5. Domingos, P., Richardson, M.: Mining the network value of customers. In: Proceedings of the Seventh ACM SIGKDD International Conference on Knowledge Discovery and Data Mining, pp. 57–66. ACM (2001)
6. Esmaeili, M., Aryanezhad, M.B., Zeephongsekul, P.: A game theory approach in seller-buyer supply chain. Eur. J. Oper. Res. **195**(2), 442–448 (2009)
7. Friedman, L.: Game-theory models in the allocation of advertising expenditures. Oper. Res. **6**(5), 699–709 (1958)
8. Lasaulce, S., Tembine, H.: Game Theory and Learning for Wireless Networks: Fundamentals and Applications. Academic Press, Cambridge (2011)

9. Masucci, A.M., Silva, A.: Strategic resource allocation for competitive influence in social networks. In: 2014 52nd Annual Allerton Conference on Communication, Control, and Computing (Allerton), pp. 951–958. IEEE (2014)
10. Singh, N., Vives, X.: Price and quantity competition in a differentiated duopoly. RAND J. Econ. 546–554 (1984)
11. Varma, V., Morarescu, I.C., Lasaulce, S., Martin, S.: Opinion dynamics aware marketing strategies in duopolies. In: 56th IEEE Conference on Decision and Control, CDC 2017 (2017)

Opinion Dynamics with Multi-body Interactions

Leonie Neuhäuser[1,2,4(✉)] ⓘ, Michael T. Schaub[3,4] ⓘ, Andrew Mellor[2] ⓘ, and Renaud Lambiotte[2] ⓘ

[1] Hertie School, Berlin, Germany
[2] Mathematical Institute, University of Oxford, Oxford, UK
{mellor,renaud.lambiotte}@maths.ox.ac.uk
[3] Department of Engineering Science, University of Oxford, Oxford, UK
[4] RWTH Aachen University, Aachen, Germany
{neuhaeuser,schaub}@cs.rwth-aachen.de

Abstract. We introduce and analyse a three-body consensus model (3CM) for non-linear consensus dynamics on hypergraphs. Our model incorporates reinforcing group effects, which can cause shifts in the average state of the system even in if the underlying graph is complete (corresponding to a mean-field interaction), a phenomena that may be interpreted as a type of peer pressure. We further demonstrate that for systems with two clustered groups, already a small asymmetry in our dynamics can lead to the opinion of one group becoming clearly dominant. We show that the nonlinearity in the model is the essential ingredient to make such group dynamics appear, and demonstrate how our system can otherwise be written as a linear, pairwise interaction system on a rescaled network.

Keywords: Consensus · Diffusion · Non-linear dynamics · Networks · Group dynamics · Multi-body interactions · Opinion formation

1 Introduction

Networks provide a powerful framework for the modelling of dynamical systems. Many networked dynamical systems can be described by a set of differential equations describing *pairwise* interactions between the nodes:

$$\dot{x}_i = \sum_j A_{ij} f(x_i, x_j) \qquad \text{for} \quad i \in \{1, \dots, N\}. \tag{1}$$

where x_i is the state of node i, $A \in \mathbb{R}^{N \times N}$ is the adjacency matrix of the underlying graph ($A_{ij} = 1$, if node i connects to node j, and 0 otherwise), and $f(x_i, x_j)$

Parts of this work have been presented in [12]. L. Neuhäuser was financially supported by the Hertie School. M. Schaub was supported by the European Union's Horizon 2020 research and innovation programme under the Marie Sklodowska-Curie grant agreement No 702410.

S. Lasaulce et al. (Eds.): NetGCooP 2021, CCIS 1354, pp. 261–271, 2021.
https://doi.org/10.1007/978-3-030-87473-5_23

is a function describing the interactions between nodes i and j. Important examples of the above type of dynamics include diffusion [11] or oscillator dynamics [1]. In particular, the dynamics of opinion formation have been considered in the context of dynamical processes on networks [15,16], including opinion formation models such as the de Groot model [6], bounded-confidence models [5] and threshold models [20].

However, it is increasingly realized that such pairwise interaction models may not be sufficient to describe a range of important phenomena, ranging from collaborations of authors [14] to neuronal activity [7]. Accordingly, various models that focus on the importance of group interactions, i.e., situations when the basic unit of interaction involves more than two nodes have been proposed in the literature [2,8,17].

These *multi-body* interaction models are particularly relevant for social dynamics, which have long been argued to be emergent phenomena that are not merely based on pairwise interactions between members of a community, but often include complex mechanisms of peer influence and reinforcement. Such group dynamics, which may lead to 'higher-order' dynamical effects, may indeed be essential to better understand phenomena such as hate communities, echo chambers and polarisation in society.

2   A Multi-body Interaction Model for Non-linear Consensus

Motivated by the above discussed scenarios, we here introduce a simple multi-body interaction model for opinion formation within social systems. As a first step towards studying the higher-order effects of multi-body dynamics, we concentrate on a three-body consensus model (3CM), in which the interactions between triplets of nodes are governed by the following differential equations:

$$\dot{x}_i = \sum_{j,k=1}^{N} \mathcal{A}_{ijk} g_i^{\{j,k\}}(x_i, x_j, x_k) \quad \text{for } i \in \{1, \dots, N\}. \tag{2}$$

Here $\mathcal{A}_{ijk}$ describes the adjacency tensor of node triplets $\{i, j, k\}$, where $\mathcal{A}_{ijk} = 1$ if the group of nodes interact and $\mathcal{A}_{ijk} = 0$ otherwise. We further model the group (multi-body) interaction function $g_i^{\{j,k\}}(x_i, x_j, x_k)$ as:

$$g_i^{\{j,k\}}(x_i, x_j, x_k) = s(\|x_j - x_k\|) \left[(x_j - x_i) + (x_k - x_i) \right]. \tag{3}$$

For each triplet $\{i, j, k\}$, this function comprises (a) the joint influence of the node-pair j, k on node i, modeled by the linear term $[(x_j - x_i) + (x_k - x_i)]$, which is (b) modulated by an influence function $s(\|x_j - x_k\|)$ of their state differences. In the following we assume $g_i^{\{j,k\}}$ is the same for each interacting node triplet, for the sake of simplicity.

Note that if the modulation function $s(x)$ is monotonically decreasing, nodes j and k reinforce their influence on i if they have similar states x_j, x_k, whereas the influence of nodes j, k on node i is diminished if their states are very different. This property is reminiscent of non-linear voter models for discrete dynamics [9], where voters change opinion with a probability depending non-linearly on the fraction of disagreeing neighbours.

In addition to the ability to describe a reinforcing dynamics, the functional form of our model has some further desirable symmetry properties. In particular, we remark that (2) is invariant to translation and rotation of all node states. This is a desirably property for many opinion formation process, as it ensures that the opinion formation is only influenced by the relative position of the node states x_i and independent of a specific global reference frame. This property can be shown by observing that any rotation is norm preserving, and thus $s(\|x_j - x_k\|)$ is rotational and translational invariant. Since the term $[(x_j - x_i) + (x_k - x_i)]$ is translation invariant and linear, any translation and rotation applied to all states will leave (2) invariant. Note that this 'quasi-linearity' of the interaction function $g_i^{\{j,k\}}(x_i, x_j, x_k)$ is in close correspondence to the necessary and sufficient conditions for translation and rotational invariance for pairwise interaction systems [19]. In the following we will restrict our scope to scalar states x_i. In this case, the above described invariance simply implies an invariance under a change of signs or a global shift of all states.

3   Results

3.1   Reduction to Network Model and Higher-Order Effects

Interestingly, it can be shown that the above dynamics can be rewritten in terms of a (in general) *time-varying and state dependent* weighted adjacency matrix $\mathfrak{W}(t, x, \mathcal{A})$, whose entries describe the three-body influence on node i exerted over the 'pairwise link' (i, j):

$$(\mathfrak{W})_{ij} = \sum_k \mathcal{A}_{ijk} s(\|x_j - x_k\|) = \sum_{k \in \mathcal{I}_{ij}} s(\|x_j - x_k\|). \tag{4}$$

Here $\mathcal{I}_{ij}$ is the index-set of nodes that interact in a triplet with nodes i, j; and for simplicity we have written $\mathfrak{W} = \mathfrak{W}(t, x, \mathcal{A})$, omitting the dependencies of $\mathfrak{W}$. Accordingly, we can rewrite the dynamics (2) as:

$$\dot{x}_i = 2 \sum_j \mathfrak{W}_{ij}(x_j - x_i) =: -2 \sum_j L_{ij}^{\mathfrak{W}} x_j, \tag{5}$$

where we have defined the Laplacian $L^{\mathfrak{W}}$ of the 3CM model via $L_{ij}^{\mathfrak{W}} = -\mathfrak{W}_{ij}$ for $i \neq j$, and $L_{ii}^{\mathfrak{W}} = \sum_j \mathfrak{W}_{ij}$.

As discussed above, the entries of the weighted adjacency matrix $\mathfrak{W}$ (and thus of $L^{\mathfrak{W}}$) are in general time-varying, state and topology dependent for a general modulation function $s(t)$, and so the above rewriting does not imply that the system can be understood via pairwise interaction of the form (1). There is one important exception, though. If $s(x)$ is constant, the group interaction function

g is linear and the three-body dynamical system can therefore be rewritten as a rescaled pairwise dynamical system defined on a graph. The weighted adjacency matrix $\mathfrak{W}$ and corresponding graph Laplacian $L^{\mathfrak{W}}$ are then constant in time. In fact, in this case $L^{\mathfrak{W}}$ becomes the so-called motif Laplacian proposed by [3] for community detection in higher-order networks: the entries of $L^{\mathfrak{W}}$ count the nodes involved in interaction triplets (triangles on the corresponding graph). This emphasizes that multi-body dynamical effects beyond rescaled pairwise interactions can only appear for non-linear interaction functions, regardless of topology of the multi-body interactions encoded in $\mathcal{A}$.

3.2   Convergence to Consensus and Average-Opinion Dynamics

From our above rewriting (5), it is easy to see that a global consensus, in which $x_i(t) = x_j(t)$ for all i, j, is a fixed point of our model. Using standard arguments, it can be shown that convergence to consensus is guaranteed as long as the scaling $s(\|x_j - x_k\|)$ is positive. For generic initial conditions this is only the case if the modulation function $s(x)$ is positive definite. We will therefore focus on this scenario in the following.

Despite the fact that in our model the 3-body interactions as undirected, i.e., the adjacency tensor $\mathcal{A}$ is completely symmetric in all pairs of indices, the average opinion $\bar{x} = (\sum_i x_i)/N$ is however not invariant over time, in general. To see this observe that

$$\dot{\bar{x}}(t) = \frac{1}{N} \sum_{i=1}^{N} \dot{x}_i(t) = -\frac{2}{N} \sum_{i,j=1}^{N} L_{ij}^{\mathfrak{W}} x_j(t), \tag{6}$$

which is zero only (i) when there is global consensus, or (ii) when the Laplacian $L^{\mathfrak{W}}$ (interpreted as a graph Laplacian of a directed graph) corresponds to a balanced graph, i.e., the in-degree equals the out-degree for every node and thus $\sum_i L_{ij}^{\mathfrak{W}} = \mathbf{0}$. While the former condition is dynamically trival, the latter condition will in general depend on a complex interplay between the states x_i, the structure of the node-triplet interaction and the form of the interaction function. One exception here is again the case in which the modulation function $s(x)$ is constant and the dynamics therefore becomes linear. In this case $L^{\mathfrak{W}}$ will be a symmetric matrix and therefore trivially correspond to a balanced (undirected) graph. Note that if at any instance of time t_0 the Laplacian $L^{\mathfrak{W}}$ becomes balanced, the average opinion will be conserved for all $t \geq t_0$.

3.3   Shifts Towards Majority Opinions on Complete Graphs

To exhibit how the non-linear reinforcing diffusion dynamics (2) can lead to a shift of the average state over time, we first study the dynamics on a structurally featureless 'complete' hypergraph in which all triplet interactions are present. We split the nodes into two factions with binary opinions $x_i = 0$ or $x_j = 1$, respectively. In this case, for initial distributions with average $\bar{x}(0) = 0.5$, the

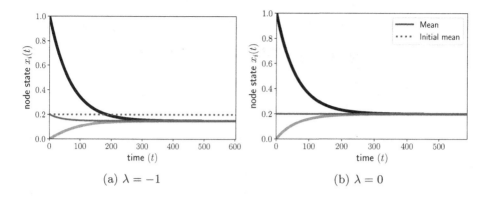

Fig. 1. 3CM dynamics on a hypergraph with all-to-all connectivity. We display simulations for a modulation function $s(x) = \exp(\lambda x)$ for different values of λ, and an unbalanced initial condition in which 80% of the nodes have opinion 0 and 20% have opinion 1 ($\bar{x}(0) = 0.2$) Dotted red lines indicate the initial value of the average node state. Black (grey) solid lines represent the evolution of the state of nodes whose initial configuration is one (or zero). (a) The setting $\lambda < 0$ will results in a final consensus value that is shifted away from the initial average. (b) In contrast, the average state is conserved for $\lambda = 0$ no matter what the initial average was (as the dynamics is linear). (Color figure online)

Laplacian $L^{\mathfrak{W}}$ will correspond to a balanced graph for which in-degree equals out-degree for all nodes, and accordingly the final consensus value will be the average of the initial opinions, which is invariant in this case. In contrast, initial distributions with $\bar{x}(0) \neq 0.5$, necessarily lead to an unbalanced graph Laplacian $L^{\mathfrak{W}}$. We remark that these conditions depend on the regular topology of the system and will not hold for systems with more general interaction structure.

Note that if $\bar{x} \neq 0.5$ the initial groups must have different size, with one group being a relative majority or minority, respectively. Let us now consider a decreasing modulation function $s(x) = \exp(\lambda x)$ with $\lambda < 0$, such that the opinions of similar nodes reinforce each other. In this case, any deviation from an initial average $\bar{x}(0) = 0.5$ grows in time, with a drift towards the majority. This is shown in Fig. 1. In the context of opinion dynamics, this type of dynamics may be interpreted as a kind of peer-pressure, which causes the average opinion in the system to shift towards the initial majority opinion.

3.4   Opinion Dynamics in Clustered Systems

To gain some insight into the interplay between the system structure and our dynamics, we consider a system defined on simple modular hypergraph as displayed in Fig. 2. Here the system consists of two fully connected equally sized clusters of 10 nodes, i.e., $\mathcal{A}_{ijk} = 1$ for all triplets (i, j, k) that are in the same cluster and $\mathcal{A}_{ijk} = 0$, otherwise. In addition, the clusters are connected by 80 randomly chosen triplet interactions, of which a fraction $p \in [0, 1]$ is oriented

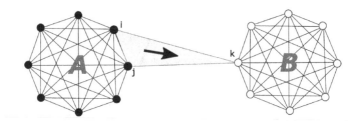

Fig. 2. Schematic: 3CM dynamics on a modular hypergraph. We consider a binary initialisation of two clusters, above indicated as black and white node colours. For a triplet interaction (i, j, k) oriented towards cluster B (top), the fact that there is a close consensus in cluster A between nodes i, j accelerates the rate of change of node k in B. In contrast, the influence on the nodes i, j in A is small, since the nodes in the pairs (i, k) and (j, k), belong to different clusters and are thus in relative disagreement, which decreases the effect of cluster B on A.

towards cluster B, and the remainder towards cluster A. Here we say a triplet interaction (i, j, k) is oriented towards a specific cluster if two of its nodes are in the oppositve cluster (see Fig. 2). For instance, if nodes i, j are in cluster A and node k is in cluster B, the triplet interaction is oriented towards cluster B (recall, however, that all nodes of the triplet interact, i.e., there is no 'directionality' in the coupling tensor $\mathcal{A}$). We consider an initially polarized state of opinions such that nodes in cluster A have initial state $x_A(0) = 0$ and nodes in cluster B have initial state $x_B(0) = 1$.

In contrast to the fully connected system described in the previous section, here an initial average of $\bar{x}(0) = 0.5$ does not guarantee that the average is invariant over time (i.e., the induced graph is not balanced). This is due to the reinforcing group effects which result in asymmetric interactions as encoded by the induced graph described by $\mathfrak{W}$. To see this, consider a triplet interaction oriented towards cluster B (see Fig. 2. Since there is (local) consensus in cluster A, for monotonically decreasing $s(x)$ the influence of the nodes from A onto the node in B is increased, whereas the opposite influence is decreased. Following this reasoning, the relative influence of cluster A on B (and vice versa) thus depends on the relative number of triplet interactions oriented towards each cluster, which is determined by parameter p: for small p most of the 'cross-cluster' triplet interactions will be oriented towards A, for large p most interactions will be oriented towards B.

In Fig. 3, we show how the relative number of oriented triplet interactions measured by p affect the final consensus value and the convergence towards consensus. For these results we averaged our 20 simulations with varying p. As seen in Fig. 3(a), we observe a shift in the final consensus value towards the initial value in cluster A or in cluster B, depending on the percentage of triplet interactions oriented away from that cluster. The asymmetry also influences the rate of convergence towards consensus, as shown in Fig. 3(b), i.e., a relative increase in oriented triplets leads to a faster rate of convergence. Our simulations also reveal higher fluctuations in the asymptotic state for values close to $p = 0.5$.

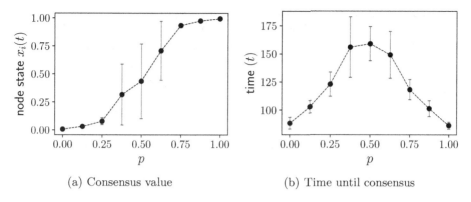

(a) Consensus value   (b) Time until consensus

Fig. 3. 3CM dynamics on a modular hypergraph. Simulations of 3CM on two inter-connected clusters of 10 nodes, with the modulation function $s(x) = \exp(-100x)$. All results are averaged over 20 simulations, where the error bars denote one standard deviation. (a) Final consensus value as a function of the directionality parameter p. As the fraction of triplet interactions oriented from cluster A to cluster B increases, so does the consensus value towards the initial state in cluster A. (b) The rate of convergence is significantly faster when the triplet interactions are mostly oriented towards one cluster, i.e., for extreme values of p.

This result indicates that the process is sensitive to small deviations from balance in the initial topology, which can lead to comparably large differences in the consensus value.

Note that similar observations can also be made if one cluster forms a "minor-ity" and is comparably smaller than the other cluster (the majority). Indeed, depending on the relative number of triplets oriented towards the majority, the opinion of the minority cluster may have a much stronger influence on the final consensus value than the majority. An example of this situation is shown in Fig. 4. Here the opinion of the minority cluster A with 5 nodes 'dominates' the opinion of the majority cluster B with 10 nodes for a non-linear reinforcing modulation function $s(x)$ (left). In the context of opinion dynamics, this type of behavior is akin to a "minority influence", in which small groups can dominate the formation of an opinion not because of their size, but due to their inter-nal cohesion. In contrast, if we remove the effect of opinion reinforcement and consider a linear coupling the initial opinion of majority will have the strongest effect on the final consensus state (as expected from a distributed averaging).

3.5   Time-Scale Separations in Clustered Systems and Multi-body Interactions

Finally, we investigate interplay between the topology in a clustered hypergraph and our multi-body interaction dynamics for initial conditions that are not piece-wise constant. Specifically, we are interested in examining the different time

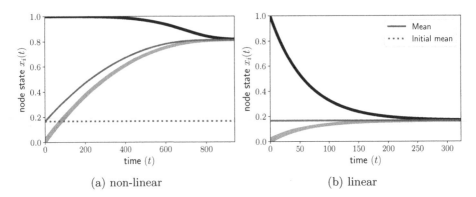

(a) non-linear   (b) linear

Fig. 4. Minority influence through reinforcing opininon. We display simulations for a modulation function $s(x) = \exp(\lambda x)$ for $\lambda = -10$ and two biased clusters of different sizes, which are connected with a single triplet interaction oriented towards cluster A. Cluster A comprises the majority of nodes (10 nodes) whereas cluster B consists only of 5 nodes. (a) While intuition may suggest a final consensus that is leaning towards the initial opinion 0 of the majority cluster A, we observe the opposite behavior due to opinion reinforcing effect of the nonlinear coupling, which leads to an (effectively) oriented connection between B and A. (b) If the dynamics are linear (right), the initial average opinion is conserved and therefore the majority opinion dominates the final consensus value.

scales in the dynamics induced by the clustered topology (as is also well known for pairwise interaction systems in the fixed [4] and time-varying case [10]). The different time-scales are here associated to a fast convergence of states inside clusters, followed by a slower convergence towards global consensus.

For simplicity, we consider here again the setting of a clustered hypergraph with 2 clusters described above for $p = 1$ (cf. Fig. 2 and Fig. 3). This time, however, the nodes in each cluster may have different states initially. For our experiments, we initialise nodes in different clusters uniformly at random over disjoint intervals, such that nodes of cluster A have random initial states in the interval $I_A = [0, 0.5]$ and nodes in cluster B have random initial states in $I_B = [0.5, 1]$ (see Fig. 5). The initial cluster averages of the node states are thus far apart.

Now two effects lead to a fast multi-body consensus inside each cluster. First, each of the clusters are internally fully-connected. Second, the inter-cluster-dynamics will generally have a weaker effect since the difference of the initial conditions means that $s(\|x_i - x_j\|)$ will be small if nodes i and j are in different clusters. As a result, we first observe a fast dynamics within the clusters in which nodes approach the cluster-average state (Fig. 5, bottom) and then a slower dynamics between the two clusters (Fig. 5, top).

However, the final outcome of this process critically depends on the modulation function $s(x)$. For $s(x) = \exp(\lambda x)$, with $\lambda = -100$, we observe an asymmetric shift towards cluster B for $p = 1$, the final opinion as shown in Fig. 5 (left).

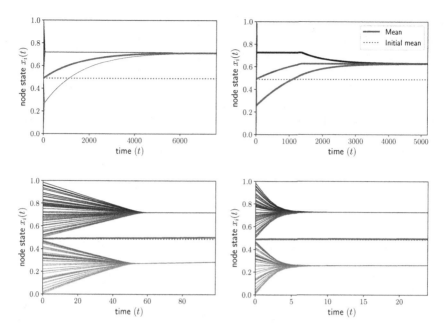

Fig. 5. Time-scale separation in clustered systems and the influence of the modulation function $s(x)$. We simulate the dynamics of two clusters A and B connected with 80 (random) triplet interactions all oriented towards nodes in A. The nodes states x_i are initialised uniformly at random over two separate intervals I_A, I_B with $I_A \cap I_B = \emptyset$, such that $x_i(0) \in I_A$ if i is in cluster A and $x_i(0) \in I_B$ if i is in cluster B. The left figures correspond to the exponential modulation function $s(x) = \exp(\lambda x)$ for $\lambda = -100$. The right figures correspond to a Heaviside modulation function with threshold $\phi = 0.2$. In both cases, we observe a timescale separation with a fast, symmetric dynamics inside the clusters, followed by a slow dynamics in which cluster B exhibits a disproportionate influence compared to its size (both cluster have the same size). For the Heaviside function, the process becomes linear when the values in the two clusters are less separated than the Heaviside-threshold. The fast transient inter-cluster dynamics is shown in the bottom figures, with qualitatively similar results for both modulation functions.

If we consider other modulation functions, however, the results can be different. For instance we may consider a coupling via a (shifted) heaviside function of the form:

$$s(\|x_j - x_k\|) = H(\|x_j - x_k\| - \phi) = \begin{cases} 0 & \text{if } \|x_j - x_k\| < \phi \\ 1 & \text{otherwise,} \end{cases} \qquad (7)$$

which switches between a zero interaction and linear diffusion when the difference of the neighbouring triangle nodes becomes smaller than a threshold $\phi \in (0,1)$.

If we consider the Heaviside function with threshold $\phi = 0.2$ as the modulation function, the dynamics between the clusters behaves initially very similar to the exponential case (see Fig. 5). This behavior continues until the two cluster means are less separated than ϕ. As shown in Fig. 5 (right), the dynamics between the clusters then become linear and therefore the average opinion remains constant from then onwards.

We remark that the Heaviside function is not positive-definite, so that the above dynamics do not necessarily converge to a global consensus asymptotically. Indeed, consider a setting with three clusters that are connected by a triplet interaction with exactly one node in each cluster. If the initial states are separated more than the Heaviside constant, all inter-cluster interactions would be zero and therefore the system would at most converge to a decoupled 'polarised' state with three independent opinions, one for each cluster.

4   Conclusion

We have explored a model for opinion formation with multi-body interactions, defined on hypergraphs, in order to identify the impact of reinforcing opinions on the dynamics. We found that these non-linear multi-body interactions can cause dynamical phenomena such as shifts in the average opinion that would not appear in a corresponding pairwise system. In situations with two connected, polarised groups the dynamics can lead to the opinion of one group clearly dominating the other. These findings are important to better understand processes governed by reinforcing group effects in society.

In standard linear opinion formation models with two-body dynamical system such as the de-Groot model [6] or consensus dynamics [13,18], it is known that the asymptotic behavior of the dynamics is dominated by the networks structure. For instance, the mixing time is determined by the spectral gap. In contrast, in our model knowing the structure alone is not sufficient to understand the asymptotic behavior. Indeed, we have shown the initial node states can lead to an effectively oriented flow in the dynamics and thus lead to an opinion formation process that can be dominated, even by relatively small groups of nodes—provided, they have a coherent opinion. In real-world setings, such an insight could be exploited to steer the opinion towards a desirable outcome by modifying the balance in the dynamical system by seeding (i.e., changing the initial states) or eliminating components (i.e. changing the topology). Future work will consider such issues in more detail. A specific challenge in this context will be the derivation of simple, computable (heuristic) strategies that would enable for such a control, without having to assees all minute details of the hypergraph and initial conditions. For instance, deriving some form of higher-order generalization of network centralities, related to the dynamical properties of such multi-body interaction systems, would be an interesting avenue to pursue. As a first attempt one may make use of the matrix representation $\mathfrak{W}$ of the 3CM system to derive appropriate centrality values. However, this would assume a fixed initialisation is known—one thus would need to generalize this notion to identify the important actors and connections in the 3CM system.

Acknowledgements. Michael T. Schaub and Leonie Neuhäuser acknowledge funding by the Ministry of Culture and Science (MKW) of the German State of North Rhine-Westphalia ("NRW Rückkehrprogramm").

References

1. Arenas, A., Díaz-Guilera, A., Kurths, J., Moreno, Y., Zhou, C.: Synchronization in complex networks. Phys. Rep. **469**(3), 93–153 (2008)
2. de Arruda, G.F., Petri, G., Moreno, Y.: Social contagion models on hypergraphs (2019)
3. Benson, A.R., Gleich, D.F., Leskovec, J.: Higher-order organization of complex networks. Science **353**(6295), 163–166 (2016). https://doi.org/10.1126/science. aad9029
4. Chow, J., Kokotovic, P.: Time scale modeling of sparse dynamic networks. IEEE Trans. Autom. Control **30**(8), 714–722 (1985). https://doi.org/10.1109/TAC.1985. 1104055
5. Deffuant, G., Neau, D., Amblard, F., Weisbuch, G.: Mixing beliefs among inter-acting agents. Adv. Complex Syst. **3**(01n04), 87–98 (2000)
6. DeGroot, M.H.: Reaching a consensus. J. Am. Stat. Assoc. **69**(345), 118–121 (1974)
7. Giusti, C., Pastalkova, E., Curto, C., Itskov, V.: Clique topology reveals intrinsic geometric structure in neural correlations. In: Proceedings of the National Academy of Sciences, vol. 112, no. 44, pp. 13455–13460 (2015). https://doi.org/10.1073/pnas. 1506407112
8. Iacopini, I., Petri, G., Barrat, A., Latora, V.: Simplicial models of social contagion. Nat. Commun. **10**(1), 1–9 (2019). https://doi.org/10.1038/s41467-019-10431-6
9. Lambiotte, R., Redner, S.: Dynamics of non-conservative voters. EPL (Europhys. Lett.) **82**(1), 18007 (2008). https://doi.org/10.1209/0295-5075/82/18007
10. Martin, S., Morărescu, I., Nešić, D.: Time scale modeling for consensus in sparse directed networks with time-varying topologies. In: 2016 IEEE 55th Conference on Decision and Control (CDC), pp. 7–12 (2016). https://doi.org/10.1109/CDC. 2016.7798238
11. Masuda, N., Porter, M.A., Lambiotte, R.: Random walks and diffusion on networks. Phys. Rep. **716–717**, 1–58 (2017). https://doi.org/10.1016/j.physrep.2017.07.007
12. Neuhäuser, L., Mellor, A., Lambiotte, R.: Multi-body interactions and non-linear consensus dynamics on networked systems (2019)
13. Olfati-Saber, R., Fax, J.A., Murray, R.M.: Consensus and cooperation in networked multi-agent systems. Proc. IEEE **95**(1), 215–233 (2007)
14. Patania, A., Petri, G., Vaccarino, F.: The shape of collaborations. EPJ Data Sci. **6**(1), 18 (2017). https://doi.org/10.1140/epjds/s13688-017-0114-8
15. Proskurnikov, A.V., Tempo, R.: A tutorial on modeling and analysis of dynamic social networks. Part i. Ann. Rev. Control **43**, 65–79 (2017)
16. Proskurnikov, A.V., Tempo, R.: A tutorial on modeling and analysis of dynamic social networks. Part ii. Ann. Rev. Control **45**, 166–190 (2018)
17. Schaub, M.T., Benson, A.R., Horn, P., Lippner, G., Jadbabaie, A.: Random walks on simplicial complexes and the normalized Hodge 1-Laplacian. SIAM Rev. **62**(2), 353–391 (2020)
18. Tsitsiklis, J.N.: Problems in decentralized decision making and computation. Ph.D. thesis (1984)
19. Vasile, C.I., Schwager, M., Belta, C.: SE(N) invariance in networked systems. In: 2015 European Control Conference (ECC), pp. 186–191. IEEE (2015). https://doi. org/10.1109/ECC.2015.7330544
20. Watts, D.J.: A simple model of global cascades on random networks. In: Proceed-ings of the National Academy of Sciences, vol. 99. no. 9, pp. 5766–5771 (2002)

Modeling Limited Attention in Opinion Dynamics by Topological Interactions

Francesca Ceragioli[1], Paolo Frasca[2(✉)], and Wilbert Samuel Rossi[3]

[1] DISMA, Politecnico di Torino, Turin, Italy
`francesca.ceragioli@polito.it`
[2] Univ. Grenoble Alpes, CNRS, Inria, Grenoble INP, GIPSA-lab,
38000 Grenoble, France
`paolo.frasca@gipsa-lab.fr`
[3] University College Groningen, University of Groningen,
Groningen, The Netherlands
`w.s.rossi@rug.nl`

Abstract. This work explores models of opinion dynamics with opinion-dependent connectivity. Our starting point is that individuals have limited capabilities to engage in interactions with their peers. Motivated by this observation, we propose an opinion dynamics model such that interactions take place with a limited number of peers: we refer to these interactions as topological, as opposed to metric interactions that are postulated in classical bounded-confidence models.

Keywords: Opinion dynamics · Limited attention · Nonsmooth dynamical systems

1 Introduction

Driven by the evolution of digital communication and social networking services, there is an increasing interest for mathematical models of opinion dynamics in social networks. Among the many models proposed in the literature, a few have become popular in the control community [26,27]. In the perspective of the control community, opinion dynamics distinguish themselves from consensus dynamics because consensus is prevented by some other feature of the dynamics. In many popular models, this feature is an opinion dependent limitation of the connectivity. Chief examples are bounded confidence models [16,22], where social agents influence each other iff their opinions are closer than a threshold.

This way of defining influence assumes that agents have always access to the opinions of all fellow agents and may lead to agents being influenced by a large number of their fellows, possibly the whole population. Instead, the number of possible interactions is bounded in practice by the limited time and efforts

Supported in part by MITI CNRS via 80 PRIME grant DOOM and by ANR via project HANDY, number ANR-18-CE40-0010.

S. Lasaulce et al. (Eds.): NetGCooP 2021, CCIS 1354, pp. 272–281, 2021.
https://doi.org/10.1007/978-3-030-87473-5_24

that individuals can devote to social interactions. Similar *limitations of attention* are well documented in psychology and sociology, for instance by the notion of Dunbar number [18], and become evermore crucial in today's age of information bonanza [21]. Indeed, in online social media, natural limitations of attention interplay with the way the online platforms are designed. Users interact via the contents they share: out of the pool of fresh contents, the online platform selects for each user the best contents in order to maximize engagement, mainly based on similarities between users [23]. As the notion of Dunbar number was originally defined with reference to primates, the reader will not find surprising that similar ideas have also been fundamental in the study of flocking in animal groups, as testified by numerous theoretical and experimental works [3,5,15,20,24]. The importance of considering networks where the number of neighbors is limited has also been understood by graph theorists, who have studied the properties of what they call k-nearest-neighbors graph: for instance, it is known that k must be logarithmic in n to ensure connectivity [4].

However, few works have incorporated this important observation in suitable models of opinion dynamics. Before surveying these important references, we briefly describe the contribution of this paper. In our effort to make the case for limited attention in opinion dynamics, we study a simple continuous-time model (first appeared in the survey paper [2]) in which every agent is influenced by her closest k nearest neighbors. In this paper, we provide some preliminary results about this continuous-time dynamics. Our results concentrate on two axes: studying the main properties of its equilibria, including their robustness to disruptions, and proving convergence results in special cases. We describe the equilibria of the dynamics, distinguishing a special type of *clusterization* equilibria that are constituted of separate clusters, and we discuss the robustness of clustered equilibria to disruptions, such as the addition of new agents. Regarding the question of convergence, we are able to provide a proof in two cases: when the total number of agents n is small enough compared to number of neighbors k, namely $n \leq 2k + 1$, and when $k = 1$, that is, agents are only influenced by one "best friend".

The difficulties in studying k-nearest-neighbors dynamics originate from two key features: (1) interactions are not reciprocal; (2) whether two agents interact does not only depend on their two states, but also on the states of all the other agents. In the literature, models with any of these features are still relatively few. In classical bounded confidence models [16,22], interactions are reciprocal as long as the interaction thresholds are equal for all agents [7–10,22], and any lack of reciprocity makes the analysis much more delicate [12,14,25]. In our model, not only interactions are non-reciprocal, but they are also non-metric: whether two agents interact is not solely determined by the distance between their two opinions. For this reason, we follow a consolidated tradition [5] and refer to our connectivity model where agents can interact with their k nearest neighbors as *topological*.

Topological interactions are becoming increasingly popular in the applied mathematics community, especially for second order models [13]. Kinetic and continuum models with topological interactions are also actively studied [6,17, 31]. Among first order "opinion" models, [1] has recently used Petri nets to define a class of models where interactions depend on the opinions of multiple agents: despite some similarities, our model does not belong to this class. In our recent papers [28,29], we have studied two dynamics with asynchronous updates (with and without sub-sampling, respectively) that are discrete-time counterparts of the model we propose here. Finally, our contribution here differs from the one of [2] as the latter focuses on specific properties of the equilibria, such as the distribution of their clusters' sizes, studied by extensive simulations, whereas we are interested in analytical results about dynamical properties like convergence to the equilibria and about their robustness to perturbations. Our robustness analysis is inspired by the approach taken in [7] for bounded confidence models.

The rest of this paper has the following structure. Section 2 introduces the model, Sect. 3 develops its analysis, and Sect. 4 discusses our results.

2 Mathematical Model

Let n and k be two integers with

$$1 \le k < n,$$

and let $V = \{1, \ldots, n\}$ be the set of agents. Each agent is endowed with a scalar opinion $x_i \in \mathbb{R}$. For every agent $i \in V$, her neighborhood N_i is defined in the following way. The elements of $V \setminus \{i\}$ are ordered by increasing values of $|x_j - x_i|$; then, the first k elements of the list (i.e. those with smallest distance from i) form the set N_i of current neighbors of i. Should a tie between two or more agents arise, priority is given to agents with lower index. Note that N_i depends on the state, namely one should write $N_i(x(t))$: nevertheless, we omit to explicitly write the dependence of N_i on the state. Based on the current definition of N_i, agent i's opinion evolves according to

$$\dot{x}_i = \sum_{\ell \in N_i} (x_\ell - x_i) \tag{1}$$

We denote by $F(x)$ the righthand side of (1). In order to describe the inter-agent interactions allowed by a state $x \in \mathbb{R}^n$, it is convenient to define the directed graph

$$G(x) = (V, E(x)) \quad \text{with} \quad E(x) = \bigcup_{i \in V} \{(i,j), j \in N_i\},$$

where N_i is the set of neighbors of i. Clearly, if $k = n - 1$ the graph $G(x)$ is complete (up to self-loops). In using some simple graph theory in this paper, we take some background and standard jargon for granted: a concise summary can be found in [19, Ch. 1]. The chosen tie-breaking rule makes the right-hand side $F(x)$ well defined for any $x \in \mathbb{R}^n$. The neighborhoods depend on the current state and, therefore, on time. This fact makes dynamics (1) a piecewise-continuous system [11]. Its solutions shall be intended in a *semi-classical* sense, that is, as piecewise-smooth solutions $\phi(t)$ such that the right-derivative of the solution is equal to the right-hand side at all times, that is,

$$\lim_{h \to 0^+} \frac{\phi(t+h) - \phi(t)}{h} = F(\phi(t)) \quad \text{for all times } t.$$

We conjecture that a forward complete and unique solution exists from every initial condition: a rigorous verification of this fact, which is assumed to hold true in the rest of this paper, is left to future work. Note that choosing a more general notion of solutions, e.g. Caratheodory's, would prevent unicity and the produced multiple solutions would make the subsequent analysis more delicate.

3   Analysis

This section details our results on dynamics (1). We first study equilibria, then convergence properties, and finally reconsider equilibria to study their stability.

3.1   Equilibria

A *cluster* is a subset of agents that have the same opinion: $C \subset V$ such that $x_i = x_j$ for all $i, j \in C$. A state x is called *clusterization* if every agent belongs to a cluster of at least $k + 1$ elements. Finally, a clusterization with only one cluster is said to be a *consensus*.

A state $x \in \mathbb{R}^n$ is said to be an *equilibrium* for (1) when the right-hand side $F(x)$ is zero. For any $i \in \{1, \ldots, n\}$, it is immediate to see that $\dot{x}_i = 0$ if i belongs to a cluster of at least $k + 1$ elements. This condition is also necessary when i is the index of the smallest or of the largest component. Therefore, all clusterizations are equilibria and all non-consensus equilibria have at least two clusters of at least $k + 1$ elements, but not all equilibria are clusterizations. It is possible to obtain a simple counterexample by considering $k = 2$ and $n = 7$ with

$$x_1 = x_3 = x_5 = 0, \quad x_7 = \frac{1}{2}, \quad x_2 = x_4 = x_6 = 1. \tag{2}$$

Note that this example exploits the tie-breaking rule. However, this is not necessary, as the following example shows: consider $k = 4$ and $n = 14$ with

$$x_1 = x_2 = x_3 = x_4 = x_5 = 0,$$
$$x_6 = x_7 = \frac{2}{5}, \quad x_8 = x_9 = \frac{3}{5},$$
$$x_{10} = x_{11} = x_{12} = x_{13} = x_{14} = 1.$$

3.2   Dynamical Properties

We can readily observe that, for any two agents i and j,

$$\frac{d}{dt}(x_i - x_j) = \sum_{\ell \in N_i \setminus N_j} (x_\ell - x_i) - \sum_{m \in N_j \setminus N_i} (x_m - x_j) - |N_i \cap N_j|\,(x_i - x_j). \tag{3}$$

This formula allows us to derive a few consequences. First, we observe that if $N_i(t) = N_j(t)$ for all $t \geq t_0$, then $x_i - x_j \to 0$. Second, we can deduce that the dynamics preserves the order of the agents.

Proposition 1 (Order preservation). *If $x_i(t_0) > x_j(t_0)$, then $x_i(t) > x_j(t)$ for all $t \geq t_0$.*

Proof. Observe that (3) can be rewritten as

$$\frac{\mathrm{d}}{\mathrm{d}t}(x_i - x_j) = \sum_{\ell \in N_i \setminus N_j} x_\ell - \sum_{\ell \in N_j \setminus N_i} x_m - k(x_i - x_j)$$

$$\geq -k(x_i - x_j),$$

where the inequality holds because $|N_i \setminus N_j| = |N_j \setminus N_i|$ and $x_\ell \geq x_m$ for any $\ell \in N_i \setminus N_j$ and $m \in N_j \setminus N_i$. By this bound and Gronwall lemma, $x_i - x_j$ cannot reach zero in finite time. ☐

As a consequence of this property, we can assume from now on with no loss of generality that the agents are *sorted in ascending order* of opinions, that is, $x_i(t) \leq x_{i+1}(t)$ for all $i \in \{1, \ldots, n-1\}$ and all $t \geq 0$. The following proposition formally justifies this fact.

Proposition 2 (Re-ordering agents). *Let $x(t)$ be a solution and σ be a permutation on the index set $\{1, \ldots, n\}$. Assume that for all pairs of distinct indices i, j the permutation satisfies $\sigma(i) < \sigma(j)$ if either $x_i(0) < x_j(0)$ or $x_i(0) = x_j(0)$ and $i < j$. Then, the following facts hold true:*

1. *$x_{\sigma(i)}(t) = x_i(t)$ for all $i \in \{1, \ldots, n\}$ and for all $t \geq 0$;*
2. *if $\sigma(i) < \sigma(j)$, then $x_{\sigma(i)}(t) \leq x_{\sigma(j)}(t)$ for all $t \geq 0$.*

Proof. To verify the first claim, notice that the definition of σ does not interfere with the tie-breaking rule that is used in the definition of the neighborhoods, therefore the dynamics of the agents is unchanged.

To verify the second claim, observe the following facts. If $x_{\sigma(i)}(0) < x_{\sigma(j)}(0)$, then $x_{\sigma(i)}(t) < x_{\sigma(j)}(t)$ for $t > 0$ by Proposition 1. If $x_{\sigma(i)}(0) = x_{\sigma(j)}(0)$, then $N_{\sigma(i)} = N_{\sigma(j)}$ and therefore $x_{\sigma(i)}(t) = x_{\sigma(j)}(t)$ also for $t > 0$ by (3). ☐

From now on we will assume that agents are sorted in ascending order. We can now deduce a convergence result for small groups.

Proposition 3 (Consensus for small groups). *If $n \leq 2k + 1$, then $x(t)$ converges to a consensus.*

Proof. Since $n \leq 2k + 1$, the two agents with lowest and highest opinion share at least one neighbor. Therefore, their difference evolves according to

$$\frac{\mathrm{d}}{\mathrm{d}t}(x_n - x_1) = \sum_{\ell \in N_n \setminus N_1}(x_\ell - x_n) - \sum_{\ell \in N_1 \setminus N_n}(x_\ell - x_1) - |N_1 \cap N_n|(x_n - x_1)$$

$$\leq -(x_n - x_1),$$

which implies exponential convergence to zero by Gronwall lemma. ☐

Simulations suggest that the dynamics converge also for larger groups, though not necessarily to consensus; see Fig. 1.

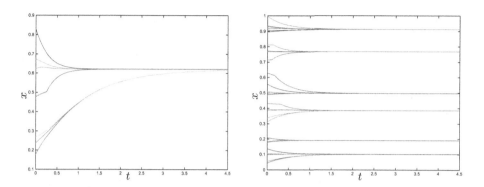

Fig. 1. Two typical evolutions of the dynamics with $k = 3$ from random initial conditions. We observe convergence to consensus for $n = 7$ (left) and to a clusterization for $n = 30$ (right). The non-smooth nature of the trajectories is also very visible.

3.3 Special Case $k = 1$

In the case $k = 1$, the dynamics takes the form

$$\dot{x}_i = x_{\mathrm{cl}[i]} - x_i \quad i \in \{1, \ldots, n\},$$

where $\mathrm{cl}[i]$ denotes the index of the closest agent to i. This specific form has three important consequences.

Lemma 1. *If $k = 1$, then the following facts hold true.*

1. *All equilibria are clusterizations.*
2. *For every $x \in \mathbb{R}^n$, the interaction graph $G(x)$ is the union of weakly connected components, such that each component contains exactly one circuit of length 2 and the two nodes of the circuit can be reached from all nodes of the component.*
3. *Two disconnected components cannot become connected in the evolution.*

Proof. Claim 1: We observe that the only possibility for the right-hand side to be zero is that $x_{\mathrm{cl}[i]} = x_i$ for all i.

Claim 2: Observe that $\mathrm{cl}[i]$ can only be equal to either $i - 1$ or $i + 1$, except for the extreme agents, for which necessarily $\mathrm{cl}[1] = 2$ and $\mathrm{cl}[n] = n - 1$. Therefore, the sequence $\delta_i = \mathrm{cl}[i] - i$ is such that $\delta_1 = 1$ and $\delta_n = -1$ and must therefore change sign an odd number of times. Where it changes from positive to negative, there is a pair of reciprocal edges; where it changes from negative to positive, there is a disconnection. Therefore, every connected component has a pair of nodes that are connected to each other and that can be reached through a directed path from all other nodes. See Fig. 2 for an illustrative example.

Claim 3: Let there be a disconnection between j and $j+1$. Then,

$$\frac{\mathrm{d}}{\mathrm{d}t}(x_{j+1} - x_j) = (x_{j+2} - x_{j+1}) + (x_j - x_{j-1}) \geq 0,$$

implying that the distance $x_{j+1} - x_j$ cannot decrease. Moreover,

$$\frac{\mathrm{d}}{\mathrm{d}t}(x_j - x_{j-1}) = -(x_j - x_{j-1}) - (x_{\mathrm{cl}[j-1]} - x_{j-1}) \leq 0,$$

because the second term either is negative or, if positive, must be smaller or equal in magnitude than $x_j - x_{j-1}$. Therefore, $x_j - x_{j-1}$ cannot increase. Since an analogous reasoning implies that $x_{j+2} - x_{j+1}$ cannot increase, the two components cannot become connected in the future.   □

Fig. 2. Example of weakly connected component of graph $G(x)$.

These facts allow to draw a conclusion about convergence.

Proposition 4 (Clusterization). *If $k = 1$, then $x(t)$ converges to a clusterization.*

Proof. The third statement of Lemma 1 implies that weakly connected component can only split. Since the number of individuals is finite, the splitting process terminates with a finite number of constant weakly connected components. After that termination time, the topology does not change. Since each connected component has a globally reachable node, then each group of agents is guaranteed to converge to consensus [19, p. 61], therefore producing a clusterization.   □

Unfortunately, the idea of the proof of Proposition 4 does not extend to $k > 1$, because in general disconnected components can become connected.

3.4 Stability and Robustness of Equilibria

It is easy to see that non-clusterization equilibria are not stable in general. For instance, consider example (2) with a small perturbation on agent 7: the ensuing dynamics leads to a clusterization with two clusters. Instead, clusterizations exhibit several stability properties.

We shall begin by considering small perturbations of the opinions. We say that a clusterization state is *structurally stable* if, after a perturbation, the dynamics converges to another clusterization that has the same clusters (though not necessarily taking on the same opinion values). More formally, the clusterization $\bar{x}$ is said to be structurally stable if there exists a neighborhood of $\bar{x}$ such that, for every y' in that neighborhood, any solution issuing from y' converges to a clusterization $\bar{y}$ that has the following property: for any pair i, j of individuals, $\bar{x}_i = \bar{x}_j$ if and only if $\bar{y}_i = \bar{y}_j$.

Proposition 5 (Structural stability of small clusters). *A clusterization is structurally stable if and only if all of its clusters have cardinality not larger than $2k + 1$.*

Proof. If all clusters have cardinality not larger than $2k+1$, then after the perturbation Proposition 3 can be applied. To prove the opposite implication, observe that if one cluster has cardinality at least $2k+2$, then a suitable perturbation can split it into two separate clusters of cardinality at least $k + 1$, thereby creating a clusterization with different structure.   □

We shall also consider different kinds of disruptions, namely the addition or removal of one agent. We say that a clusterization is stable to these disruptions if, after the addition or removal of an agent, the other agents do not change their opinion.

Proposition 6 (Stability to removals). *A clusterization is stable to removals if and only if all of its clusters have cardinality larger than $k + 1$.*

Proof. It is clear that agents in a cluster remain at equilibrium after the removal, unless the cluster size goes below the threshold $k + 1$.   □

Proposition 7 (Stability to additions). *Every clusterization is stable to additions.*

Proof. Agents within a cluster of size at least $k + 1$ will not be influenced by any new arrival.   □

4   Conclusion

The stability properties of the equilibria of dynamics (1) should be contrasted with the lack thereof shown by the equilibria of the corresponding metric bounded confidence model, which reads as

$$\dot{x}_i = \sum_{\ell : |x_\ell - x_i| < d} (x_\ell - x_i), \tag{4}$$

where $d > 0$ is an interaction radius. It is well-known [8,10] that this dynamics converges to clusterizations. If a new agent is added to such a clusterization, either the new agent is too far apart from the original agents and nothing happens, or the new agent falls within the visibility radius from a cluster. In the latter case, the new agents and the agents in the cluster influence each other and therefore change their opinions, converging to a common intermediate value. Actually, if the new agent falls within the visibility radius of two clusters, the two clusters eventually merge.

In contrast, clusters produced by (1) are much more stable. In our opinion, this stability intriguingly reminds the stability that is exhibited by norms and organizations in societies. Indeed, sociologists and ethologists have observed since a long time [30, 32, 33] that social norms and social structures are typically rather stable across time, despite the fact that the composition of the social groups evolve, notably with the arrival of new members. Our insights about k-neighbor interactions suggest that limitations of attention can have stabilizing effects in societies.

Acknowledgements. The authors are grateful to Emiliano Cristiani, Julien Hendrickx, Samuel Martin, Benedetto Piccoli and Tommaso Venturini for fruitful discussions that, along the years, have shaped their point of view on the topic of this paper.

References

1. Angeli, D., Manfredi, S.: A Petri net approach to consensus in networks with joint-agent interactions. Automatica **110**, 108466 (2019)
2. Aydoğdu, A., et al.: Interaction network, state space, and control in social dynamics. In: Bellomo, N., Degond, P., Tadmor, E. (eds.) Active Particles, Volume 1. MSSET, pp. 99–140. Springer, Cham (2017). https://doi.org/10.1007/978-3-319-49996-3_3
3. Aydoğdu, A., et al.: Modeling birds on wires. J. Theor. Biol. **415**, 102–112 (2017)
4. Balister, P., Bollobás, B., Sarkar, A., Walters, M.: Connectivity of random k-nearest-neighbour graphs. Adv. Appl. Probab. **37**(1), 1–24 (2005)
5. Ballerini, M.: Interaction ruling animal collective behavior depends on topological rather than metric distance: evidence from a field study. Proc. Natl. Acad. Sci. **105**(4), 1232–1237 (2008)
6. Blanchet, A., Degond, P.: Topological interactions in a Boltzmann-type framework. J. Stat. Phys. **163**(1), 41–60 (2016). https://doi.org/10.1007/s10955-016-1471-6
7. Blondel, V.D., Hendrickx, J.M., Tsitsiklis, J.N.: On Krause's multi-agent consensus model with state-dependent connectivity. IEEE Trans. Autom. Control **54**(11), 2586–2597 (2009)
8. Blondel, V.D., Hendrickx, J.M., Tsitsiklis, J.N.: Continuous-time average-preserving opinion dynamics with opinion-dependent communications. SIAM J. Control. Optim. **48**(8), 5214–5240 (2010)
9. Canuto, C., Fagnani, F., Tilli, P.: An Eulerian approach to the analysis of Krause's consensus models. SIAM J. Control. Optim. **50**(1), 243–265 (2012)
10. Ceragioli, F., Frasca, P.: Continuous and discontinuous opinion dynamics with bounded confidence. Nonlinear Anal. Appl. B **13**(3), 1239–1251 (2012)
11. Ceragioli, F., Frasca, P.: Discontinuities, generalized solutions, and (dis)agreement in opinion dynamics. In: Tarbouriech, S., Girard, A., Hetel, L. (eds.) Control Subject to Computational and Communication Constraints: Current Challenges. LNCS, vol. 475, pp. 287–309. Springer, Heidelberg (2018). https://doi.org/10.1007/978-3-319-78449-6_14
12. Chazelle, B., Wang, C.: Inertial Hegselmann-Krause systems. IEEE Trans. Autom. Control **62**(8), 3905–3913 (2017)
13. Chen, C., Chen, G., Guo, L.: On the minimum number of neighbors needed for consensus of flocks. Control Theory Technol. **15**(4), 327–339 (2017). https://doi.org/10.1007/s11768-017-7097-7
14. Chen, G., Su, W., Mei, W., Bullo, F.: Convergence properties of the heterogeneous Deffuant-Weisbuch model. Automatica **114**, 108825 (2020)

15. Cristiani, E., Frasca, P., Piccoli, B.: Effects of anisotropic interactions on the structure of animal groups. J. Math. Biol. **62**(4), 569–588 (2011). https://doi.org/10.1007/s00285-010-0347-7
16. Deffuant, G., Neau, D., Amblard, F., Weisbuch, G.: Mixing beliefs among interacting agents. Adv. Complex Syst. **03**(01n04), 87–98 (2000)
17. Degond, P., Pulvirenti, M.: Propagation of chaos for topological interactions. Ann. Appl. Probab. **29**(4), 2594–2612 (2019)
18. Dunbar, R.: Neocortex size as a constraint on group size in primates. J. Hum. Evol. **22**(6), 469–493 (1992)
19. Fagnani, F., Frasca, P.: Introduction to Averaging Dynamics over Networks. Lecture Notes in Control and Information Sciences, Springer, Heidelberg (2017). https://doi.org/10.1007/978-3-319-68022-4
20. Giardina, I.: Collective behavior in animal groups: theoretical models and empirical studies. HFSP J. **2**(4), 205–219 (2008). pMID: 19404431
21. Gonçalves, B., Perra, N., Vespignani, A.: Modeling users' activity on Twitter networks: validation of Dunbar's number. PloS ONE **6**(8), e22656 (2011)
22. Krause, U.: A discrete nonlinear and non-autonomous model of consensus formation. In: Communications in Difference Equations, pp. 227–236 (2000)
23. Lazer, D.: The rise of the social algorithm. Science **348**(6239), 1090–1091 (2015)
24. Martin, S.: Multi-agent flocking under topological interactions. Syst. Control Lett. **69**, 53–61 (2014)
25. Mirtabatabaei, A., Bullo, F.: Opinion dynamics in heterogeneous networks: convergence conjectures and theorems. SIAM J. Control. Optim. **50**(5), 2763–2785 (2012)
26. Proskurnikov, A., Tempo, R.: A tutorial on modeling and analysis of dynamic social networks. Part II. Ann. Rev. Control **45**, 166–190 (2018)
27. Proskurnikov, A.V., Tempo, R.: A tutorial on modeling and analysis of dynamic social networks. Part I. Ann. Rev. Control **43**, 65–79 (2017)
28. Rossi, W.S., Frasca, P.: Asynchronous opinion dynamics on the k-nearest-neighbors graph. In: IEEE Conference on Decision and Control, pp. 3648–3653 (2018)
29. Rossi, W.S., Frasca, P.: Opinion dynamics with topological gossiping: asynchronous updates under limited attention. IEEE Control Syst. Lett. **4**(3), 566–571 (2020)
30. Sethi, R.: Evolutionary stability and social norms. J. Econ. Behav. Organ. **29**(1), 113–140 (1996)
31. Shvydkoy, R., Tadmor, E.: Topological models for emergent dynamics with short-range interactions. arXiv preprint (2018)
32. Simmel, G.: The persistence of social groups. Am. J. Sociol. **3**(5), 662–698 (1898)
33. Van de Waal, E., Borgeaud, C., Whiten, A.: Potent social learning and conformity shape a wild primate's foraging decisions. Science **340**(6131), 483–485 (2013)

Electrical Networks

Optimal Pricing Approach Based on Expected Utility Maximization with Partial Information

Chao Zhang[1]([✉])(ⓘ), Hang Zou[1], Samson Lasaulce[2], Vineeth S. Varma[2], Lucas Saludjian[3], and Patrick Panciatici[3]

[1] L2S, CNRS-CentraleSupelec-Univ. Paris Saclay, Gif-sur-Yvette, France
`chao.zhang@centralesupelec.fr`
[2] CRAN, CNRS & Univ. Lorraine, Nancy, France
[3] RTE, Paris, France

Abstract. Real-time pricing is considered as a promising strategy to flatten the power consumption provided with perfect knowledge of consumers' demand level. However, the gathering of full information of demand levels might be cumbersome or even impossible for the provider in practical scenarios. In this paper, instead of assuming the perfectly known demand levels, we investigate the problem where the provider has the sole knowledge of the probabilistic distribution of the demand levels. Furthermore, a penalty term caused by the prediction error of the consumption prediction is introduced due to the incomplete information. By solving the stochastic optimization problem, the optimal consumption prediction and optimal price to maximize the expected social welfare is derived analytically. Numerical results show that the degradation on the social welfare brought by the partial information can be less than 1% when the price and consumption prediction are well designed.

Keywords: Real-time pricing · Demand response · Consumption prediction

1 Introduction

An essential goal of smart grids is to create reliable communications between many components. the exchange and control of information can provide more effective generation and transmission of electricity, resulting in flattening the power consumption. To reduce the power consumption during the peak time, demand side management (DSM) has been proposed and shown to be a promising strategy in certain scenarios [1,2].

Pricing-based demand response (DR) is one of the most widely used DSM methods, where the electricity price designed by the provider is related to the overall demand (or aggregate load) of the served consumers. In a smart grid

© Springer Nature Switzerland AG 2021
S. Lasaulce et al. (Eds.): NetGCooP 2021, CCIS 1354, pp. 285–293, 2021.
https://doi.org/10.1007/978-3-030-87473-5_25

system, the energy provider can send the tariff information to the energy consumption controller (ECC) unit located at the consumer's devices (e.g., the smart meters), and thus the consumers can schedule their activities to low price periods. To shift the power consumption from the rush hours, the dynamic pricing is in accordance with the overall demand levels. Several different pricing have been proposed recently. For instance, the time of use pricing has on-peak tariff and off-peak tariff [3], the day-ahead pricing predicts the following day's consumption and propose a tariff according to its consumption prediction [4]. Due to the large deployment of smart meters, the real-time pricing can be implemented by communicating the demand of current period and feeding back the real-time tariff to the consumers. Under the ideal communication environment, the real-time price can be designed to maximize the social welfare or the benefit of the provider [5,6].

However, due to the limitation of available resources in the communication channels, the perfect observation of the demand levels can be prohibitive. In this paper, we consider the case where only partial information can be acquired by the provider. More precisely, the provide has the sole knowledge of the demand level statistics instead of knowing its instantaneous realizations. Moreover, the information shortage about the demand levels in provider's side leads to the uncertainty of total power consumption of the system. As a consequence, the procurement (or generation) of electricity in advance will be affected and thus a penalty term has been introduced in this paper to model this impact. Knowing the statistics of demand and the penalty term to the provider's cost function, the stochastic optimization problem has been studied here. The rest of the paper is organized as follows. The system model is introduced in Sect. 2. The problem is formulated and The optimal price and optimal load prediction are derived in Sect. 3 and Sect. 4 respectively. The paper ends by numerical results.

2 System Model

In this paper, we consider a smart power system consisting of an unique energy provider and several consumers. It is assumed that there is an ECC unit embedded in each consumer's smart meter. The role of the ECC is to control the power consumption such that the consumer's utility can be maximized.

Let $\mathcal{N} = \{1, \ldots, N\}$ denote the set of all the consumers. For each consumer $i \in \mathcal{N}$, denote x_i^k as the power consumption of consumer i at time slot k. In fact, considering the problems during several time slots $\mathcal{K} = \{1, \ldots, K\}$, the solutions can be found separately in each time-slot. Without loss of generality, we consider our problem for one given time-slot and remove the index k for all the definitions. For each consumer i. the available power consumption interval I_i is defined as

$$I_i = [m_i, M_i] \tag{1}$$

and thus $m_i \leq x_i \leq M_i$.

2.1   Utility Function of Consumers

The energy demand of each consumer depends on several parameters, e.g., the climate condition, tariff, the variation of the energy demand over different time of the day. For all consumers, we denote the utility function as $U(x, \omega)$, where x represents the power consumption and ω is a parameter representing the satisfaction level (based on climate, time and so on). Here we choose the utility function proposed in [5] defined as:

$$
U(x, \omega) = \begin{cases} \omega x - \frac{\alpha}{2} x^2, & \text{if } 0 \leq x \leq \frac{\omega}{\alpha} \\ \frac{\omega^2}{2\alpha}, & \text{if } x > \frac{\omega}{\alpha}. \end{cases} \tag{2}
$$

2.2   The Cost Function of the Provider

Denoting the total power consumption L as $L = \sum_{i=1}^{N} x_i$, the cost function $C_1(L)$ for the provider, representing the cost of providing L units of energy, is chosen as [5]:

$$
C_1(L) = aL^2 + bL + c \tag{3}
$$

Different from most of the existing works, we also consider the cost caused by the prediction error of the power consumption. Note that it is assumed that the provider has imperfect knowledge of ω_i. The imperfect information can be induced by privacy issue, or the bad quality of communication channels. Indeed, in the worst case (all consumer want to keep its private information), it is assumed that the provider knows solely the probability density function (p.d.f.) of the ω_i, which is possible to be acquired by knowing the past realizations (values) of the ω_i. As a consequence, the provider needs to predict the total consumption by $\tilde{L}$ and thus brings a penalty term depending on $L - \tilde{L}$. For instance, the provider purchased $\tilde{L}$ units of energy in advance from the energy generator. If $\tilde{L} > L$, the provider will sell the superfluous energy to the Transmission System Operator (TSO) with a *lower* price. If $\tilde{L} < L$, the provider needs to purchase more energy from the TSO with a *higher* price to satisfy the energy need by all the consumers. Define the penalty term as $C_2(L - \tilde{L})$, with $C_2(\cdot)$ fulfilling the following properties:

1) The penalty function is non-negative.

$$
C_2(x) \geq 0 \tag{4}
$$

2) The penalty function is non-decreasing when $L - \tilde{L} > 0$ and non-increasing otherwise.

$$
\frac{\partial C_2(x)}{\partial x} \geq 0 \text{ if } x > 0 \tag{5}
$$

$$
\frac{\partial C_2(x)}{\partial x} \leq 0 \text{ if } x < 0 \tag{6}
$$

In this paper, we choose the absolute value function to describe the penalty term as follows:

$$C_2(L - \tilde{L}) = d|L - \tilde{L}| \tag{7}$$

Hence, the total cost of the provider can be expressed as:

$$C(L, \tilde{L}) = C_1(L) + C_2(L - \tilde{L}) \tag{8}$$

3   Problem Formulation

3.1   Demand Side Response of Consumers

For user $i \in \{1, \ldots, N\}$, consuming x_i kW electricity with a tariff P_i dollars per kWh needs to pay $P_i x_i$ dollars to the provider, its utility can be expressed as:

$$W_i(x_i, \omega_i) = U(x_i, \omega_i) - P_i x_i \tag{9}$$

To maximize its own utility, the optimal power consumption can be written as:

$$x_i^\star = \begin{cases} m_i, & \text{if } \frac{\omega_i - P_i}{\alpha} \leq m_i \\[2mm] \frac{\omega_i - P_i}{\alpha}, & \text{if } m_i \leq \frac{\omega_i - P_i}{\alpha} \leq M_i \\[2mm] M_i, & \text{if } \frac{\omega_i - P_i}{\alpha} \geq M_i \end{cases} \tag{10}$$

For the sake of clarify, we assume the power consumption interval is sufficient large so that the condition $m_i \leq \frac{\omega_i - P_i}{\alpha} \leq M_i$ is always met. Hence, the optimal consumption can be simplified to

$$x_i^\star = \frac{\omega_i - P_i}{\alpha} \tag{11}$$

3.2   Expected Social Welfare Maximization Problem

When the communication channel between consumers and the provider is assumed to be perfect (lossless information can be exchanged), the social welfare can be optimized by using the algorithm proposed in [5]. However, for the scenario with no available communication channels unreliable communication channels, the maximization problem becomes totally different. In our case, it is assumed that the provider has the sole knowledge of the p.d.f. of all the ω_i. Hence, due to the lack of information, it is difficult to optimize the instantaneous social welfare. Consequently, we propose to maximize the expected social welfare as follows:

$$\mathbb{E}_{x_i, \omega_i} \left[\sum_{i \in \mathcal{N}} U(x_i, \omega_i) - C(L, \tilde{L}) \right] \tag{12}$$

Note that the power consumption x_i has been determined by (11) in the consumer's side. Plug (11) into (12), the expected social welfare can be rewritten as:

$$
\begin{aligned}
&\overline{W}(P_1, \ldots, P_N, \widetilde{L}) \\
&= \mathbb{E}_{\omega_i}[\sum_{i \in \mathcal{N}} U(x_i^\star, \omega_i) - C(L^\star, \widetilde{L})] \\
&= \mathbb{E}_{\omega_i}[\sum_{i \in \mathcal{N}} U(\frac{\omega_i - P_i}{\alpha}, \omega_i) - C(L^\star, \widetilde{L})]
\end{aligned}
\tag{13}
$$

where $L^\star = \sum_{i \in \mathcal{N}} x_i^\star$. To maximize the expected social welfare defined by (13), the provider can optimize the tariff P_i and the prediction of total power consumption $\widetilde{L}$ as follows:

$$
\max_{P_1, \ldots, P_N, \widetilde{L}} \quad \mathbb{E}_{\omega_i}[\sum_{i \in \mathcal{N}} U(\frac{\omega_i - P_i}{\alpha}, \omega_i) - C(L^\star, \widetilde{L})]
\tag{14}
$$

Additionally, from the following proposition, it can be seen that the expected welfare can be optimized by using a common tariff P for all consumers rather than a different tariff for each consumer.

Proposition 1. *For any given tariff $\widetilde{P}_1, \ldots, \widetilde{P}_N$, the following inequality always hold:*

$$
\overline{W}(\widetilde{P}_1, \ldots, \widetilde{P}_N, \widetilde{L}) \leq \overline{W}(P_1^c, \ldots, P_N^c, \widetilde{L})
\tag{15}
$$

where $P_i^c = \frac{1}{N} \sum_{j=1}^N \widetilde{P}_j$ for every $i \in \mathcal{N}$.

Proof. It can be calculated that

$$
\begin{aligned}
&\overline{W}(\widetilde{P}_1, \ldots, \widetilde{P}_N, \widetilde{L}) - \overline{W}(P_1^c, \ldots, P_N^c, \widetilde{L}) \\
&= \frac{1}{2\alpha}(\sum_{i=1}^N (P_i^c)^2 - \sum_{i=1}^N \widetilde{P}_i^2)
\end{aligned}
\tag{16}
$$

Note that $\sum_{i=1}^N \widetilde{P}_i = \sum_{i=1}^N P_i^c$. Furthermore, it can be found that

$$
(P_1^c, \ldots, P_N^c) = \underset{\sum_{i=1}^N P_i = C}{\arg\min} \sum_{i=1}^N P_i^2
\tag{17}
$$

As a consequence, it can be seen that

$$
\sum_{i=1}^N (P_i^c)^2 - \sum_{i=1}^N \widetilde{P}_i^2 \leq 0.
\tag{18}
$$

Our claim is proved.

According to Prop. 3.1, the optimization problems (14) can be further simplified to the following problem:

$$\max_{P,\widetilde{L}} \quad \mathbb{E}_{\omega_i}[\sum_{i \in \mathcal{N}} U(\frac{\omega_i - P}{\alpha}, \omega_i) - C(L^\star, \widetilde{L})] \tag{19}$$

The provider will find the optimal price and the optimal power consumption prediction to maximize its expected social welfare. The approach to derive them will be presented in the subsequent section.

4   Optimal Consumption Prediction and Price

In this section, we proposed one approach to derive the optimal power consumption prediction and the optimal price. One can easily observe that these two variables are correlated. Without loss of optimality loss, the optimization problem (19) can be solved in two steps. Firstly, we focus on find the optimal consumption prediction for a given price P, i.e.,

$$\widetilde{L}^\star(P) \in \arg\max_{\widetilde{L}} \quad \mathbb{E}_{\omega_i}[\sum_{i \in \mathcal{N}} U(\frac{\omega_i - P}{\alpha}, \omega_i) - C(L^\star, \widetilde{L})] \tag{20}$$

When the $\widetilde{L}^\star(P)$ has been derived, the optimal price can be obtained by solving the following problem:

$$P^\star \in \arg\max_{P} \quad \mathbb{E}_{\omega_i}[\sum_{i \in \mathcal{N}} U(\frac{\omega_i - P}{\alpha}, \omega_i) - C(L^\star, \widetilde{L}^\star(P))] \tag{21}$$

4.1   Optimal Prediction of the Total Power Consumption

We notice that only the term $C_2(L^\star - \widetilde{L})$ is related to the prediction $\widetilde{L}$, thus the optimization problem (22) is equivalent to the following problem:

$$\widetilde{L}^\star(P) \in \arg\max_{\widetilde{L}} \quad \mathbb{E}_{\omega_i}[|\sum_{i=1}^{N} \frac{\omega_i - P}{\alpha} - \widetilde{L}|] \tag{22}$$

Proposition 2. *For a given price P, the optimal prediction of the total power consumption can be written as*

$$\widetilde{L}^\star(P) = \text{MED}(\omega) - \frac{NP}{\alpha} \tag{23}$$

where MED(.) represents the median of the variable and $\omega = \sum_{i=1}^{N} \frac{\omega_i}{\alpha}$.

Proof. The proof is omitted because of the lack of space.

According to the Prop. 4.1, the optimal prediction decreases when the price rises. This can be explained by the fact that the rise of price will bring a degradation of the power consumption, and further the prediction of total power consumption will decline.

4.2   Optimal Price

Knowing the optimal prediction for a given price, the second step can be done by solving the optimization problem (21). Interestingly, according to (23), the problem can be further simplified. We notice that the term $C(L^\star, \widetilde{L}^\star(P))$ can be rewritten as:

$$
\begin{aligned}
& C(L^\star, \widetilde{L}^\star(P)) \\
&= C_1(L^\star) + d \left| \sum_{i=1}^{N} \frac{\omega_i - P}{\alpha} - \widetilde{L}^\star(P) \right| \\
&= C_1(L^\star) + d \left| \sum_{i=1}^{N} \frac{\omega_i}{\alpha} - \mathrm{MED}(\omega) \right|
\end{aligned}
\tag{24}
$$

Since $d | \sum_{i=1}^{N} \frac{\omega_i}{\alpha} - \mathrm{MED}(\omega)|$ is independent of P, the optimization problem (21) can be further simplified as:

$$
P^\star \in \arg\max_{P} \quad \mathbb{E}_{\omega_i} \left[\sum_{i \in N} U(\frac{\omega_i - P}{\alpha}, \omega_i) - C_1(L^\star) \right]
\tag{25}
$$

Therefore, when the prediction has been optimized, the pricing problem is independent of the optimal prediction $\widetilde{L}^\star$.

Proposition 3. *The optimal price to maximize the expected social welfare can be derived by solving (25) and written as:*

$$
P^\star = \frac{2a\alpha\mathbb{E}[\omega] + \alpha b}{\alpha + 2aN}
\tag{26}
$$

Proof. The proof is omitted because of the lack of space.

Knowing the optimal price $P^\star$, the optimum prediction of the total power consumption can be calculated. Plug (26) into (22), the optimal consumption prediction can be written as:

$$
\widetilde{L}^\star(P^\star) = \mathrm{MED}(\omega) - \frac{2Na\alpha\mathbb{E}[\omega] + N\alpha b}{\alpha^2 + 2Na\alpha}
\tag{27}
$$

5   Simulation Results

In this section, numerical results are shown to evaluate the performance of our approach. We consider a system with an unique provider and $N = 10$ consumers. For the sake of simplicity, we assume each ω_i is identically independent distributed (i.i.d.) and remains fixed during one time-slot. Each ω_i is uniformly distributed over the interval $[\omega_{\min}, \omega_{\max}]$. The scenario with asymmetric ω_i can be treated in the same way. The parameter d to define the penalty term is set to be 0.1. For other parameters, they are chosen same as [5], i.e., $\alpha = 0.5$, $a = 0.01$, $b = 0$, and $c = 0$.

Firstly, we assess the performance degradation by using our approach compared with the algorithm proposed in [5]. The reason for the degradation is twofold: the imperfect knowledge of w_i in our scenario leads to the sub-optimal power consumption solution, and also the deviation between the real power consumption and predicted power consumption brings the penalty to the provider. Assume $w_{min} = 2.5 - \sigma$ and $w_{max} = 2.5 + \sigma$, Fig. 1 represents the expected social welfare against σ. It can be observed that the performance degradation induced by the imperfect knowledge is quite small. Even with largest σ, the optimality loss is close to 1%. Furthermore, the performance degradation rises when σ increases. Indeed, larger σ leads to higher variance of w_i. When w_i changes faster, knowing the real value of w_i becomes more important. As a consequence, when the provider has limited resource to communicate with the consumers, it is better to communicate with the consumer which has higher variance of its w_i.

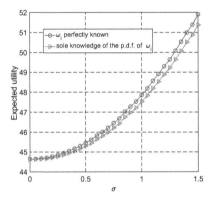

Fig. 1. Even have the sole knowledge of the p.d.f. of w_i, the relative optimality loss is less than 1% under typical scenarios.

In Fig. 2, the optimal price against the expectation of w_i is shown. Assume $w_{min} = \mathbb{E}[w_i] - 1.5$ and $w_{max} = \mathbb{E}[w_i] + 1.5$. The optimal price is derived by computing the expected utility defined in (25) with exhaustive search. From this figure, it can be seen that the optimal price is linear to $\mathbb{E}[w_i]$, which verified our analytical result shown in Prop. 4.2. When w_i increases, the consumer prefers to respond with a higher power consumption to maximize its individual welfare. Therefore, the provider needs to increase the price to avoid the high load for the system. Therefore, it is better to have a rise in price in the evening since consumers are more demanding and have a drop in price after midnight since consumers are much less demanding at that time.

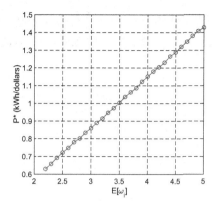

Fig. 2. When consumers are more probably to have higher power consumption, the price of the electricity designed by the provider will rise.

Acknowledgement. This work is funded by the RTE-CentraleSupelec Chair on "Digital Transformation of Electricity Networks".

References

1. Mohsenian-Rad, A., Wong, V.W., Jatskevich, J., Schober, R., LeonGarcia, A.: Autonomous demand-side management based on game-theoretic energy consumption scheduling for the future smart grid. IEEE Trans. Smart Grid **1**(3), 320–331 (2010)
2. Palensky, P., Dietrich, D.: Demand side management: demand response, intelligent energy systems, and smart loads. IEEE Trans. Ind. Inform. **7**(3), 381–388 (2011)
3. Hatami, A., Seifi, H., Sheikh-el-Eslami, M.K.: A stochastic-based decision-making framework for an electricity retailer: time-of-use pricing and electricity portfolio optimization. IEEE Trans. Power Syst. **26**(4), 1808–1816 (2011)
4. Joe-Wong, C., Sen, S., Ha, S., Chiang, M.: Optimized day-ahead pricing for smart grids with device-specific scheduling flexibility. IEEE J. Sel. Areas Commun. **30**(6), 1075–1085 (2012)
5. Samadi, P., Mohsenian-Rad, A.H., Schober, R., Wong, V.W., Jatskevich, J.: Optimal real-time pricing algorithm based on utility maximization for smart grid. In: IEEE SmartGridComm 2010 (2010)
6. Roozbehani, M., Dahleh, M.A., Mitter, S.K.: Volatility of power grids under real-time pricing. IEEE Trans. Power Syst. **27**(4), 1926–1940 (2012)

Price Formation and Optimal Trading in Intraday Electricity Markets

Olivier Féron[1(✉)], Peter Tankov[2(✉)], and Laura Tinsi[1,2(✉)]

[1] EDF Lab, Palaiseau, France
olivier-2.feron@edf.fr
[2] ENSAE, Institut Polytechnique de Paris, Palaiseau, France
{peter.tankov,laura.tinsi}@ensae.fr

Abstract. We study price formation in intraday electricity markets in the presence of asymmetric information and intermittent generation. We use stochastic control theory to identify optimal strategies of agents with market impact and exhibit the Nash equilibrium in closed form for a finite number of agents as well as in the asymptotic setting of Mean field games. We show that our model is able to reproduce some empirical facts observed in the market (price impact, volatility), and allows producers to deal with risks and costs related to intermittent renewable generation.

Keywords: Stochastic games · Renewable energies · Electricity markets

1 Introduction

The world electricity markets are presently undergoing a major transformation driven by the transition towards a carbon-free energy system. The intraday markets are increasingly used by the renewable producers to compensate forecast errors. This improves market liquidity and at the same time creates feedback effects of the renewable generation on the market price, leading to negative correlations between renewable infeed and prices, and, negative impact on the revenues of renewable producers.

The aim of this paper is to build an equilibrium model for the intraday electricity market, to understand the price formation and identify optimal strategies for renewable producers in the setting where renewable generation forecasts may affect market prices. We consider renewable producers, optimizing their revenues based on imperfect forecasts of terminal production. The actions of each agent impact the market, leading to a stochastic game where players interact through the market price. We exhibit a closed-form Nash equilibrium for this game in the linear-quadratic setting, first for a finite number of agents with perfect information, and then in the asymptotic Mean field game setting, with imperfect information. We then show by numerical simulations that our model reproduces the observed stylized features of the market price, such as the volatility patterns and the market impact.

© Springer Nature Switzerland AG 2021
S. Lasaulce et al. (Eds.): NetGCooP 2021, CCIS 1354, pp. 294–305, 2021.
https://doi.org/10.1007/978-3-030-87473-5_26

Correlations between renewable infeed and intraday market prices have been studied by a number of authors. Kiesel and Paraschiv [10] perform an econometric analysis of the German intraday market and show that a deeper penetration of renewable energies increases market liquidity and price-infeed correlations. Karanfil and Li [9] draw similar conclusions on the Danish market, and exhibit the impact of renewable energies on price and volatility. Gruet, Rowińska and Veraart [14] establish a negative correlation between the wind energy penetration and the day ahead market prices. Jonsson et al. [8] show not only prices are negatively correlated with the penetration of intermittent energies but also that the latter modifies significantly the spot price distribution.

Optimal strategies in the intraday market for a single wind energy producer have already been studied. In the price-taker setting, Garnier and Madlener [6] solve a discrete optimal trading problem to arbitrate between immediate and delayed trading when price and production forecast are uncertain. In [13], Morales et al. consider a multimarket setting to derive an optimal bidding strategy in the day ahead and adjustment markets while minimizing the cost in the balancing market. This approach has been enhanced by Madsen et al., [16] and then by Delikaraoglou et al. [5] , where the wind energy producer is first a price maker in the balancing market , and then in both the spot and balancing markets. Still in the price-maker setting, continuous approaches have also been developed. Aïd, Gruet and Pham [1], consider the optimal trading rate and power generation for producer with uncertain terminal residual demand. Tan and Tankov [15] develop an optimal trading model with a quantified evolution of forecast uncertainty and exhibit optimal strategies depending on forecast updates. We differ from the latter by considering an equilibrium setting with many agents and determining the market price as the result of their interactions.

Explicit results for dynamic equilibria are often difficult to obtain. In particular, Nash equilibria often lead to coupled Partial Differential Equation (PDE) systems. In the imperfect information setting, the problem may be simplified by assuming a continuum of agents and using the Mean field game (MFG) approach.

The Mean field games are stochastic differential games with a large number of symmetric agents, which were originally studied by Lions and Lasry [12] and Huang, Caines, and Malhamé [7]. The equilibrium of such a game is characterized through a coupled system of a Hamilton-Jacobi-Bellman and a Fokker-Planck equation. Carmona and Delarue [3] proposed an alternative way to formalize the system inspired by the Pontryagin principle and relating the Mean field game solution to a McKean-Vlasov Forward Backward Stochastic Differential Equation (FBSDE). From the Mean field game solution one can derive an ε-Nash equilibrium of the corresponding N-player game.

Financial markets and energy systems are a natural domain of applications of MFG. Alasseur, Tahar and Matoussi [2] develop a model for the optimal management of energy storage and distribution in a smart grid system through an extended MFG. Casgrain and Jaimungal [4] apply it to optimal trade execution with price impact and deterministic terminal liquidation condition. They dealt with incomplete information and heterogeneous sub-populations of agents.

The paper is organized as follows. Section 2 describes the main elements of the model. Section 3 is devoted to the setting of complete information, where each agent observes the production forecast of all other agents. In Sect. 4, we consider a more realistic setting, where each agent observes only its own production forecast as well as the national production forecast. In Sect. 5 we perform empirical analysis of intraday market and confront it to the theoretical results obtained. Section 6 concludes the paper.

2   The Model

We place ourselves in the intraday market for a given delivery hour starting at time T, where time 0 corresponds to the opening time of the market. We assume that market participants can trade during the entire period $[0, T]$.

To model the price and the forecasts, we introduce a filtered probability space $(\Omega, \mathcal{F}, \mathbb{F} := (\mathcal{F}_t)_{t \in [0,T]}, \mathbb{P})$ to which all processes are adapted. As the agents' strategies may impact the market price, we distinguish the price without price impact or fundamental price $S := (S_t)_{t \in [0,T]}$ from the market price $P := (P_t)_{t \in [0,T]}$, where the market impact is included.

We consider small renewable energy producers that use the intraday market to manage the volume risk associated to the imperfect production forecast. They observe a common national production forecast. In addition, each agent has access to the individual production forecast, which may or may not be observed by other agents. We assume that the forecast process of ith renewable producer at time t is given by $X_t^i := \overline{X}_t + \check{X}_t^i$, where $\overline{X} := (\overline{X}_t)_{t \in [0,T]}$ is common for all agents (one can see this component as the national forecast), and the processes $\check{X}^i := (\check{X}_t^i)_{t \in [0,T]}$ for $i = 1, \ldots, N$ represent the individual production forecast of each renewable producer. Each small renewable producer aims to maximise her gain from trading in the market where they take their positions denoted $\phi^i := (\phi_t^i)_{t \in [0,T]}$ for $i = 1, \ldots, N$. The agents control their positions by choosing the trading rate, denoted by $\dot{\phi}_t^i$, $i = 1, \ldots, N$, at time t. They also face a terminal volume constraint $\phi_T^i = X_T^i$, which is enforced as a penalty. In Sects. 4, we use a generic agent to model the renewable producers in the Mean field setting. The agent has the same characteristics and goals as the ones of the small renewable energy producers above. The forecast process of this producer at time t is denoted $X_t = \overline{X}_t + \check{X}_t$, her state process position is given by $\phi := (\phi_t)_{t \in [0,T]}$ and controlled by $\dot{\phi} := (\dot{\phi}_t)_{t \in [0,T]}$.

In the following sections, we say that a strategy $\dot{\phi}^i$, $i = 1, \ldots, N$ (resp $\dot{\phi}$), is admissible if it is $\mathbb{F}$-adapted and square integrable.

3   A Complete Information Game

In this section we place ourselves in a complete information setting to find the unique Nash equilibrium. We assume that there are N identical agents in the market and they all observe the individual production forecast of the other agents. The filtration $\mathbb{F} := (\mathcal{F}_t)_{t\in[0,T]}$ models the information available to all of them. Without loss of generality, we assume that the initial position is $\phi_0^i = 0$ for all $i = 1, \ldots, N$, so that the position of the ith agent at time t is given by $\phi_t^i = \int_0^t \dot{\phi}_s^i ds$. The strategies impact the market price P as follows:

$$P_t^N = S_t + a\bar{\phi}_t^N, \quad \forall t \in [0, T], \tag{1}$$

where $\bar{\phi}_t^N = \frac{1}{N}\sum_{i=1}^N \phi_t^i$ is the average position of the agents and a is a constant. The agents trading in the market at time t incur an instantaneous cost,

$$\dot{\phi}_t^i P_t^N + \frac{\alpha(t)}{2}(\dot{\phi}_t^i)^2, \quad \forall t \in [0, T]$$

for i^{th} agent. The first term represents the actual cost of buying the electricity, and the second term represents the cost of trading, where $\alpha(.)$ is a continuous strictly positive function on $[0, T]$ reflecting the variation of market liquidity at the approach of the delivery date.

The processes S and $(X^i)_{i=1}^N$ satisfy the following assumption.

Assumption 1. *The processes S and $(X^i)_{i=1}^N$ are square integrable martingales with respect to the filtration $\mathbb{F}$.*

Each producer wishes to maximize the objective function:

$$J^{N,i}(\phi^i, \phi^{-i}) := -\mathbb{E}\left[\int_0^T \left\{\frac{\alpha(t)}{2}(\dot{\phi}_t^i)^2 + \dot{\phi}_t^i(S_t + a\bar{\phi}_t^N)\right\} dt + \frac{\lambda}{2}(\phi_T^i - X_T^i)^2\right], \tag{2}$$

where λ determines the strength of the imbalance penalty and $\phi^{-i} := (\phi^1, \ldots, \phi^{i-1}, \phi^{i+1}, \ldots, \phi^N)$ is the vector of all positions except i^{th} agent's one.

The optimal strategy of each player depends on other players' actions and we want to describe the resulting dynamical equilibrium, which we define formally below.

Definition 1 (Nash equilibrium). *We say that $(\dot{\phi}^{i*})_{i=1}^N$ is a Nash equilibrium for the N-player game if it is a vector of admissible strategies, and for each $i = 1, \ldots, N$,*

$$J^{N,i}(\phi^i, \phi^{-i*}) \leq J^{N,i}(\phi^{i*}, \phi^{-i*}) \tag{3}$$

for any other admissible strategy $\dot{\phi}^i$ of player i in the market.

The following theorem characterizes the Nash equilibrium of the N-player game. In the theorem, we denote the average forecast process by $\overline{X}_t^N := \frac{1}{N}\sum_{i=1}^N X_t^i$ and use the following shorthand notation:

$$\Delta_{s,t} := \int_s^t \frac{\eta(u,t)}{\alpha(u)} du \quad \text{with} \quad \eta(s,t) = e^{-\int_s^t \frac{a}{\alpha(u)} du} \quad \text{and} \quad \widetilde{\Delta}_{s,t} := \int_s^t \alpha^{-1}(u) du. \tag{4}$$

298   O. Féron et al.

Theorem 2. *Under Assumption 1, the unique Nash equilibrium in the complete information N-player game is given by*

$$\phi_t^{i*} = \int_0^t \Delta_{s,t} \frac{\lambda d\overline{X}_s^N - dS_s}{1 + \left(\frac{a}{N} + \lambda\right)\Delta_{s,T}} + \Delta_{0,t} \frac{\lambda \overline{X}_0^N - S_0}{1 + \left(\frac{a}{N} + \lambda\right)\Delta_{0,T}}$$
$$+ \int_0^t \widetilde{\Delta}_{s,t} \frac{\lambda d(X_s^i - \overline{X}_s^N)}{1 + \left(\frac{a}{N} + \lambda\right)\widetilde{\Delta}_{s,T}} + \widetilde{\Delta}_{0,t} \frac{\lambda(X_0^i - \overline{X}_0^N)}{1 + \left(\frac{a}{N} + \lambda\right)\widetilde{\Delta}_{0,T}}.$$

The equilibrium price has the following shape:

$$P_t^{N*} = S_t + a\overline{\phi}_t^{N*}, \qquad \overline{\phi}_t^{N*} = \int_0^t \Delta_{s,t} \frac{\lambda d\overline{X}_s^N - dS_s}{1 + \left(\frac{a}{N} + \lambda\right)\Delta_{s,T}} + a\Delta_{0,t} \frac{\lambda \overline{X}_0^N - S_0}{1 + \left(\frac{a}{N} + \lambda\right)\Delta_{0,T}}$$

$$(5)$$

From the expression of the equilibrium price (5) in Theorem 2, we observe that the price impact is composed of a deterministic part, a path dependent stochastic part relying on the past values of S and $\overline{X}$. These processes both play a symmetric role in the price impact up to some coefficients.

Let us consider a finite number of players and let the penalization parameter λ go to infinity. Then, from the expression (5) of $\overline{\phi}^{N*}$, we can derive, that the price impact depends only on the aggregate production forecast and the final aggregate position matches exactly with the final production forecast:

$$\overline{\phi}_t^{N*} \xrightarrow[\lambda \to \infty]{\mathbb{P}} \int_0^t \frac{\Delta_{s,t}}{\Delta_{s,T}} d\overline{X}_s^N + \frac{\Delta_{0,t}}{\Delta_{0,T}} \overline{X}_0, \qquad \overline{\phi}_T^{N*} \xrightarrow[\lambda \to \infty]{\mathbb{P}} \overline{X}_T^N$$

The complete information setting is questionable since one could argue that, in practice, the producers do not observe the individual forecasts of other players. The complexity of determining a Nash equilibrium in a partial information setting motivates us to consider the partial information problem in the Mean field setting.

4 An Incomplete Information Game

In this section we assume that the agents do not observe each other's individual production forecasts and consider the associated Mean field game. We then investigate on the existence of an ε-Nash equilibrium approximation in the N-player incomplete information game.

We consider a generic agent, and the filtration $\mathbb{F}$ contains the information available to this agent. In addition we introduce a smaller filtration, containing the common noise and denoted by $\mathbb{F}^0$. This filtration contains the information about the fundamental price and common part of the production forecast of the generic agent.

Throughout the paper and for any $\mathbb{F}$-adapted process $(\zeta_t)_{t \in [0,T]}$, we will denote $\bar{\zeta}_t = \mathbb{E}[\zeta_t | \mathcal{F}_t^0]$. The game is now modeled by the interaction of agents through the conditional distribution flow $\mu_t^\phi := \mathcal{L}(\phi_t | \mathcal{F}_t^0)$ of the state process. In the price impact function defined in the previous section, expectation with respect to the empirical measure, will be replaced by an integral with respect to the measure flow such that the market price is now given by:

$$P_t = S_t + a\bar{\phi}_t. \tag{6}$$

The generic agent wants to maximize the objective function:

$$J^{MF}(\phi, \bar{\phi}) := -\mathbb{E}\left[\int_0^T \frac{\alpha(t)}{2}\dot{\phi}_t^2 + \dot{\phi}_t(S_t + a\bar{\phi}_t)dt + \frac{\lambda}{2}(\phi_T - X_T)^2\right], \tag{7}$$

We now define what is meant be a Mean field equilibrium and make some additional assumptions:

Definition 2 (Mean field equilibrium). *An admissible strategy ϕ^* is a Mean field equilibrium if it maximizes (7) and $\bar{\phi} = \bar{\phi}^*$.*

Assumption 3.

– *The process X is a square integrable martingale with respect to the filtration $\mathbb{F}$.*
– *The process S and the process $\overline{X}$, defined by $\overline{X}_t := \mathbb{E}[X_t | \mathcal{F}_t^0]$ for $0 \leq t \leq T$, are square integrable martingales with respect to the filtration $\mathbb{F}^0$.*

Note that if X is an $\mathbb{F}$-martingale, then $\overline{X}$ is by construction an $\mathbb{F}^0$-martingale, but it may not necessarily be a martingale in the larger filtration $\mathbb{F}$.

The following theorem characterizes the Mean field equilibrium in our setting. In the theorem we use the same shorthand notation (4) as before.

Theorem 4. *Under Assumption 3, the unique Mean field equilibrium is given by*

$$\phi_t^* = \int_0^t \Delta_{s,t}\frac{\lambda d\overline{X}_s - dS_s}{1 + \lambda\Delta_{s,T}} + \widetilde{\Delta}_{s,t}\frac{\lambda d\check{X}_s}{1 + \lambda\widetilde{\Delta}_{s,T}} + \Delta_{0,t}\frac{\lambda\overline{X}_0 - S_0}{1 + \lambda\Delta_{0,T}} + \widetilde{\Delta}_{0,t}\frac{\lambda\check{X}_0}{1 + \lambda\widetilde{\Delta}_{0,T}}.$$

The equilibrium price has the following shape:

$$P_t = S_t + a\int_0^t \Delta_{s,t}\frac{\lambda d\overline{X}_s - dS_s}{1 + \lambda\Delta_{s,T}} + a\Delta_{0,t}\frac{\lambda\overline{X}_0 - S_0}{1 + \lambda\Delta_{0,T}}.$$

We now would like to construct an approximate equilibrium in the N-player setting and quantify how close this approximate equilibrium will be to the true solution. To address this question, both the N-agent problem and the Mean field problem must be defined on the same probability space.

Assumption 5.

– *The process S is a square integrable martingale with respect to the filtration $\mathbb{F}^0$.*

– *The process $\bar{X}$ and the processes $(X^i)_{i=1}^N$ are square integrable martingale with respect to the filtration $\mathbb{F}$.*
– *For $i = 1, \ldots, N$, almost surely, $\mathbb{E}[X_t|\mathcal{F}_t^0] = \mathbb{E}[X_t^i|\mathcal{F}_t^0] := \overline{X}_t$, $i = 1, \ldots, N$.*
– *The processes $(\check{X}^i)_{i=1}^N$ defined by $\check{X}_t^i = X_t^i - \overline{X}_t$ for $t \in [0,T]$, are orthogonal square integrable martingales with respect to the filtration $\mathbb{F}$, which are identically distributed.*

Definition 3 (ε-Nash equilibrium). *We say that a strategy $(\dot{\phi}^{i*})_{i=1}^N$ is an ε-Nash equilibrium for the N-player game if it is admissible and:*

$$J^{N,i}(\phi^i, \phi^{-i*}) - \varepsilon \le J^{N,i}(\phi^{i*}, \phi^{-i*}), \quad \forall i \in \{1, \ldots, N\}, \forall t \in [0,T]$$

for any other admissible control $\dot{\phi}^i$.

Proposition 1. *Under assumption 5, we define an admissible strategy for the N-player game as follows.*

$$\phi_t^{i*} = \int_0^t \Delta_{s,t} \frac{\lambda d\overline{X}_s - dS_s}{1 + \lambda \Delta_{s,T}} + \widetilde{\Delta}_{s,t} \frac{\lambda d\check{X}_s^i}{1 + \lambda \widetilde{\Delta}_{s,T}} + \Delta_{0,t} \frac{\lambda \overline{X}_0 - S_0}{1 + \lambda \Delta_{0,T}} + \widetilde{\Delta}_{0,t} \frac{\lambda \check{X}_0^i}{1 + \lambda \widetilde{\Delta}_{0,T}}$$

Then, for any $\varepsilon > 0$, there exists N_ε with $N_\varepsilon = \mathcal{O}(\frac{1}{\varepsilon^2})$, such that for all N with $N \ge N_\varepsilon$, this strategy is an ε-Nash equilibrium of the N-player game.

To complete the analysis led in the previous part, we would like to numerically confront the model results to some empirical findings.

5   Numerical Illustration

Volatility and price impact showed to be characteristics of interest in empirical studies. As they affect strategies of the agents, we want the model to capture these patterns in order to manage the risks and costs related to renewable production uncertainty.

5.1   Data Presentation

We use the limit order book data from the intraday EPEX market of January 2017 for the Germany delivery zone to perform the analyses on the market price. To exhibit a linear price impact and calibrate the volatility of the market price P defined in the theoretic model, we used the observed midquote price that we denote $\tilde{P}$. The volatility of the production forecast that we denote $\tilde{X}$, is also calibrated from empirical wind energy forecasts over January 2015.

In order to conduct these empirical studies we assume the following dynamics for $\tilde{P}$ and $\tilde{X}$:

$$d\tilde{P}_t = \mu_t dt + \sigma_t dW_t^{\tilde{P}}, \qquad d\tilde{X}_t = \tilde{\sigma}_t dW_t^{\tilde{X}} \quad \forall t \in [0,T], \tag{8}$$

where $W^{\tilde{P}}, W^{\tilde{X}}$ are $\mathbb{F}$-Brownian motions, and μ, σ, $\tilde{\sigma}$ are $\mathbb{F}$-adapted processes. For the simulated data, we assume all through Sect. 5 the following dynamics:

$$dS_t = \sigma_t^s dW_t^0, \qquad d\overline{X}_t = \sigma_t^0 dB_t^0, \qquad d\check{X}_t^i = \sigma_t^x dB_t^i, \quad i = 1, \ldots, N, \quad \forall t \in [0,T]$$

where σ^s, σ^0 and σ^x are deterministic functions and W^0, B^0, B^i are $\mathbb{F}$-Brownian motions in the complete information setting (respectively $\mathbb{F}^0$ for W^0, B^0 and $\mathbb{F}^i$ for B^i in the incomplete setting). We also assume all through Sect. 5 the liquidity function $\alpha(.)$ is given by:

$$\alpha(t) = \alpha \times (T - t) + \beta, \quad \forall t \in [0,T], \quad \alpha > 0, \beta > 0. \tag{9}$$

The model parameters are specified in Table 1.

5.2   Price Impact

In this section, we compare the simulated price impact of the model to the empirical one. In the empirical data, the market price is identified to the observed midquote price, and we make the simplifying assumption that we can derive the price impact from this single variable. We regress the midquote price shift just before and just after a transaction on the traded volumes:

$$\Delta \tilde{P}_t = \tilde{P}_{t+} - \tilde{P}_{t-} = a * \text{sign}(\text{Volume}_t) * \text{Volume}_t, \quad \forall t \in [0,T],$$

as the empirical analogous of the linear price impact in the model given by expression (1). The volume sign corresponds to the side of the order that trigger the transaction: if it is sell order then the sign is $-$, if it is a buy order it is $+$.

The results of this study are available in Fig. 1. We present the price impact for several delivery hours, the regression was made over more than 4750 data points for each of them. The price shifts are heteroscedastic and seem to be less significant for small volumes. Despite the small volume effect, the p-value and the R^2 indicate a significant regression coefficient and are coherent with the price impact assumption. In Fig. 2, we draw a trajectory of the fundamental price S starting from $t = 0$ an hour before the delivery time, to the delivery time T. We also draw the market price P associated with the different homogeneous settings studied in the paper: the N-player Nash equilibrium, the ε-Nash equilibrium, and the Mean field one.

The price impact matches with the production forecast changes. If producers think they have underestimated their production forecast with respect to their supply commitment (negative values of the production forecast changes process), there will be an excess of sell positions in the market, thus the price impact is negative. On the contrary, if they think they overestimated the final production forecast, there will be a lack of supply in the market and a negative price impact.

5.3   Volatility

We want to investigate whether or not the uncertain production forecast has an impact on the market price variations, and show that the volatility observed empirically in the intraday market can be explained by this phenomenon.

We are interested in estimating the instantaneous volatility σ of $\tilde{P}$, introduced in the dynamics (8). Following [11], we use a kernel-based non parametric estimator of the instantaneous volatility:

$$\hat{\sigma}_t^2 = \frac{\sum_{i=1}^n K_h(t_{i-1} - t)\Delta \tilde{P}_{t_{i-1}}^2}{\sum_{i=1}^n K_h(t_{i-1} - t)(t_i - t_{i-1})},$$

where $K_h(x) = \frac{1}{h}K(\frac{x}{h})$ and $K(.)$ is the Epanechnikov kernel. We used a generic choice: $h = 0.1$ hour for all the delivery dates and hours of January. We also estimate the volatility $\tilde{\sigma}$ of the empirical wind production forecast $\tilde{X}$ introduced in the dynamic (8) using the same method with $h = 1$ hour, and use it to calibrate the volatility of the production forecasts σ^0 and σ^x in the model.

In Fig. 3, the first graph and the second graph represent respectively the empirical volatility of the midquote $\tilde{P}$ and the estimated variations of the Nash equilibrium market price P, for different hours in function of the time to delivery. During peak hours, activity and thus liquidity in the market is more important. In order to adapt the liquidity to the delivery hour considered in the model, we chose different levels of the liquidity coefficients α and β for the function $\alpha(.)$ defined in (9), available in Table 2. Apart from these coefficients, all the other parameters are the same as in Table 1. The model reproduces the increasing shape of the empirical market price volatility when we approach the delivery time. Moreover, by adapting the liquidity coefficients, the model also captures the different levels of volatility according to the delivery hour.

6   Conclusion

We developed a linear quadratic model and derived a dynamic price equilibrium in the intraday electricity market. We focused on the integration of renewable energies in the energy supply system. We considered intermittent energy producers first in a complete information setting, then, in a more realistic incomplete information one.

The model provides closed form optimal strategies for agents taking into account their own incertitude. It leads to a dynamic equilibrium on the market, and reproduces some important empirical patterns such as the price impact and the volatility. For these reasons, a practical use of this mathematical tool might help to better optimize the renewable furniture system.

Acknowledgements. Financial support from LABEX ECODEC (ANR-11-IDEX-0003/Labex Ecodec/ANR-11-LABX-0047), FIME Research Initiative and Agence Nationale de Recherche (ANR project EcoREES) is gratefully acknowledged.

A   Appendix

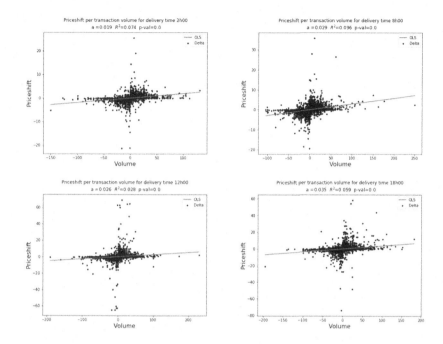

Fig. 1. Price impact over January 2017 for different delivery times

Table 1. Parameters of the model

Parameter	Value	Parameter	Value
S_0	40€/MWh	a	1 €/MWh^2
σ^s	10 €/MWh.h$^{\frac{1}{2}}$	λ	100€/MWh^2
$\overline{X}_0, \check{X}_0^i$	0 MWh	N	100
σ^x	73 MWh/h$^{1/2}$	α	0.3 €/s.MW^2
σ^0	73 MWh/h$^{1/2}$	β	0.1 €/MW^2

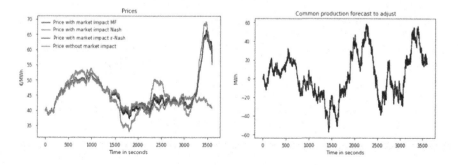

Fig. 2. Theoretical price impact and common production forecast changes associated

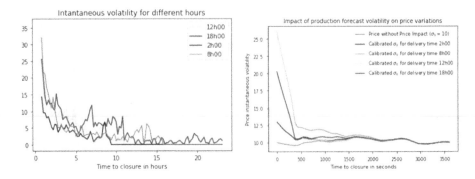

Fig. 3. Instantaneous market volatility and wind energy production forecast impact

Table 2. Liquidity coefficients in function of the hour trading activity

Hours	Coefficients	
	α (€/s.MW^2)	β (€/MW^2)
2h00	1.2	0.5
8h00	0.5	0.2
12h00	0.7	0.3
18h00	0.3	0.1

References

1. Aïd, R., Gruet, P., Pham, H.: An optimal trading problem in intraday electricity markets. Math. Financ. Econ. **10**(1), 49–85 (2016). https://doi.org/10.1007/s11579-015-0150-8
2. Alasseur, C., Ben Tahar, I., Matoussi, A.: An extended mean field game for storage in smart grids. J. Optim. Theory Appl. **184**(2), 644–670 (2020)
3. Carmona, R., Delarue, F.: Probabilistic Theory of Mean Field Games with Applications I. Springer, Cham (2018)
4. Casgrain, P., Jaimungal, S.: Mean field games with partial information for algorithmic trading. Technical report, arXiv.org (2019)
5. Delikaraoglou, S., Papakonstantinou, A., Ordoudis, C., Pinson, P.: Price-maker wind power producer participating in a joint day-ahead and real-time market. In: 2015 12th International Conference on the European Energy Market (EEM), pp. 1–5. IEEE (2015)
6. Garnier, E., Madlener, R.: Balancing forecast errors in continuous-trade intraday markets. Energy Syst. **6**(3), 361–388 (2015). https://doi.org/10.1007/s12667-015-0143-y
7. Huang, M., Malhamé, R.P., Caines, P.E., et al.: Large population stochastic dynamic games: closed-loop mckean-vlasov systems and the nash certainty equivalence principle. Commun. Inf. Syst. **6**(3), 221–252 (2006)
8. Jónsson, T., Pinson, P., Madsen, H.: On the market impact of wind energy forecasts. Energy Econ. **32**(2), 313–320 (2010)

9. Karanfil, F., Li, Y.: Wind up with continuous intraday electricity markets? The integration of large-share wind power generation in denmark. Chaire European Electricity Markets working paper (2015)
10. Kiesel, R., Paraschiv, F.: Econometric analysis of 15-minute intraday electricity prices. Energy Econ. **64**, 77–90 (2017)
11. Kristensen, D.: Nonparametric filtering of the realized spot volatility: a kernel-based approach. Econom. Theor. **26**(1), 60–93 (2010)
12. Lasry, J.M., Lions, P.L.: Mean field games. Japan. J. Math. **2**(1), 229–260 (2007)
13. Morales, J.M., Conejo, A.J., Pérez-Ruiz, J.: Short-term trading for a wind power producer. IEEE Trans. Power Syst. **25**(1), 554–564 (2010)
14. Rowińska, P.A., Veraart, A., Gruet, P.: A multifactor approach to modelling the impact of wind energy on electricity spot prices (2018). SSRN 3110554
15. Tan, Z., Tankov, P.: Optimal trading policies for wind energy producer. SIAM J. Financ. Math. **9**(1), 315–346 (2018)
16. Zugno, M., Morales, J.M., Pinson, P., Madsen, H.: Pool strategy of a price-maker wind power producer. IEEE Trans. Power Syst. **28**(3), 3440–3450 (2013)

Distributed Static State Feedback Control for DC Microgrids

Sifeddine Benahmed[1]([envelope]) [iD], Pierre Riedinger[1] [iD], and Serge Pierfederici[2] [iD]

[1] Centre de Recherche en Automatique de Nancy (CRAN) UMR 7036,
University of lorraine and CNRS, Nancy, France
{sifeddine.benahmed,pierre.riedinger}@univ-lorraine.fr
[2] Laboratoire d'Energétique et de Mécanique Thèorique et Appliquée (LEMTA)
UMR 7563, University of lorraine and CNRS, Nancy, France
serge.pierfederici@univ-lorraine.fr

Abstract. This paper addresses the problem of Current Sharing (CS) and Average Voltage Regulation (AVR) in Direct Current (DC) microgrids composed of several interconnected Distributed Generation Units (DGUs), power lines and loads. To achieve the control objectives (CS and AVR), the system is augmented with distributed integral actions. A distributed-based static state feedback control architecture is proposed. This latter guarantees the global asymptotic convergence of the system state to the set of all equilibrium points for which the control objectives are achieved, thanks to the passivity property of the DGU with local controller. Simulation results are provided to illustrate the effectiveness of the proposed methodology.

Keywords: Distributed control · Multi-agent systems · DC microgrids

1 Introduction

Microgrids (MG) are a novel concept of distributed electrical network that can be composed of several interconnected power supplies and loads. This concept represents an efficient key component to simplify the integration of renewable energy sources. Moreover, Direct Current (DC) MGs have received an increasing interest in power system control engineering community. This growing interest is due to its efficiency, simplicity and wide range of applicability [2,4].

Effective control strategies are needed to achieve high performance operation and ensure the stability of the MG. These objectives require not only local management but also cooperation between the interconnected Distributed Generation Units (DGUs) and loads [4].

This work was supported by the French PIA project Lorraine Université d'Excellence, reference ANR-15-IDEX-04-LUE and by the ANR under the grant HANDY ANR-18-CE40-0010.

S. Lasaulce et al. (Eds.): NetGCooP 2021, CCIS 1354, pp. 306–314, 2021.
https://doi.org/10.1007/978-3-030-87473-5_27

Common problems in the control of DC MG are Current Sharing (CS) and Voltage Regulation (VR). CS or, equivalently, load sharing aims to share proportionally the current demand between the different DGUs taking into consideration their power capacity. VR aims to guarantee a certain voltage level for the loads since this latter must be supplied by a nominal value of voltage under any perturbation [9]. Achieving these goals is an arduous and generally impossible task as the CS requires a voltage deviation from its reference values. Therefore, an alternative is to provide an average voltage regulation (AVR), i.e., the average values of voltages at the Points of Common Coupling (PCC) is equal to the average of its references [1].

Many works are presented in the literature to control DC MGs. Generally, the main difficulty is to guarantee global stability when CS and AVR objectives are simultaneously considered. In [7] and [9], only voltage stabilization is considered. Moreover, in [10] and [5], the aforementioned objectives are considered but without proof of global convergence.

In this paper, a new distributed methodology to control DC-MG is proposed where each element of the MG has its controller and exchanges information with its neighbors over a communication network. The novelty of this work is the use of two distributed integral actions to achieve both AVR and CS objectives. In addition, the proposed control approach is LMI-based which makes it attractive numerically. Finally, the use of passivity of interconnected systems to prove the global asymptotic convergence allows to extend the result to more general MG-problems, e.g., MG with Storage Units, etc.

The paper is organized as follows: in Sect. 2, some notation and preliminaries are given. In Sect. 3, the general framework of the studied DC-MG model is presented. The control objectives are detailed in Sect. 4. In Section sec5, integral actions are considered to deal with the control objectives, the design of the proposed distributed control is presented. In Sect. 6, the simulation results are presented. Finally, Sect. 7 concludes the paper.

2   Notation and Preliminaries

Notation: The symbols, $\mathbb{R}$ and $\mathbb{R}_{>0}$ stand respectively for the set of real and positive real numbers. To simplify notation we denote a column vector as an n-tuple $(x_1, x_2, \cdots, x_n)$ whose entries x_i can be also column vectors or equivalently $(x_1, x_2, \cdots, x_n) = \begin{bmatrix} x_1^T & x_2^T & \cdots & x_n^T \end{bmatrix}^T$. The notation $\mathbf{I}_n$ is used to denote the identity matrix of the size $(n \times n)$. The transpose of a matrix A is denoted by A^T. The vector of dimension n with all components equal 1 is denoted $\mathbb{1}_n$. $\mathbb{0}_{m \times p}$ stands for the zero matrix of the size $(m \times p)$. The empty set is represented by $\emptyset$. The symbol $\otimes$ represents the Kronecker product. The notation $diag(A_1, ..., A_n)$ denotes the block diagonal matrix having the matrices A_1 to A_n on the diagonal and 0 every where else.

Convergence to a Set: If $d(x, y)$ denotes a distance in a metric space, the distance of a point x to a set S is defined by: $d(x, S) = \inf_{y \in S} d(x, y)$. A trajectory $x(\cdot)$ is said to converge asymptotically to a set S if $\lim_{t \to +\infty} d(x(t), S) = 0$.

Passivity Theory: A linear system (A, B, C) is strictly passive [3] if there exists a matrix $P = P^T > 0$ and a scalar $\epsilon > 0$ s.t.: $A^T P + PA < -\epsilon P$, $PB = C^T$.

3   DC Microgrid Model

In this work we consider, a DC-MG composed of N distributed Generation Units (DGUs) connected through q resistive power lines. A simple electrical scheme example of the considered model is shown in Fig. 1. The generic energy source of each DGU is modeled as a DC voltage source that supplies a local load through a DC-DC converter. The local load is connected to the Point of Common Coupling (PCC) through an RLC (low-pass) filter. Furthermore, two types of local load are considered, Resistive load R_{Li} and unknown constant current source I_{Li}. The model of the DGU$_i$ is described by the following dynamic equations:

$$\text{DGU}_i \begin{cases} L_i \dot{I}_i = -R_i I_i - V_i + u_i, \\ C_i \dot{V}_i = I_i - I_{Li} - \dfrac{V_i}{R_{Li}} - \sum_{j \in \mathcal{N}_i^{pow}} \dfrac{1}{R_{l_{ij}}} (V_i - V_j), \end{cases} \tag{1}$$

where I_i is the generated current, V_i is the voltage at the PCC near the DGU$_i$, $I_{l_{ij}}$ is the power line current, L_i and R_i are, respectively, the output filter inductance and resistance, C_i is the output shunt capacitor, R_{Li} is the local resistive load, $R_{l_{ij}}$ is the power line resistance and $\mathcal{N}_i^{pow}$ denotes the set of nodes connected, respectively, by power lines to the i-th DGU.

The DC power network is represented by a connected and undirected graph $\mathcal{G}^{pow} = (\mathcal{V}^{pow}, \mathcal{L}^{pow})$ (see [6] for more details about graph theory). The nodes, $\mathcal{V}^{pow} = \{1, ..., N\}$, represent the DGUs. The topology of the power network is represented by a weighted Laplacian matrix $\mathcal{L}^{pow} \in \mathbb{R}^{N \times N}$ whose elements are related to the coupling term $\sum_{j \in \mathcal{N}_i^{pow}} \frac{1}{R_{l_{ij}}} (V_i - V_j)$ in (1).

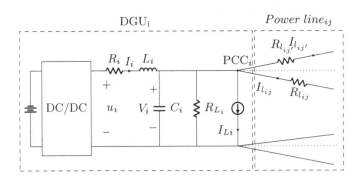

Fig. 1. The considered electrical scheme of the DC Microgird with DGUs and power lines.

The overall microgrid system for all the DGUs can be written, compactly, as:

$$\text{MG} \begin{cases} L\dot{I} = -RI - V + u, \\ C\dot{V} = I - R_L^{-1}V - \mathcal{L}^{pow}V - I_L, \end{cases} \tag{2}$$

where $I, V, I_L, u \in \mathbb{R}^N$. As well, $C, R_L, R, L \in \mathbb{R}^{N \times N}$ are positive definite diagonal matrices, e.g., $L = \text{diag}(L_1, \cdots, L_N)$.

4   Motivation and Problem Formulation

First, we present the considered control objectives. When sharing current between several supplies, the current demand should be shared proportionally, but not necessarily equally.

Objective 1 *(Current Sharing)*. *At steady state, currents need to fulfill the following requirement*

$$\lim_{t \to \infty} \omega_i I_i = \omega_i I_i^e = \omega_j I_j^e \quad \forall i, j \in \mathcal{V}^{pow},$$

where the weight ω_i, $i = 1, \cdots, N$ are given parameters.

In fact, ω_i^{-1} can be chosen as the corresponding DGU$_i$ rated current. Hence, a relatively small value of ω_i corresponds to a relatively large generation capacity of DGU$_i$.

Generally, achieving Objective 1 does not permit to attend an equilibrium voltage $V^e = V^{ref}$ at the same time. Hence, as in [8] an average voltage regulation is considered, where the aim of the controller is to have the weighted average value of V^e equal to the weighted average value of the desired reference voltages V^{ref}. Assuming that there exists a reference voltage V_i^{ref} at the PCC, for all DGU$_i$, the second control objective can be stated as

Objective 2 *(Average Voltage Regulation)*

$$\lim_{t \to \infty} \mathbb{1}_N^T W^{-1} V(t) = \mathbb{1}_N^T W^{-1} V^e = \mathbb{1}_N^T W^{-1} V^{ref},$$

where $W = \text{diag}(\omega_1, \cdots, \omega_N)$, $\omega_i > 0$, for all DGU$_i$.

The choice of the weights for voltages as ω_i^{-1} is motivated by the fact that the DGU$_i$ with the highest capacity should impose the voltage of the MG [8].

Now, we are able to state the control problem as:

Control Problem: *For a given reference voltage V^{ref} and an unknown load current I_L, design a distributed-based control scheme s.t. the state of system (2) in closed-loop converges globally and asymptotically to a set of equilibrium points S^e whose elements satisfy Objectives 1–2.*

5   Distributed Controller Design

In this section, a solution to the control problem defined in Sect. 4 is provided. First, we assume the following:

Assumption 1 *(communication network). A communication network modeled as a connected and undirected graph* $\mathcal{G}^{com} = (\mathcal{V}^{com}, \mathcal{L}^{com})$ *where* $\mathcal{V}^{pow} = \{1, ..., N\}$ *represent the DGUs and* $\mathcal{L}^{com} \in \mathbb{R}^{N \times N}$ *is a symmetric positive semidefinite Laplacian matrix, allows to exchange voltage* V_i *and current* I_i *measured at each* DGU$_i$, $i = 1, ..., N$.

Assumption 2 *(Nominal Model). All the DGUs have the same nominal values of parameters, i.e.,* $L_i = L^*$, $C_i = C^*$, $R_i = R^*$ *and* $R_{Li} = R_L^*$ $\forall i = 1, \cdots, N$ *with* L^*, C^*, R^*, $R_L^* \in \mathbb{R}_{>0}$ *represent the nominal values. Thus,* $L = L^* \boldsymbol{I}_N$, $C = C^* \boldsymbol{I}_N$, $R = R^* \boldsymbol{I}_N$ *and* $R_L = R_L^* \boldsymbol{I}_N$.

Our aim is to determine a controller including N integral actions in order to achieve Objectives 1–2. Consider system (2), let us introduce an augmented state $\mathcal{X} = (I, V, \phi, \gamma)$ whose dynamics is given by the following equations:

$$\Sigma \begin{cases} L\dot{I} = -RI - V + u, & (3a) \\ C\dot{V} = I - (R_L^{-1} + \mathcal{L}^{pow})V - I_L, & (3b) \\ \tau_\phi \dot{\phi} = W^T \mathcal{L}^{com} W I, & (3c) \\ \tau_\gamma \dot{\gamma} = -W^T \mathcal{L}^{com} W \gamma + (V - V^{ref}), & (3d) \end{cases}$$

where τ_ϕ, $\tau_\gamma \in \mathbb{R}_{>0}$ and where $\mathcal{L}^{com}$ is defined in Assumption 1.

Definition 1 *(Set of Equilibrium Points). For a given reference voltage* V^{ref} *and an unknown load current* I_L, *the set of all the equilibrium points is defined by* $S^e(I_L, V^{ref}) = \{\mathcal{X}^e = (I^e, V^e, \phi^e, \gamma^e) \in \mathbb{R}^{4N}$ *and* $u^e \in \mathbb{R}^N$ *s.t.:* $0 = -RI^e - V^e + u^e$, $0 = I^e - (R_L^{-1} + \mathcal{L}^{pow})V^e - I_L$, $0 = W^T \mathcal{L}^{com} W I^e$, *and* $0 = -W^T \mathcal{L}^{com} W \gamma^e + (V^e - V^{ref}).\}$

For a given reference voltage V^{ref} and an unknown load current I_L, one can easily prove that the set $S^e(I_L, V^{ref})$ is not empty and that Objectives 1–2 are always achieved in this set. The next part concerns the design of a state feedback controller of the form

$$u = -K(I, V, \phi, \gamma), \qquad (4)$$

with $K \in \mathbb{R}^{N \times 4N}$ and s.t. the state $\mathcal{X} = (I, V, \phi, \gamma)$ converges asymptotically to the set S^e.

5.1   Local Controllers Design

Since the controller u should be distributed, the local controllers u_i $i = 1, ..., N$ should depend only on local variables $x_i = (I_i, V_i, \phi_i, \gamma_i)$. Hence, the gain matrix K (see (4)) should be restricted to the form:

$$K = (K_I, K_V, K_\phi, K_\gamma), \qquad (5)$$

where K_I, K_V, K_ϕ, $K_\gamma \in \mathbb{R}^{N \times N}$ are diagonal matrices. The main difficulty to find a gain matrix of this form for system (3) is the existence of physical ($\mathcal{L}^{pow}$) and communication ($\mathcal{L}^{com}$) coupling terms. Hence, to simplify the design let us introduce the following change of coordinates:

$$(\tilde{I}, \tilde{V}, \tilde{\phi}, \tilde{\gamma}) = (\mathbf{I}_4 \otimes U^T)(I, V, \phi, \gamma), \tag{6}$$

where $U \in \mathbb{R}^{N \times N}$ is a unitary matrix s.t.:

$$\tilde{\mathcal{L}}^{com} = U^T W^T \mathcal{L}^{com} W U = \mathrm{diag}(0, \lambda_2, \ldots, \lambda_N)$$

where $\lambda_i < \lambda_j \ \forall \ i < j$. The matrix U exists because $W^T \mathcal{L}^{com} W$ is a symmetric matrix and $\lambda_1 = 0$ since the graph $\mathcal{G}^{com}$ is connected. In this new basis, system (3) can be rewritten as follows:

$$\tilde{\Sigma} \begin{cases} L\dot{\tilde{I}} = -R\tilde{I} - \tilde{V} + \tilde{u}, & (7a) \\[4pt] C\dot{\tilde{V}} = \tilde{I} - (R_L^{-1} + \tilde{\mathcal{L}}^{pow})\tilde{V} - \tilde{I}_L, & (7b) \\[4pt] \tau_\phi \dot{\tilde{\phi}} = \underbrace{\begin{bmatrix} 0 & 0 & \cdots & 0 \\ 0 & \lambda_2 & \cdots & 0 \\ \vdots & \vdots & \ddots & \vdots \\ 0 & 0 & \cdots & \lambda_N \end{bmatrix}}_{\tilde{\mathcal{L}}^{com}} \tilde{I}, & (7c) \\[4pt] \tau_\gamma \dot{\tilde{\gamma}} = -\tilde{\mathcal{L}}^{com}\tilde{\gamma} + (\tilde{V} - \tilde{V}^{ref}), & (7d) \end{cases}$$

where $\tilde{\mathcal{L}}^{pow} = U^T \mathcal{L}^{pow} U$ and $(\tilde{u}, \tilde{I}_L, \tilde{V}^{ref}) = (\mathbf{I}_3 \otimes U^T)(u, I_L, V^{ref})$. Note that the matrices L, C, R and R_L remain unchanged by Assumption 2. Consider a controller $\tilde{u}$ of the form:

$$\tilde{u} = -\tilde{K}(\tilde{I}, \tilde{V}, \tilde{\phi}, \tilde{\gamma}), \tag{8}$$

where $\tilde{K} = (\mathcal{K} \otimes \mathbf{I}_N)$ and $\mathcal{K} = \begin{bmatrix} k_I & k_V & k_\phi & k_\gamma \end{bmatrix} \in \mathbb{R}^{1 \times 4}$. Let us remark that system $\tilde{\Sigma}$ with the control law (8) is composed of N interconnected subsystems which can be written using a permutation matrix as follows, for $i = 1, \cdots, N$:

$$\tilde{\Sigma}_i : \dot{\tilde{x}}_i = Acl_i \tilde{x}_i + d_i - B_p \sum_{j \in \mathcal{N}_i^{pow}} l_{i,j} C_p(\tilde{x}_i - \tilde{x}_j), \tag{9}$$

where $\tilde{x}_i = (\tilde{I}_i, \tilde{V}_i, \tilde{\phi}_i, \tilde{\gamma}_i)$, $d_i = -(0, C^{*-1}\tilde{I}_{Li}, 0, \frac{1}{\tau_\gamma}\tilde{V}_i^{ref})$, $B_p = (0, C^{*-1}, 0, 0)$, $C_p = \begin{bmatrix} 0 & 1 & 0 & 0 \end{bmatrix}$, $l_{i,j}$ for $1 \le i,j \le N$ denotes the elements of $\tilde{\mathcal{L}}^{pow}$ and

$$Acl_i = \begin{bmatrix} A & \mathbb{0}_{2 \times 2} \\ \frac{\lambda_i}{\tau_\phi} 0 & 0 \ 0 \\ 0 \ \frac{1}{\tau_\gamma} \ 0 & -\frac{\lambda_i}{\tau_\gamma} \end{bmatrix} - \begin{bmatrix} L_t^{*-1} \\ \mathbb{0}_{3 \times 1} \end{bmatrix} \mathcal{K}, \tag{10}$$

and where $A = \begin{bmatrix} -R^* L^{*-1} & -L^{*-1} \\ C^{*-1} & -(C^* R_L^*)^{-1} \end{bmatrix}.$

In this basis, it can be noticed that the local variables $\tilde{x}_i = (\tilde{I}_i, \tilde{V}_i, \tilde{\phi}_i, \tilde{\gamma}_i)$ $i = 1, ..., N$ are only coupled by the term $B_p \sum_{j \in \mathcal{N}_i^{pow}} l_{i,j} C_p(\tilde{x}_i - \tilde{x}_j)$ (related to the matrix $\tilde{\mathcal{L}}^{pow}$). The next theorem shows how it is possible to determine the unique gain matrix $\mathcal{K}$ in (10) for all the subsystems by removing the last coupling terms in the right member of (9) and using some passivity arguments.

Theorem 1 *(**Main result**). If there exists a static state feedback* $\mathcal{K} = \begin{bmatrix} k_I & k_V & k_\phi & k_\gamma \end{bmatrix}$ *s.t. the triples* (Acl_i, B_p, C_p) *for* $i = 2, ..., N$ *and* $(\breve{A}cl_1, \breve{B}_1, \breve{C}_1)$ *are strictly passive where:*

$$\breve{A}cl_1 = \begin{bmatrix} A & \mathbb{0}_{2\times 1} \\ 0 & \frac{1}{\tau_\gamma} & 0 \end{bmatrix} - \begin{bmatrix} L^{*-1} \\ \mathbb{0}_{2\times 1} \end{bmatrix} \begin{bmatrix} k_I & k_V & k_\gamma \end{bmatrix},$$

$$\breve{B}_1 = (0, C^{*-1}, 0), \quad \breve{C}_1 = \begin{bmatrix} 0 & 1 & 0 \end{bmatrix}, \tag{11}$$

and where Acl_i, B_p, C_p *and* A *are given with subsystems (9), then the state of the augmented system (3) in closed-loop with*

$$u = -(\mathcal{K} \otimes \boldsymbol{I}_N)(I, V, \phi, \gamma)$$

converges asymptotically to an equilibrium $\mathcal{X}^e \in S^e(V^{ref}, I_L)$ *for which the control objectives 1–2 are satisfied.*

6   Simulation

In this section we aim to validate the proposed controller by simulation. We consider a MG composed of 4 DGUs with non-identical electrical parameters and communication links (see Fig. 2). The controller was designed using the nominal parameter of the MG and then applied on the MG model with the real parameters. The system is initially at a steady state with load current $I_L(0) = [5 \ 10 \ 30 \ 20]$ A. Then, at the time instant $t = t_1$ the load current is stepped up with $\Delta I_L = [10 \ 15 \ 20 \ 30]$ A. As we can see in Fig. 3 and 4, the weighted average voltage converges to the weighted average value of the reference voltages (see Objective 2). Furthermore, the voltages at the PCC converge, without oscillations, to a steady state near to the reference voltage

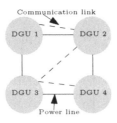

Fig. 2. MG with 4 DGUs, power lines, and communication links.

$V^{ref} = 380V$. Moreover, Fig. 4 shows clearly that the weighted currents converge to the same consensus value achieving Objective 1 and the generated currents converge asymptotically to the desired steady state, asymptotically.

The results illustrate the robust performance of the proposed controllers under the change in the load current and the presence of parametric discrepancies from the nominal values.

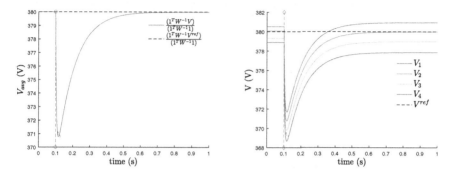

Fig. 3. From the left: weighted average voltage at the PCC and the weighted average reference voltage value (dashed line); voltage at the PCC of each DGU together with the reference value (dashed line).

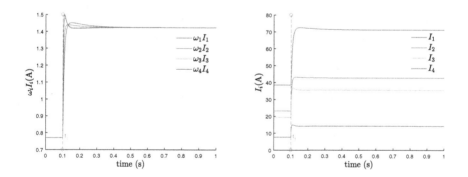

Fig. 4. From the left: the weighted generated currents of the DGUs; generated currents.

7   Conclusion

A distributed-based Static-State-feedback control scheme, including integral actions to achieve both proportional Current Sharing and Average Voltage Regulation in DC power-networks has been proposed. Distributed integral actions have been used to achieve the control objectives by exploiting a communication network. The simulation results clearly show that the control objectives are achieved with unknown load and even with significant discrepancies between nominal and real parameters of the DGUs.

References

1. Cucuzzella, M., Trip, S., De Persis, C., Cheng, X., Ferrara, A., van der Schaft, A.: A robust consensus algorithm for current sharing and voltage regulation in DC microgrids. IEEE Trans. Control Syst. Technol. **27**(4), 1583–1595 (2019)

2. Hirsch, A., Parag, Y., Guerrero, J.: Microgrids: a review of technologies, key drivers, and outstanding issues. Renew. Sustain. Energy Rev. **90**, 402–411 (2018)

3. Khalil, H.K.: Nonlinear Systems, 3rd edn. Prentice-Hall, Upper Saddle River (2002)

4. Kumar, J., Agarwal, A., Agarwal, V.: A review on overall control of dc microgrids. J. Energy Storage **21**, 113–138 (2019)

5. Lu, X., Guerrero, J. M., Sun, K.: Distributed secondary control for dc microgrid applications with enhanced current sharing accuracy. In: 2013 IEEE International Symposium on Industrial Electronics, pp. 1–6, May 2013

6. Mesbahi, M., Egerstedt, M.: Graph Theoretic Methods in Multiagent Networks. Princeton University Press, Princeton (2010)

7. Soloperto, R., Nahata, P., Tucci, M., Ferrari-Trecate, G.: A passivity-based approach to voltage stabilization in dc microgrids. In: 2018 Annual American Control Conference (ACC), pp. 5374–5379, June 2018

8. Trip, S., Cucuzzella, M., Cheng, X., Scherpen, J.: Distributed averaging control for voltage regulation and current sharing in dc microgrids. IEEE Control Syst. Lett. **3**(1), 174–179 (2019)

9. Tucci, M., Riverso, S., Ferrari-Trecate, G.: Line-independent plug-and-play controllers for voltage stabilization in DC microgrids. IEEE Trans. Control Syst. Technol. **26**(3), 1115–1123 (2018)

10. Zhang, X., Dong, M., Ou, J.: A distributed cooperative control strategy based on consensus algorithm in dc microgrid. In: 2018 13th IEEE Conference on Industrial Electronics and Applications (ICIEA), pp. 243–248, May 2018

Design of a Combinatorial Double Auction for Local Energy Markets

Diego Kiedanski[1]([✉]), Daniel Kofman[1][iD], and Ariel Orda[2][iD]

[1] Télécom Paris, Paris, France
`diego.kiedanski@telecom-paristech.fr`
[2] Technion, Haifa, Israel

Abstract. Local energy markets allow neighbours to exchange energy among them. Their traditional implementation using sequential auctions has proven to be inefficient and even counterproductive in some cases. In this paper we propose a combinatorial double auction for the exchange of energy for several time-slots simultaneously. We suppose that participants have a flexible demand; flexibility being obtained, for example, by the usage of a battery. We show the benefits of the approach and we provide an example of how it can improve the utility of all the participants in the market.

Keywords: Auction · Smart grids · Local energy markets

1 Introduction

Local energy markets (LEMs) have been proposed as a paradigm to better exploit the benefits of distributed local energy generation [3]. The various proposed market mechanisms target to encourage neighbours to exchange energy locally - within the same low voltage distribution grid, for example - in order to reduce their energy bill or even to generate revenue. In most implementations, the market mechanisms consist of a sequence of auctions that allow the participants to trade energy for the next time-slot (usually 15 or 30 min long). For a review of different proposals and implementations, the reader is referred to [12], [8,9] and the references therein. LEMs are usually implemented as double auctions, with players (households) submitting both buying and selling bids. In particular, a house with renewable generation can be a buyer or a seller, depending on the time-slot.

In addition, if households have flexibility in their consumption profiles (for example, thanks to energy storage systems), they will schedule their load to obtain the most out of the market. In spite of this, it is known that the system architecture involving sequential auctions does not fully exploit the available flexibility and can even be counterproductive in some cases [6]. For example, when players are subject to Time-of-Use tariffs (ToU), their beliefs about future market prices can lead them to postpone their demand, only to follow it with

© Springer Nature Switzerland AG 2021
S. Lasaulce et al. (Eds.): NetGCooP 2021, CCIS 1354, pp. 315–320, 2021.
https://doi.org/10.1007/978-3-030-87473-5_28

a huge peak in consumption before a change in the price from the cheap ToU period to the most expensive one [5].

In this paper, we put forward the design of an approach based on a combinatorial double auction [7,11,13] that improves the utility of all players and increases the total traded energy. Even though combinatorial auctions have already been proposed [1,10] the design presented here is the first to exploit the structure of flexible demand derived from energy storage.

2 Mathematical Model of Players

Let $\mathcal{N} = \{1, \ldots, N\}$ denote the set of players and $\mathcal{T} = \{1, \ldots, T\}$ the set of time-slots in a given day. Each player can consume energy by using appliances (water heater, A/C, charging electric vehicles, TV, etc.) and might produce energy (e.g. photovoltaic generation). Let x_t^i denote the demand of player i at time-slot t, where a positive value of x_t^i represents excess of consumption while a negative value stands for a surplus of renewable energy (the definition of x is independent of possible flows with a battery, those flows will be introduced through additional variables). The demand profile $x^i = (x_1^i, \ldots, x_T^i)$ of player i is assumed fixed and known.

To simplify the presentation, we suppose that the flexibility of each player is introduced only by batteries (for example, the demand of the appliances is not shifted in time). Let $\mathbf{S}^i$ denote the total capacity of player i's battery (possibly 0), $\mathbf{S}_0^i$ the initial state of charge and s_t^i the amount of charged ($s_t^i \geq 0$) or discharged ($s_t^i < 0$) energy at time-slot t. The feasible set of charging/discharging decisions $\mathcal{F}^i$ is given by Eq. (1).

$$\mathcal{F}^i = \left\{ s^i : \mathbf{S}_0^i + \sum_{t=1}^{j} s_t^i \in [0, \mathbf{S}^i], \ \forall j \in \mathcal{T}; s^i \in \mathbb{R}^T \right\} \tag{1}$$

The state of the battery at time-slot t is precisely: $\mathbf{S}_0^i + \sum_{j=1}^{t} s_j^i$.

Furthermore, we will denote by $n^i = s^i + x^i$, with $s^i \in \mathcal{F}^i$, the net consumption of player i as seen from the grid.

In addition to trading in the market, households can trade with their traditional electricity company (TEC). Each player's contract with the TEC stipulates a price for buying energy β_t^i and a price for selling energy back to the grid ζ_t^i at time-slot t. Consequently, the cost faced by player's i at time-slot t when consuming a load of w_t^i is given by:

$$C_t^i(w_t^i) = \beta_t^i \max\{w_t^i, 0\} - \zeta_t^i \max\{-w_t^i, 0\}.$$

As it is with most tariffs that allow for injecting back into the main grid, we will assume that the price of buying $\beta_t^i > \zeta_j^i$, $\forall t, j$ so that buying and re-selling to the TEC is never optimal.

2.1   Utility of Players Trading in the Market and with the TEC

We introduce here the definition we use of the utility of any given player when trading in the local market.

At time-slot t, player i might be able to trade a fraction $\lambda_t^i \in [0, 1]$ of her net load n_t^i in the local market. If player i trades $\lambda_t^i n_t^i$ in the local market, then it will have to trade the quantity $(1 - \lambda_t^i)n_t^i$ with the TEC. Denoting $\mathcal{P}^i$ the payment of player i associated with the total quantity traded in the local market (positive if buying, negative if selling) among all time-slots, the utility of player i is given by:

$$u^i(x^i, s^i, \lambda^i, \mathcal{P}^i) = \begin{cases} -\mathcal{P}^i - \sum_{t=1}^{T} C_t^i((1 - \lambda_t^i)n_t^i) & \text{if } (n^i - x^i \in \mathcal{F}^i) \\ -\infty & \text{otherwise} \end{cases} \quad (2)$$

An interpretation of the above is as follows: if players do not manage to consume their desired energy consumption profile x^i, then they are dissatisfied beyond repair. Otherwise, their utility is simply their total cost, which is given as the cost associated with the market and the cost associated with trading with the TEC.

The maximum utility that a player can obtain without participating in the market and only trading with the TEC is given by:

$$\alpha^i = \max_{s^i \in \mathcal{F}^i} u^i(s^i + x^i, 0, 0) \quad (3)$$

The optimization problem specified in (3) coincides with the optimal control of a battery subject to a fixed price tariff and it is equivalent to a linear programming problem [2].

3   Auction Model

We put forward the design of a combinatorial double auction that exploits the flexibility available for players. Unlike the traditional auctions used for LEMs in which players bid the quantity they want to buy or sell for a single time-slot, we allow players express in their bids their desire to acquire specific profiles of energy spanning multiple periods. We proceed to explain the bidding format, the allocation and the pricing rules.

3.1   Bidding Format and Allocation Rule

In the proposed auction, each player expresses all her acceptable trading profiles and the utility associated with each one of them. To do so, each player bids a feasible set of consumption profiles $\hat{\mathcal{F}}^i$ (this can be done by bidding the battery capacity, initial state of charge and the player's demand $\hat{x}^i$) and her utility function u^i, such as the one defined in Eq. (2). Here, we use the $\hat{h}$ notation to emphasize that the bid needs not to be truthful. From the bids, we can obtain

$\hat{\alpha}^i$, the maximum utility that player i can guarantee without trading in the local market, according to her reported information.

Observe that to bid the utility function u^i, it suffices to bid the set of buying and selling prices β^i, ζ^i.

Regarding the allocation rule, it will be derived from the optimal solution of optimization problem (4a)–(4e). As the objective function of the allocation problem, we decided to use Eq. (4a), which maximizes the value of all the local trades. The value is defined as the price that players would have to pay to the TEC to buy (sell) the same amount of energy. That way, the maximum amount of profit can be distributed among the market participants. This is analogous to finding the clearing price in a double auction such as [4].

$$\max_{n^i, \lambda^i, \mathcal{P}^i} \quad \sum_{i \in \mathcal{N}} \sum_{t \in \mathcal{T}} C_t^i \left(\lambda_t^i n_t^i \right) \tag{4a}$$

$$\text{subject to:} \qquad \sum_{i \in \mathcal{N}} \mathcal{P}^i \quad \geq 0 \tag{4b}$$

$$\mathcal{P}^i + \sum_{t \in \mathcal{T}} C_t^i \left[1 - \lambda_t^i n_t^i \right] \quad \leq -\hat{\alpha}^i \quad \forall i \in \mathcal{N} \tag{4c}$$

$$\sum_{i \in \mathcal{N}} \lambda_t^i n_t^i \quad = 0 \quad \forall t \in \mathcal{T} \tag{4d}$$

$$n^i \in \hat{\mathcal{F}}^i + \hat{x}^i \quad \forall i \in \mathcal{N} \tag{4e}$$

The first constraint (4b) ensures that if the equality holds, all the money is redistributed among the participants according to the market decisions, while if the inequality is strict, the market maker obtains a profit. Constraint (4c) guarantees that the auction is individually rational, i.e., each players is at least as good as if she had not participated in the local market. It's important to note that encoding all of the $\mathcal{N}$ constraints (4c) requires a total of $\mathcal{NT}$ additional binary variables. Equalities (4d) ensure that the amount of sold energy is equal to the energy bought in every time-slot. The last constraint guarantees that only feasible net consumption profiles are used. Finally, the amount of energy traded by player i at time-slot t is given by $\lambda^{i*} n^{i*}$, where λ^{i*} and n^{i*} are the optimal solutions of optimization problem (4a).

3.2   Payment Rule

As a payment rule, one alternative is to use the value of $\mathcal{P}^{i*}$ in the optimal solution of (4a). For the cases in which the values of $\mathcal{P}^{i*}$ will not be unique, a predefined rule can be used to choose among the possible values. One such could be to select the values of $\mathcal{P}^{i*}$ that maximize a given fairness criterion.

We proceed to illustrate our proposal with an example.

3.3   A Simple Example

Consider two players 1 and 2 such that: $x^1 = (0, -1, 0)$, $x^2 = (0, 0, 1)$, $\beta^1 = \beta^2 = (2, 3, 3)$, $\zeta^1 = \zeta^2 = (1, 1, 1)$, $\mathbf{S}^1 = \mathbf{S}^2 = 1$, $\mathbf{S}_0^1 = \mathbf{S}_0^2 = 0$.

If player 1 does not trade in the market, she will sell all her energy at price 1, for a total utility of $\alpha^1 = 1$, net consumption profile $n^1 = (0, -1, 0)$ and no need to user her battery $s^1 = (0, 0, 0)$. Analogously, player's 2 utility is -2 as she charges her battery during the first time-slot and discharges it in the last one ($s^2 = (1, 0, -1)$) to obtain a net consumption profile $n^2 = (1, 0, 0)$.

We will now assume that the two players decide to participate in the auction and they do so truthfully. In the optimal solution of the allocation problem defined by their bids, it holds that $n^1 = (0, -1, 0) = -n^2$, $\lambda^1 = \lambda^2 = (0, 1, 0)$. Furthermore, the maximum value is attained at: $3 \times (1) + 1 \times (-1) = 2$.

Regarding the payments, we have that for player 1: $\mathcal{P}^{1*} \leq -1$ and for player 2: $\mathcal{P}^{2*} \leq 2$. Consequently, any payment from player 2 to player 1 in the interval $\mathcal{P}^{2*} \in (1, 2)$ will leave both players better off than before.

3.4 General Properties of the Solution

First, observe that in (4a), the scenario without trades ($\mathcal{P}_t^i = \lambda_t^i = 0, \forall i \in \mathcal{N}, \forall t \in \mathcal{T}$) is always feasible and therefore, a solution exists. This solution needs not to be unique, as discussed in Subsect. 3.2. Secondly, when all players bid truthfully, the proposed auction obtains the consumption and trading profiles that maximize the value of the trades. The obtained allocation outperforms the results obtained when players maximize their individually utility and attempt to trade later using sequential auctions. An example of this was given in the previous subsection. There, the total utility of players went from -1, had they tried to trade in sequential auctions using the net profiles that maximized their individual utilities, to 0 by trading in the proposed auction.

4   Conclusion

In this paper we introduced a combinatorial double auction to be used in local energy markets as a replacement to run several sequential auctions in the same day, one for each time-slot. The proposed model maximizes the value of the trades in the local market by exploiting the latent flexibility of the players, given that players bid truthfully. Future lines of research include variations of the proposed mechanism that are strategy-proof or that require less binary variables.

References

1. Carlsson, P., Andersson, A.: A flexible model for tree-structured multi-commodity markets. Electron. Commer. Res. **7**(1), 69–88 (2007). https://doi.org/10.1007/s10660-006-0063-y
2. Hashmi, M.U., Mukhopadhyay, A., Bušić, A., Elias, J., Kiedanski, D.: Optimal storage arbitrage under net metering using linear programming. In: 2019 IEEE International Conference on Communications, Control, and Computing Technologies for Smart Grids (SmartGridComm), pp. 1–7. IEEE (2019)

3. Horta, J.L.: Innovative paradigms and architecture for future distribution electricity networks supporting the energy transition. Theses, Télécom ParisTech, April 2018. https://pastel.archives-ouvertes.fr/tel-01998249
4. Huang, P., Scheller-Wolf, A., Sycara, K.: Design of a multi-unit double auction e-market. Comput. Intell. **18**(4), 596–617 (2002)
5. Kiedanski, D., Kofman, D., Maillé, P., Horta, J.: Misalignments of objectives in demand response programs: a look at local energy markets. In: IEEE SmartGrid-Comm 2020, Tempe, USA, October (Submitted)
6. Kok, J.K., Warmer, C.J., Kamphuis, I.: PowerMatcher: multiagent control in the electricity infrastructure. In: Proceedings of the Fourth International Joint Conference on Autonomous Agents and Multiagent Systems, pp. 75–82 (2005)
7. Li, L., Liu, Y., Ming Liu, K., Lei Ma, X., Yang, M.: Pricing in combinatorial double auction-based grid allocation model. J. China Univ. Posts Telecommun. **16**(3), 59–65 (2009). https://doi.org/10.1016/S1005-8885(08)60228-9, http://www.sciencedirect.com/science/article/pii/S1005888508602289
8. Lopez-Rodriguez, I., Hernandez-Tejera, M., Lopez, A.L.: Methods for the management of distributed electricity networks using software agents and market mechanisms: a survey. Electr. Power Syst. Res. **136**, 362–369 (2016)
9. Mengelkamp, E., Diesing, J., Weinhardt, C.: Tracing local energy markets: a literature review. it - Inf. Technol. **61**(2–3), 101–110 (2019). https://doi.org/10.1515/itit-2019-0016
10. Penya, Y., Jennings, N.: Combinatorial markets for efficient energy management. In: IEEE/WIC/ACM International Conference on Intelligent Agent Technology. IEEE. https://doi.org/10.1109/iat.2005.54
11. Samimi, P., Teimouri, Y., Mukhtar, M.: A combinatorial double auction resource allocation model in cloud computing. Inf. Sci. **357**, 201 – 216 (2016). https://doi.org/10.1016/j.ins.2014.02.008, http://www.sciencedirect.com/science/article/pii/S0020025514001054
12. Weinhardt, C., et al.: How far along are local energy markets in the DACH+ Region? A comparative market engineering approach. In: Proceedings of the Tenth ACM International Conference on Future Energy Systems, pp. 544–549 (2019)
13. Xia, M., Stallaert, J., Whinston, A.B.: Solving the combinatorial double auction problem. Eur. J. Oper. Res. **164**(1), 239 – 251 (2005). https://doi.org/10.1016/j.ejor.2003.11.018, http://www.sciencedirect.com/science/article/pii/S0377221703008981

Author Index

Printed in the United States
by Baker & Taylor Publisher Services